MARE Publication Series

Volume 22

The MARE Publication Series is an initiative of the Centre for Maritime Research (MARE). MARE is an interdisciplinary social-science network devoted to studying the use and management of marine resources. It is based jointly at the University of Amsterdam and Wageningen University (www.marecentre.nl).

The MARE Publication Series addresses topics of contemporary relevance in the wide field of 'people and the sea'. It has a global scope and includes contributions from a wide range of social science disciplines as well as from applied sciences. Topics range from fisheries, to integrated management, coastal tourism, and environmental conservation. The series was previously hosted by Amsterdam University Press and joined Springer in 2011.

The MARE Publication Series is complemented by the Journal of Maritime Studies (MAST) and the biennial People and the Sea Conferences in Amsterdam.

More information about this series at http://www.springer.com/series/10413

Petter Holm • Maria Hadjimichael
Sebastian Linke • Steven Mackinson
Editors

Collaborative Research in Fisheries

Co-creating Knowledge for Fisheries Governance in Europe

Springer

Editors
Petter Holm
Norwegian College of Fishery Science
UiT The Arctic University of Norway
Tromsø, Norway

Sebastian Linke
School of Global Studies
University of Gothenburg
Gothenburg, Sweden

Maria Hadjimichael
Innovative Fisheries Management,
Department of Planning
Aalborg University
Aalborg, Denmark

University of Cyprus
Nicosia, Cyprus

Steven Mackinson
Centre for Environment, Fisheries and
Aquaculture Science (Cefas)
Lowestoft, Suffolk, UK

Scottish Pelagic Fishermen's Association
Fraserburgh, UK

ISSN 2212-6260 ISSN 2212-6279 (electronic)
MARE Publication Series
ISBN 978-3-030-26783-4 ISBN 978-3-030-26784-1 (eBook)
https://doi.org/10.1007/978-3-030-26784-1

Cover illustration: 'All aboard' Fishermen, Scientists and Managers working together for sustainable fisheries. Copyright: GAP2 project

This Springer imprint is published by the registered company Springer Nature Switzerland AG.
The registered company address is: Gewerbestrasse 11, 6330 Cham, Switzerland

We dedicate this book to our late colleague and friend, Doug Wilson, a leading voice in the social science of European fisheries governance and enthusiastic proponent of collaboration and inclusivity. Doug was one of the catalysts and creators of the GAP programme and crafted the investigations to delve into understanding the conditions that determine the success of participatory research. Doug's thinking has influenced all the authors and work described in this book, but you will find his voice most strongly in Chaps. 2 and 17 where he worked closely with his long-time friends.

Foreword

This book completes more than 10 years of research on the role and value of stakeholder-driven science in support of effective fisheries governance. The aim of the book is to demonstrate how stakeholder knowledge can be complementary to scientific knowledge. New research practices and advice frameworks are being developed in support of the ongoing transition in fisheries governance focusing on co-creation processes and establishing a common knowledge base for fisheries management in Europe.

The European experiences can be inspirational and provide insight to how stakeholder-driven science can be implemented across the globe. In Canada, the USA and Australia, participatory research is already on the move; these countries are also searching for better ways to set up co-creation processes to integrate stakeholder-driven science into the advice process and subsequent policy-making.

This book has grown from the EU-funded FP7 research project Bridging the Gap Between Science and Stakeholders (GAP) and provides experiences on collaborative research by presenting the findings from 14 fisher-scientist partnerships across 12 European countries. The research has been conducted by interdisciplinary teams each composed of social and natural scientists working closely together. These researchers and fishers had to cross disciplinary boundaries and developed a common language to communicate and share their deep and thorough understanding to improve fisheries management in the specific settings.

The genesis of the research idea goes way back and was inspired by the ICES Fishery Systems Working Group and an increasing acknowledgement in the fisheries science community to strengthen collaboration between scientists and stakeholders in the production of knowledge to support fisheries management decision-making. I have had the pleasure to observe the development of the GAP research project in its formulation and later to be involved as co-supervisor for one of the PhD students associated with GAP. I am very impressed by the depth of the case studies and the way they present existing knowledge gaps. The case studies demonstrate great variety in types of co-creation processes and approaches to fill the gaps in understanding and applying different types of knowledge.

The underlying theme in most chapters—if not all—is the emergence of a new type of scientifically literate fisher. It would be naive to ignore that this fisher is also a political actor. Through the co-creation process, the fisher forms new alliances and enters new networks enabling him/her to influence the management discourse to pursue his/her own interest. Thus, the cases indirectly indicate a need for a knowledge broker, who on the one hand has the scientific depth and on the other hand understands the practicalities of fisheries and has the respect of the fishers. The way forward is neither to make a fisher into a scientist nor to make a scientist into a fisher, and it is impossible to clone the two breeds. The recipe to create a knowledge broker institution has not yet been found, but an interesting spinoff of the GAP project might show the way forward. Two highly respected scientists—from national research institutes, both instrumental in the formulation of the GAP project and heavily involved in science-stakeholders collaboration—have changed positions and career patterns by taking positions as chief scientific advisors for, respectively, Dutch and Scottish pelagic fisheries associations and in that capacity have undertaken a knowledge broker role.

Turning the working papers from the GAP project into a book manuscript has been a major task. It is fair to say that without the tireless effort from Petter Holm and his team of co-editors Steve Mackinson, Maria Hadjimichael and Sebastian Linke, this valuable contribution on participatory research would never have reached publication. Besides the trivial issues of preparing an edited volume, the lead editor has been challenged by the fact that the co-creation arrangements or partnerships presented in the book all have been chosen from a pragmatic perspective and present already well-established relationships between fishers and scientists in the specific case studies. However, the flip side of this approach is that to some degree—at least in earlier drafts—the case studies did not clearly illustrate the cultural gaps, which occur when establishing fisher-scientist partnerships. The cultural gaps are an important dimension when addressing existing shortcomings and challenges in setting up participatory governance structures, a concern the lead editor persistently addressed throughout the editing process with both dedication and determination—even at the expense of delaying the publication of the book! But it has been worth the wait. I cannot wait to secure my copy of the book.

Enjoy reading!

Professor, Innovative Fisheries Management (IFM) Jesper Raakjær
Aalborg University
Aalborg, Denmark
Chairman, The EU Pelagic Advisory Council
Copenhagen V, Denmark

Series Editors' Preface

As editors of the MARE Publication Series, we are proud to present yet another fine volume addressing the many challenges and opportunities for enhancing the quality of governance, this time about the fisheries of Europe. Having moved towards a participatory mode of governance, most notably with the establishment of the advisory councils for the EU's regional seas, fishermen now have a new channel for voicing concerns that breaks with the long-established hierarchical, top-down mode of governance. With fishermen directly or indirectly partaking, the decision-making process is becoming a more level playing field. By that, the established roles and relationships with other stakeholders, such as government administrators and scientists, are also changing. Fishermen are now provided with an opportunity to bring their experiential knowledge to the table, the consequences of which are worth examining. Interestingly, as pointed out by the editors of this volume, Petter Holm, Steve Mackinson, Maria Hadjimichael and Sebastian Linke, the participatory and communicative turn of fisheries science also affects the responsibilities that are put on fishermen's shoulders. From being placed at the receiving end of management, fishermen are now expected to engage in the process of knowledge production and provision. Thereby, the idea of what it means to be a fisherman alters: enter "the scientific fisherman", who is a co-producer of the knowledge base of management.

Changing the order in this manner has consequences for other stakeholders as well. For scientists, it means less detachment from the real-life struggles of fishermen, for whom livelihoods and communities are at stake. The process of science then becomes less exclusive and more transparent, which will ideally lead to a broader set of research questions investigated. But it may also bring into question the very integrity, and hence legitimacy, of science, which was always associated with the notion of objectivity and independence. However, the costs of a participatory knowledge-building exercise are expected to be outweighed by the gains of a process more open to "lay expert" knowledge. Whether this calculation is correct is an empirical question to be studied through transdisciplinary research. The operationalization of governance principles and reforms in concrete contexts may prove more challenging than anticipated. Roles and relationships between stakeholders may be harder to rearrange than one might expect. Knowledge production may meet walls when transmitted from local to higher levels of governance, which are possibly not designed to receive transdisciplinary knowledge. Therefore, as argued in this book, institutional reform at the receiving end of knowledge production must also be instigated. This may not always be easy to realize given the often entrenched modes of governance resistant to change. There is therefore a clear link between participatory research, legitimacy benefits and institutional reforms that need to be thought about in a transdisciplinary manner.

These are issues explored and arguments tested in this book, which is made up of a series of case studies from around Europe. The reader will find the questions researched to be intriguing: What are the existing knowledge gaps that need to be filled to make fisheries governance more sensitive to and inclusive of stakeholder concerns and interests? Why are current governance designs less than effective in addressing existing shortcomings in the inclusion of fishermen's knowledge? What new platforms would be needed for generating a better science generation and communication and subsequently more legitimate and effective fisheries governance? Readers of this volume may consult a previous book, Douglas Clyde Wilson's *The Paradoxes of Transparency: Science and the Ecosystem Approach to Fisheries Management in Europe*, published by the MARE Publication Series (then at Amsterdam University Press) in 2010. Holm and colleagues often refer to Wilson's work, and the reader of their book may find that they together provide a comprehensive perspective on the role of science in the new governance of fisheries transitioning towards a more participatory and transdisciplinary co-governance mode.

University of Amsterdam
Amsterdam, The Netherlands
Maarten Bavinck

UiT-The Arctic University of Norway
Tromsø, Norway
Svein Jentoft

Acknowledgements

A sincere thank you to all the partners, colleagues and friends who are committed to participatory research which has provided the bricks and mortar for this book. This book is a small reward for your continued staying power, being part of positive changes towards inclusivity that have occurred in the 16 years since the GAP project was conceived. We also thank the funders European Commission's Seventh Framework Programme Science in Society (Grant Agreement 266544) and, in particular, Philippe Galiay who shared the vision of GAP. The publisher, Springer, also deserves our thanks for their staying power and being supportive and patient with the editors and authors in our work to a finished product. We owe special thanks to Paulina Ramirez for comments and reviewing that helped us to find the thread running through the chapters, to Ajit Menon for judicious care in technical editing and to all of the external reviewers who spotted the valuable parts necessary to sharpen our focus.

Tromsø, Nicosia, Gothenburg, Fraserburgh
The Editors

Contents

Contributors

Asgeir Aglen Institute of Marine Research, Bergen, Norway

Manon Airaud IRD UMR MARBEC, Sète cedex, France

Jan I. Andersen Norwegian Fishermen's Association, Steigen, Norway

Robert Aps Estonian Marine Institute, University of Tartu, Tallinn, Estonia

Kevin Arscott South Devon and Channel Shellfishermen's Association, Devon, UK

Thomas Axenrot Department of Aquatic Resources, Institute of Freshwater Research, Swedish University of Agricultural Sciences (SLU), Drottningholm, Sweden

Victoria A. Bendall Centre for Environment, Fisheries and Aquaculture Science (Cefas), Lowestoft, Suffolk, UK

Maiken Bjørkan Norwegian College of Fishery Science, UiT The Arctic University of Norway, Tromsø, Norway
Nordland Research Institute, Bodø, Norway

Giovanni Bulian Department of Asian and North African Studies, Ca' Foscari University of Venice, Venezia, Italy

Marianna Bullo Italian National Institute for Environmental Protection and Research (ISPRA), Chioggia, Italy

Igor Celić Italian National Institute for Environmental Protection and Research (ISPRA), Chioggia, Italy
Oceanography Division (OCE), National Institute of Oceanography and Experimental Geophysics (OGS), Trieste, Italy

Lotte Worsøe Clausen DTU Aqua, National Institute of Aquatic Resources, Technical University of Denmark (DTU), Lyngby, Denmark
International Council for the Exploration of the Sea (ICES), Copenhagen, Denmark

Joan B. Company Institut de Ciències del Mar (ICM-CSIC), Barcelona, Spain

Laurent Dagorn IRD UMR MARBEC, Sète cedex, France

Barrie Deas National Federation of Fishermen's Organisations, York, UK

Mark Dubois Innovative Fisheries Management, Aalborg University Research Centre, Aalborg, Denmark
Department of Fisheries, World Fish Myanmar, Yangon, Myanmar

Mihhail Fetissov Estonian Marine Institute, University of Tartu, Tallinn, Estonia

Tomaso Fortibuoni Italian National Institute for Environmental Protection and Research, Ozzano dell'Emilia, BO, Italy
Oceanography Division (OCE), National Institute of Oceanography and Experimental Geophysics (OGS), Trieste, Italy

Gianluca Franceschini Italian National Institute for Environmental Protection and Research (ISPRA), Chioggia, Italy

Juan Freire Tecnológico de Monterrey, Business School, Mexico City, Mexico

Andrea Fusari Consorzio UNIMAR Società Cooperativa, Roma, Italy

Otello Giovanardi Italian National Institute for Environmental Protection and Research (ISPRA), Chioggia, Italy
Institute for Biological Resources and Marine Biotechnologies (IRBIM), Ancona, Italy

Giulia Gorelli Institut de Ciències del Mar (ICM-CSIC), Barcelona, Spain

Maria Hadjimichael Innovative Fisheries Management, Department of Planning, Aalborg University, Aalborg, Denmark
University of Cyprus, Nicosia, Cyprus

Paul J. B. Hart Department of Neuroscience, Psychology and Behaviour, University of Leicester, Leicester, UK

Stuart J. Hetherington Centre for Environment, Fisheries and Aquaculture Science (Cefas), Lowestoft, Suffolk, UK

Petter Holm Norwegian College of Fishery Science, UiT The Arctic University of Norway, Tromsø, Norway

Peter Hopkins Directorate-General for Maritime Affairs and Fisheries, European Commission, Brussels, Belgium

Ewan Hunter Centre for Environment, Fisheries and Aquaculture Science (Cefas), Lowestoft, UK

Reine Johansson Baltic Sea Advisory Council (BSAC), Copenhagen, Denmark

Tomas Jonsson Department of Ecology, Swedish University of Agricultural Sciences (SLU), Uppsala, Sweden
School of Bioscience, Ecological Modelling Group, University of Skövde, Skövde, Sweden

Madli Kopti Estonian Maritime Academy, Tallinn, Estonia

Marloes Kraan Wageningen Marine Research, IJmuiden, The Netherlands, & Environmental Policy Group, Wageningen University, Wageningen, The Netherlands

Sebastian Linke School of Global Studies, University of Gothenburg, Gothenburg, Sweden

Steven Mackinson Centre for Environment, Fisheries and Aquaculture Science (Cefas), Lowestoft, Suffolk, UK
Scottish Pelagic Fishermen's Association, Fraserburgh, UK

Adriano Mariani Consorzio UNIMAR Società Cooperativa, Roma, Italy

Nicholas Flores Martin Fisheries Resource Unit, Department of Fisheries and Aquaculture, Marsa, Malta

Conrad Massaguer Confraria de Pescadors de Palamós, Palamós, Catalonia, Spain

Monica Mion Italian National Institute for Environmental Protection and Research (ISPRA), Chioggia, Italy
Institute of Marine Research, Swedish University of Agricultural Sciences (SLU), Lysekil, Sweden

Gala Moreno International Seafood Sustainability Foundation (ISSF), Washington, DC, USA

Ramón Muiño Department of Biology, Faculty of Science, University of Coruña, Coruña, Spain

Jefferson Murua AZTI, Txatxarramendi Ugartea z/g, Sukarrieta, Bizkaia, Spain

Marco Nalon Italian National Institute for Environmental Protection and Research (ISPRA), Chioggia, Italy
National Institute of Aquatic Resources, Section for Ecosystem based Marine Management, Technical University of Denmark, Hirtshals, Denmark

Johnny Norrgård Lake Vättern Society of Water Conservation, Jönköping, Sweden

Verena Ohms Pelagic Advisory Council (PelAC), Zoetermeer, The Netherlands

Christian Olesen Danish Pelagic Producers Organisation (DPPO), Copenhagen, Denmark

Michael Park Scottish Whitefish Producers Association, Fraserburgh Business Centre, Fraserburgh, UK

Emma Pearson Department of Neuroscience, Psychology and Behaviour, University of Leicester, Leicester, UK

Camilla Piras Italian National Institute for Environmental Protection and Research (ISPRA), Chioggia, Italy

Pablo Pita Department of Applied Economics, Faculty of Economics and Business Administration, University of Santiago de Compostela, Santiago de Compostela, Spain

Saša Raicevich Italian National Institute for Environmental Protection and Research (ISPRA), Chioggia, Italy
Institute for Biological Resources and Marine Biotechnologies (IRBIM), Ancona, Italy

Paulina Ramírez-Monsalve Innovative Fisheries Management (IFM), Department of Planning, Aalborg University, Aalborg, Denmark

Laura Sabatini Italian National Institute for Environmental Protection and Research (ISPRA), Chioggia, Italy
Institute for Biological Resources and Marine Biotechnologies (IRBIM), Ancona, Italy

Alfred Sandström Department of Aquatic Resources, Institute of Freshwater Research, Swedish University of Agricultural Sciences (SLU), Drottningholm, Sweden

Francesc Sardà Institut de Ciències del Mar (ICM-CSIC), Barcelona, Spain

Simone Serra Consorzio UNIMAR Società Cooperativa, Roma, Italy

Malin Setzer Lake Vättern Society of Water Conservation, Jönköping, Sweden
School of Bioscience, Ecological Modelling Group, University of Skövde, Skövde, Sweden

Alan Steer South Devon and Channel Shellfishermen's Association, Devon, UK

Laurent Tezenas Montpellier SupAgro, Montpellier, France

Marieke Verweij Wing, Wageningen, The Netherlands

Duarte Vidal Department of Fisheries Socioeconomic, Centro Tecnolóxico do Mar – Fundación CETMAR, Vigo, Spain

Kai Wätjen Alfred Wegener Institute Helmholtz Centre for Polar and Marine Research, Bremerhaven, Germany

Abbreviations

AC	Advisory Council
AGCI	Associazione Generale Cooperative Italiane
BSAC	Baltic Sea Advisory Council
CEFAS	Centre for Environment, Fisheries and Aquaculture Science
CFP	Common Fisheries Policy
CFPO	Cornish Fish Producers Organisation
CPUE	Catch Per Unit Effort
CS	Case Study
CL	Cephalothorax Length
CVO	Coöperatieve Visserij Organisatie
DCF	Data Collection Framework
DEFRA	Department for Environment, Food and Rural Affairs
DIV	Division
EAF	Ecosystem Approach to Fisheries
EAFM	Ecosystem Approach to Fisheries Management
EBAF	Ecosystem-Based Approach to Fisheries
EBK	Experience-Based Knowledge
EC	European Commission
EEZ	Exclusive Economic Zone
EFH	Essential Fish Habitats
eNGO	Environmental Non-government Organization
EU	European Union
EzDK	Erzeugergemeinschaft Deutscher Krabbenfischer
FAD	Fish Aggregating Device
FAO	Food and Agriculture Organization
FDI	Fisheries Dependent Information
FHF	Norwegian Seafood Research Fund
FK	Fishermen Knowledge
FMZ	Fisheries Management Zone
FSP	Fisheries Science Partnership
FTOA	Fishing Trawler Owners Association

GAMS	General Algebraic Modelling System
GFCM	General Fisheries Commission for the Mediterranean
GIS	Geographic Information System
GFF	Galician Fishermen Federation
GPS	Global Positioning System
GSA	Geographical Sub-area
GT	Gross Tonnage
HAWG	Herring Assessment Working Group
IBM	Individual-Based Model
ICCAT	International Commission for the Conservation of Atlantic Tunas
ICES	International Council for the Exploration of the Sea
ICM-CSIC	Institute of Marine Sciences
IDESCAT	Statistical Institute of Catalonia
IFCA	Inshore Fisheries and Conservation Authority
IMR	Institute of Marine Resources
IOTC	Indian Ocean Tuna Commission
IPA	Inshore Potting Area
IRD	Institute for Research and Development
ISPRA	Istituto Superiore per la Protezione e la Ricerca Ambientale
LCA	Length Cohort Analysis
LIW	Levantine Intermediate Water
LPG	Liquid Petroleum Gas
LTMP	Long-Term Management Plan
MAMP	Multi-annual Management Plan
MAW	Modified Atlantic Water
MCZ	Marine Conservation Zone
MEDITS	Mediterranean International Trawl Survey
MLS	Minimum Landing Size
MMO	Marine Management Organisation
MPA	Marine Protected Area
MSFD	Marine Strategy Framework Directive
MS	Member States
MSARs	Monthly Shellfish Activity Returns
MSE	Management Strategy Evaluation
MSP	Marine Spatial Planning
MSY	Maximum Sustainable Yield
NCC	Norwegian Coastal Cod
NEAC	North East Arctic Cod
NFFO	National Federation of Fishermen's Organisations (UK)
NGO	Non-governmental Organization
NOK	Norwegian Kroner
NRF	Norwegian Reference Fleet
NSAS	North Sea Autumn Spawners
NSAC	North Sea Advisory Council
PCBs	Polychlorinated Biphenyls

PELAC	Pelagic Advisory Council
RAC	Regional Advisory Council
RFMO	Regional Fisheries Management Organisation
RMyP	Marine Resources and Fisheries Research Group
RRI	Responsible Research and Innovation
SD	Sub-division
SDCSA	South Devon and Channel Shellfishermen's Association
SMART	Specific, Measurable, Achievable, Realistic and Time-Bound
SSF	Small-Scale Fisheries
STECF	Scientific, Technical and Economic Committee for Fisheries
STS	Science and Technology Studies
SWAM	Swedish Agency for Marine and Water Management
TA	Technical Assistance
TAC	Total Allowable Catch
TURF	Territorial Use Rights for Fisheries
UK	United Kingdom
UNESCO	United Nations Educational, Scientific and Cultural Organization
VMS	Vessel Monitoring System
WBSS	Western Baltic Spring Spawners
WGMARS	Working Group on Marine Systems
WGMIXFISH	Working Group on Mixed Fisheries
WGSAM	Working Group on Multi-species Assessment Models
WWII	Second World War
WWF	World Wildlife Fund

Chapter 1
Bridging the Gap: Experiments in the Heart of the Transition Zone

Steven Mackinson and Petter Holm

Abstract This book is about the ongoing transition of fisheries governance and the emergence of research practices and advice frameworks that allow for the co-creation of common knowledge bases for management. This chapter introduces the context under which the GAP project ('Bridging the gap between science and stakeholders') was conceived, describes its overall approach, orientates the reader to key issues and introduces the structure of the book.

Keywords Collaborative research · Fisheries governance · European fisheries · Stakeholder involvement

1.1 Introduction

> "Our hope for the future is not only to grow the red shrimp fishery, but to grow it sustainably" Conrad Massaguer, skipper of the "Nova Gasela", Palamós, Spain.

Conrad Massaguer is a participant in the GAP project's Mediterranean red shrimp case study. A team of scientists led by Dr. Joan B. Company, fishermen, and regional policy managers have successfully brought red shrimp stocks back from the brink of collapse. The key to this accomplishment is a collaboratively produced and voluntary long-term management plan that has the approval of fishermen, Catalan regional and Spanish national government.

S. Mackinson (✉)
Centre for Environment, Fisheries and Aquaculture Science (Cefas), Lowestoft, Suffolk, UK

Scottish Pelagic Fishermen's Association, Fraserburgh, UK
e-mail: steve.mackinson@scottishpelagic.co.uk

P. Holm
Norwegian College of Fishery Science, UiT The Arctic University of Norway, Tromsø, Norway

P. Holm et al. (eds.), *Collaborative Research in Fisheries*, MARE Publication Series 22, https://doi.org/10.1007/978-3-030-26784-1_1

The plan is the result of over 5 years' work; the idea initially conceived during the first phase of the GAP project in 2008. After 15 joint meetings and a lot of talking, the final draft management plan was submitted to the Spanish Government's Fishery Ministry on 25 July 2013, where it was received with warm congratulations to those involved for having produced the first plan of its kind in the Mediterranean area.

As an example of how shared learning and collaboration on research can lead to positive outcomes for management, Massaguer's hopes for the future of the Palamós fishery reflect the overarching aspirations of the project partners in the whole GAP project: a thriving, sustainable future for European fisheries. Their story is told in detail in Chap. 10 and through a short documentary available at http://gap2.eu/launch-of-gap2themovie/.

This book is about the ongoing transition of EU fisheries governance, focusing on the emergence of research practices and advice frameworks that allow for the co-creation of common knowledge bases for management. Based on 8 years of applied research on collaborative research processes, performed in the seventh framework EU projects GAP 1 & 2, the book examines how knowledge practices in fisheries governance are changing.[1]

This chapter introduces the context under which the GAP project was conceived, describes its overall approach and introduces the structure of the book.

1.2 The Transition in Fisheries Governance and Emergence of Participatory Research

A transition is taking place in the approaches to fisheries management and the research that supports it. In the 2002 reform of the Common Fisheries Policy (CFP), pressure from many simultaneous developments stimulated the then Fisheries Commissioner, Joe Borg, to put strengthening engagement with fisheries stakeholders a priority for reform. Since then, Europe has seen progressive steps to implement this view. To many, the most visible outcome has been the creation of the Regional Advisory Councils (RACs), which are the main body for engaging with stakeholders on issues that directly or indirectly affect fisheries.[2] Prior to this, fisheries management had conventionally been done in a top-down or command-and-control form, with limited possibilities for stakeholder engagement. In particular, the space for fishermen's participation was restricted when it came to the knowledge basis for

[1] EU FP7 projects GAP1- grant agreement 217,639, GAP2 - grant agreement 266,544 www.gap2.eu

[2] Under the auspices of the Common Fisheries Policy, RACs were established by Council Decision (EC) 256/2004 with the intention to increase the participation of those affected by the CFP in the fisheries management decision-making process. In the 2013 CFP reform, they were renamed Advisory Councils (ACs). ACs are stakeholder-led organisations that provide the Commission and EU countries with recommendations on fisheries management matters. http://ec.europa.eu/fisheries/partners/advisory-councils/index_en.htm

management, which was a reserved domain for stock assessment scientists. While the original justification for this arrangement was to secure the legitimacy of management decisions, relying on the generalised trust in impartial science, this did not work as intended. Instead, the exclusion of fishermen in knowledge provision has been recognised as an important weakness of the governance regime, reinforcing important gaps between the policymakers and scientists on the one hand and fishermen and fishing communities on the other (Hind 2015).

There have also been big changes in the research arena, albeit with a time lag to the policy aspirations. When the research ideas for the GAP project were first conceived in 2003, they responded directly to the policy need, but it was not until 5 years later that the signs of change became visible in the research structures and funding mechanisms that traditionally support fisheries and environmental research actions.

In 2008, the GAP project found fertile ground in a different area, called Science in Society – a research policy area interested in society's relationship with science and its link to inclusive governance. Since then, the EU research arena has witnessed a huge shift in the expectations for collaboration in research, from traditional proposal calls that could be met by collaboration among scientific organisations alone, to calls that expected consultation with (or at least endorsement from) relevant 'stakeholders', to calls that specify a requirement for a diversity of relevant stakeholders to be partners in the project consortium and to be involved in the research framing process. GAP has been at the forefront of this transition in the management and research policy landscape, conducting applied experiments in co-creation of common knowledge bases for management, but also fostering reflection among those involved about what it takes to establish effective collaborations, and the conditions that affect the degree of success. We have explored what goes on in the transition zone between top-down management and participatory governance. That is what this book is about. It deals with the knowledge of fisheries and fisheries management and the process of (co)-production and application. It does not aim specifically to address other important issues such as power and interests, and how they affect the gaps in the interactions between 'stakeholders' and scientists and governments that they affect. Nonetheless, such issues are at play and sometimes visible in the case studies.

1.3 The GAP Approach

The GAP project aimed to reduce the tensions that arise among society, policy and science when environmental sustainability concerns appear in conflict with maintaining livelihoods. The approach was to use a process of active participation and knowledge sharing among scientists, stakeholders and policymakers to establish a common knowledge base for fisheries and then build relationships for effective governance.

In the GAP project, the term 'stakeholders' refers to all those with an interest in the science and management of fisheries and the marine environment. It is a broad term that captures many actors from society. But our main focus is on fishermen, scientists and policymakers/managers, because their knowledge and the data that they create plays a central cog in the application of scientific knowledge to fisheries management. Other stakeholders include fishing communities, dependent industries, civil society organisations (e.g. WWF, Bird Life International, Friends of the Earth, Seas at Risk), private foundations and other citizens. The different interests and responsibilities of stakeholders (from grass roots to international policy) determine the roles they play in the overall governance system. Specifically relevant to the issues addressed in this book is the fact that the term 'stakeholder' may be understood differently between its use in political parlance and in the social sciences. Political parlance does not see scientists and government bodies as stakeholders, whereas in social science, they are included. (See Chap. 2 for more discussion.)

In 2008, phase one of GAP established 13 research case studies (CS) across Europe, each one centred upon working partnerships among fishermen, scientists and policymakers. These case studies came to life in GAP2 (2011–2015), which conducted 'participatory action research' – an active, collaborative form of science involving those affected by the research outcomes, from the outset of the process, through to the implementation of any outcomes (Mackinson and Wilson 2014). The philosophy underpinning this approach is rooted in the process of shared learning: "*What I hear I forget, what I see I remember, what I do I understand*" (Xunzi340–245 BC).

The premise for this approach is based on the understanding that

- The evidence-base for management improves if the knowledge of fishermen and their experience are integrated in a meaningful way with scientific and policy knowledge.
- If knowledge is shared and co-constructed, it improves the implementation and effectiveness of management measures.
- If knowledge is shared and co-constructed among stakeholders, it improves the support for policy and societal goals to achieve responsible, sustainable, productive fisheries.

Co-constructed knowledge improves the knowledge base for fisheries with regard to credibility, legitimacy and saliency (see Chap. 2; Röckmann et al. 2015). Well-designed participatory action research is one strategy that has been shown to be effective in addressing the complex issues of knowledge, participation and decision-making in fisheries management (Reid and Hartley 2006, Johnson and van Densen 2007, Stephenson et al. 2016). In brief, well-designed means that there is an ongoing interchange based on genuine respect for participants' perspectives and contributions. Participatory action research creates not just a set of new knowledge but a social network of learning, while the action research aspect then seeks to link this network to the decision processes of marine management (Mackinson and Wilson 2014, Stephenson et al. 2016).

The GAP case studies spanned 11 different countries and covered a huge range of fisheries issues. From monitoring coastal cod populations in Norway, assessing crab stocks in the UK, modelling multispecies mixed fisheries in the North Sea, to confronting head-on the realities of a 'discard ban' in the Netherlands. While the specific research questions examined in the CS projects varied, as did the management relevance of the results, all projects were conducted with clear commitment to collaboration.

In one way, of course, the GAP project was intentionally naive. The 14 CS projects presented in this book[3] attempted to perform knowledge practices for which the appropriate institutional structures are seriously underdeveloped, producing knowledge products for which there is no ready demand. There are no accepted standards for collaborative research. There are no dedicated review processes to distinguish between acceptable and non-acceptable results. There are no formal training courses to teach best practice. While there of course is hope that the CSs succeed some way or another, this cannot be guaranteed. Instead, the nature of the CSs as experiments means that partial failure and disappointment are to be expected. One of the ways the CSs generate insight and learning is when the idealistic experience informing the collaborative projects meets up with the resistance and challenges of the established order. If transition to participatory governance and collaborative research had been easy, then it would already have happened.

1.4 Aim and Organisation of the Book

The stories of the CS projects form the basis of our journey through the subsequent chapters. Presented in individual chapters in a standardized format, each addresses questions intended to bring out their unique characteristics and experiences. How did they come about? How were research objectives negotiated among participants? If scientists took the lead, did the fishermen manage to keep up? How did the collaboration work out during the different states of the project? To what extent did the projects manage to carry through the project as planned? What difference did the collaborative research format make, in terms of the credibility and legitimacy of its results? Were the knowledge products made to count in management decisions? Collectively, the answers to these questions describe the complexity of European fisheries in some detail.

While the individual stories of the CS projects are compelling in themselves, the strength of the GAP project is the possibility to place the 14 cases alongside each other, compare them and examine them as pieces in a larger puzzle. Thus, our aim is to make them speak as a collective. In this way, the CS chapters as a collective describe the complexities and variation of European fisheries. In Chap. 2, we give

[3] Of the CS projects, 13 were formally included in GAP2. The CS on by-catch of sharks, skates and rays, reported in Chap. 16, while not part of GAP, was carried out with reference to the same perspectives and in close contact with the GAP project.

an introduction to the overall GAP approach and elaborate on three key theoretical perspectives central to this project: participation, knowledge inclusion and institutional reform. After the chapters containing the individual case studies, we return to these perspectives in Chap. 17, where we apply theory to the practical experiences of the GAP case studies and conclude what can be learned from them with respect to the three issues.

References

Hind EJ (2015) A review of the past, the present, and the future of fishers' knowledge research: a challenge to established fisheries science. ICES J Mar Sci 72(2):341–358. https://doi.org/10.1093/icesjms/fsu169

Johnson TR, van Densen WL (2007) Benefits and organization of cooperative research for fisheries management. ICES J Mar Sci 64(4):834–840

Mackinson S, Wilson DCK (2014) Building bridges among scientists and fishermen with participatory action research. In: Urquhart J, Acott TG, Symes D, Zhao M (eds) Social issues in sustainable fisheries management. Springer, Dordrecht, pp 121–139

Reid AN, Hartley TW (Eds) (2006) Partnership for a common purpose: cooperative fisheries research and management. American fisheries society symposium 52. In: Proceedings of the symposium fisheries society/sea grant symposium partnerships for a common purpose: cooperative fisheries research and management held in anchorage, Alaska, U.S.A., 13–14 September 2005. American Fisheries Society, Bethesda

Röckmann C, van Leeuwen J, Goldsborough D, Kraan M, Piet G (2015) The interaction triangle as a tool for understanding stakeholder interactions in marine ecosystem based management. Mar Policy,. Elsevier 52(C):155–162

Stephenson RL, Paul S, Pastoors M, Kraan M, Holm P, Wiber M, Mackinson S, Dankel D, Brooks K, Benson A (2016) Integrating fishers' knowledge research in science and management. ICES J Mar Sci 73(6):1459–1465. https://doi.org/10.1093/icesjms/fsw025

Chapter 2
Knowledge for Fisheries Governance: Participation, Integration and Institutional Reform

Sebastian Linke, Maria Hadjimichael, Steven Mackinson, and Petter Holm

Abstract As outlined in Chap. 1, the GAP project is situated within a transition zone from a traditional fisheries management approach that relies upon a clear separation of knowledge towards a new 'bridging perspective', which aims to establish a common knowledge base for fisheries governance. The transition builds on collaborative practices of participatory research and joint knowledge production, as will be described in the GAP case studies in subsequent chapters. Before these detailed empirical explorations, this chapter will first take a brief look at the knowledge gaps that are created by the dominant perspective of fisheries management and the resulting implications on sustainability of fisheries including the legitimacy deficits created by the traditional approach to fisheries. Second, three key domains of social science research that the GAP project connects with will be presented (participation, knowledge integration and institutional reform). Finally, some central aspects of the overall GAP approach are highlighted, and a brief overview of the GAP case studies is presented.

Keywords Collaborative research · Fisheries governance · European fisheries · Knowledge · Stakeholder involvement

---

S. Linke (✉)
School of Global Studies, University of Gothenburg, Gothenburg, Sweden
e-mail: sebastian.linke@gu.se

M. Hadjimichael
Innovative Fisheries Management, Department of Planning, Aalborg University, Aalborg, Denmark

University of Cyprus, Nicosia, Cyprus

S. Mackinson
Centre for Environment, Fisheries and Aquaculture Science (Cefas), Lowestoft, Suffolk, UK

Scottish Pelagic Fishermen's Association, Fraserburgh, UK

P. Holm
Norwegian College of Fishery Science, UiT The Arctic University of Norway, Tromsø, Norway

P. Holm et al. (eds.), *Collaborative Research in Fisheries*, MARE Publication Series 22, https://doi.org/10.1007/978-3-030-26784-1_2

2.1 Introduction: The Gaps in Traditional Fisheries Management

The traditional conceptualisation of fisheries management in the developed world established a division of knowledge between scientific expertise and policymaking on one side of the system and fisher's knowledge perspectives on the other side (Gezelius 2008; Linke et al. 2011; Holm 2003). This classic division of knowledge has been inscribed into (and is enforced by) the dominant institutional framework that established modern fisheries management after the Second World War (Nielsen 2008; Holm and Nielsen 2004). In this arrangement, which Holm and Nielsen (2004) coined the 'TAC machine', a scientific advisory system and governmental agencies are interdependently connected through an exclusive science–policy interface, with precisely defined divisions of labour for expert communities (e.g. the International Council for the Exploration of the Seas, ICES) and the clients that request their services (e.g. EU Commission and national governments) (Hegland 2012; Penas-Lado 2016). In the practical reality of fisheries management, this institutional division of labour implies that the work of fisheries scientists is constrained to react to a ready-made policy agenda by providing stock assessments and giving advice, mainly in the form of catch quotas, to policy- and decision-makers, who are equally constrained to use this knowledge as a legitimate basis for decision-making. This institutionalised practice of science–policy interactions inevitably exacerbates meaningful inputs from the fisheries sector due to specification of the science–policy 'cogs' in a finely tuned and technically inert management machinery (Schwach et al. 2007; Nielsen and Holm 2007; Wilson 2009).

However, the management system relying on these exclusive science–policy interactions has not provided desired outcomes in terms of long-term sustainable fisheries and has been criticised for creating problems with regard to the knowledge gaps and the legitimacy of the knowledge holders on either side of these gaps (Khalilian et al. 2010; Daw and Gray 2005; Nielsen and Holm 2007). By and large, fishermen have felt excluded from a management system that affects their daily businesses and livelihoods. Mistrust, frustration, noncompliance with regulations and sometimes deep-rooted conflicts between fishermen and scientists or policymakers are, at least partly, a consequence of the legitimacy deficit that this system created. Apart from the legitimacy problems, valuable knowledge from the fishing sector has also been disregarded by the TAC Machine's institutional configuration. The most dramatic example, where relevant information from inshore fishermen was not sufficiently considered, is the Canadian stock assessment and management before the collapse of the northern cod (Neis 1992; Finlayson 1994). The gaps created by the traditional top-down, science-based fisheries management system has aptly been pointed out by Hubbard (2012: 129):

> The shift away from local to remote, centralized control, mediated by scientists, marginalized those actually involved in the fisheries, who could make little input into policy decisions, resulting in growing chasms in communications between fishers and policy makers.

As a consequence of such failures and resulting legitimacy deficits, these gaps in knowledge production and use have increasingly been accepted as a key problem for conventional fisheries management and served as an important motive for reform. A common knowledge base, open to inputs from science and the fisheries sector, was envisioned both in the research community (Mackinson and Nøttestad 1998; Wilson et al. 2003; Gray 2005; Hoefnagel et al. 2006), as well as in policy discourses (CEC 2009; Penas Lado 2016). As a way to bridge the gaps for establishing a common knowledge base, participatory research practices involving scientists and stakeholders (fishermen and others) became a cornerstone of a new research and policy agenda, to which the GAP project belongs (Mackinson et al. 2011; Mackinson and Wilson 2014). This is the context in which we want to introduce the GAP case studies in this volume – as exercises exploring participatory knowledge production in practice.

As a theoretical framework, we present three themes of social science research in the following section, pertaining to collaborative knowledge production and how that knowledge is enabled by a transformed governance system. The three themes have contributed to and reflect the transition movement which the GAP project represents. As illustrated in Fig. 2.1, they can be seen as three 'pillars' of a broader governance transformation, on which the GAP project rests. A central concept by which this movement can be captured is that of the 'scientific fisherman' described by Dubois et al. (2016), which draws on the Devon crab case study described in Chap. 3 of this volume. The 'scientific fisherman' emerges from collaborative knowledge practices and allows fishermen (or their representatives) to take on a new approach by forming alliances with scientists and managers drawing on the methods, language and materials of science. As Dubois et al. highlight, the fishermen's participation in the GAP project enabled them to gain new roles and agency as knowledge actors – in the form of 'knowledge agents' rather than of 'knowledge holders'. Instead of using scientifically accredited expertise merely as providing or disproving truth claims, the fishermen used the co-created knowledge as a "political

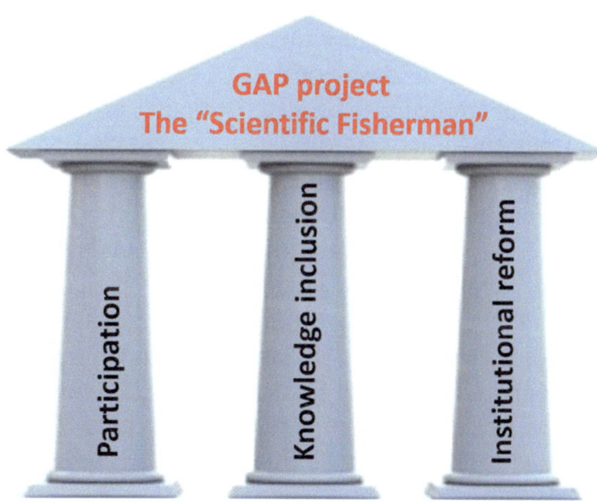

Fig. 2.1 Illustration of the three-pillared transformation of fisheries governance serving as theoretical framework for the GAP project analysis (for explication see text)

commodity to negotiate for their interests in the management arena" (ibid: 53), in relation to the sustainable use of the natural resource (Devon brown crab). However, as highlighted in Chap. 1, this book is essentially about the co-creation of knowledge among fishermen and scientist and does not explicitly address the interplay between knowledge, interests and power, despite the relevance of this nexus in several of the case studies. In Chap. 17, we will return to the scientific fisherman as a central figure emerging from the GAP project with a synthesis of lessons from the 14 case studies by using as our yardstick their contributions to the three pillars of transformation. The chapter at hand is intended to provide a background reading for Chap. 17 and aims to make the reader of this volume familiar with some key research discourses that the GAP project relates to.

2.2 Theorising GAP: Participation, Knowledge Inclusion and Institutional Reform

Fisheries management is in transition. The GAP project is both a result of that shifting policy discourse as well as an exemplification that this transition can be rendered possible in practice. Three major research issues can be identified that form tightly interconnected pillars of this transformation. First, we find that a general turn towards participation and 'principles of good governance' (COM 2001), as addressed and investigated in various social science fields, left traces also on research about fisheries governance (cf. Linke and Jentoft 2016; Griffin 2013). Second, we find a strong quest today for including knowledge from fishermen and other stakeholders in policy and management (Stephenson et al. 2016; Mackinson and Middleton 2018, Mangi et al. 2018). Third, the two first points in turn require substantial institutional reforms, the most outstanding example of these being the 2002 reform of the Common Fisheries Policy (CFP; cf. Hegland 2012; Daw and Gray 2005; Penas Lado 2016).

The three issues, or as we prefer 'pillars', are not unmated but tightly connected with each other. They are literally co-producing each other, since each of them is both dependent on developments of the other two as well as reinforcing their prospects. This mutually dependent transition of reforming fisheries management is a precondition for the improved governance situation explored in the GAP project and illustrated with the 'scientific fisherman' (Fig. 2.1). The new governance context has also been described by Röckmann et al. (2015) with the concept of an 'interaction triangle', which highlights the importance of new types and intensities of interaction between three key actor groups of fisheries management: decision-makers, scientists and other actors. Röckmann et al. argue that appropriate interactions between these three actor groups are crucial for integrating social, economic and ecological sustainability criteria for improving fisheries governance. Effective knowledge that is co-created and used in interactions among the three actor groups needs to be characterised by balancing the three knowledge criteria of credibility, legitimacy and salience. Finding the right balance (or trade-offs), to mutually improve these three criteria under recursive modes of interactions between these three actor groups is seen as an important prerequisite for more sustainable knowledge production and use in environmental governance (ibid; Cash et al. 2003; Clark et al. 2016).

2.2.1 The Turn Towards Participation

Over the last three decades, we have seen a wide-ranging turn from top-down management approaches towards more inclusive forms of 'participatory governance'. These new modes are particularly prominent in environmental governance, where new interest groups are accepted as stakeholders, and often deliberately invited to participate. The shift to participatory governance, often referred to as a 'deliberative' or 'participatory turn' (Chilvers 2009), has become a central research theme in various social sciences disciplines, particularly in Science and Technology Studies (STS) and environmental science. The turn from top-down, science-based decision-making to the novel orthodoxy of participation has been pointed out as a new 'age of participation' (Chilvers and Kearnes 2016: 2; cf. Irwin and Michael 2003). It is described as an opening-up of science and policymaking for improving democracy by including diverse forms of public engagement and stakeholder participation as well as a 'redistribution of expertise' (ibid; see also Stirling 2008; Hagendijk and Irwin 2006; Irwin 2006; Callon et al. 2009).

More recent scholarly work emphasises a need for critical investigations of the concrete methods of public participation and the implications and ambiguities that specific forms, formats, techniques and tools of participation bring about (cf. Metzger et al. 2017). Accordingly, participatory arrangements should be analysed not only in terms of their limitations and deficiencies, but also in terms of how these arrangements construct their specific subjects (Braun and Schultz 2009) and thereby enact stakeholders through particular 'performative practices' of participation (Turnhout et al. 2010).

When investigating participatory practices of collaborative research, as in the GAP case studies presented in this volume, we need to be reminded of the breadth and width of such exercises, that is the various forms, functions and objectives that participation entails (Metzger et al. 2017; Arnstein 1969; Pretty 1995). The GAP approach per se does however not engage with the broader and persisting issues of participation such as power and representation but has a more limited focus on participation of fisheries' stakeholders in the context of knowledge (co-)creation. Previous experiences from collaborative research, for example 'participatory modelling' approaches in EU fisheries research, have drawn explicit attention to the different stages of the processes: 'The appropriate stage(s) for stakeholder input in the modelling process need to be identified at an early stage. [...] To stimulate the feeling of ownership and to increase legitimacy and effectiveness, stakeholders should be involved from the *very first, the problem-framing, step*' (Röckmann et al. 2012; our emphasis). Similarly, conclude Phillipson et al. (2012, 56) from a survey on stakeholder involvement in UK research projects that 'much greater attention should be given to early processes of knowledge exchange and stakeholder engagement within the lifetime of research projects'. This sensitivity to participation in the early stages of the collaboration processes has been a key premise to the overall GAP approach. However, a particular feature of the GAP project is that it goes beyond including (early) participation in order to fix legitimacy deficits but allowing for 'real' participation in terms of joint problem definition and framing, which have been taken seriously in all case studies through close interactions between researchers and fishermen in setting up the collaborative research projects (Mackinson et al. 2015).

The GAP approach, enabling these close interactions among the participants of the collaborative research exercises in the early stages in the process, allows us to explore the performativities of these endeavours, that is the procedures that Chilvers and Kearnes (2016) refer to as 'participation in the making'. Instead of normatively embracing participation as a necessary and unproblematic step towards more democratic forms of governance and/or effective management, Chilvers and Kearnes (2016: 56) argue for 'co-productionist analyses of situated participatory experiments and practices'. Despite considerable efforts in studying how the 'participatory turn' renders science and policymaking more open, transparent and accountable (cf. Stirling 2008), 'it is striking therefore to note that the pragmatics of public participation with science and the environment have received relatively little concerted analyses in ways that deploy the tools of situated interpretive and co-productionist analyses' (Chilvers and Kearnes 2016: 5). The co-productionist type of analysis suggested by Chilvers and Kearnes implies focusing on participation as 'collective experimental practices in the making' (ibid: 15) and accepts that the outcomes of various formats, ideals, normativities and techniques of participation are not pregiven, but instead emerge within the performance of the participatory practices themselves.

What does this imply for our investigation of the collaborations in the GAP project? The perspective provided by Chilvers and Kearnes' discussion on remaking participation appears relevant for our analyses of the participatory research exercises performed in the GAP case studies. Their account of rethinking participation, and its externalised democratic norms and taken-for-granted methodological and theoretical assumptions, call for an exploration how we can approach and analyse the participatory research exercises of the GAP project. This means that from a co-productionist perspective, the cases are not merely testing how participatory research works in practice. Instead, they present examples, which in themselves shape new realities of stakeholder interaction, knowledge co-creation and how the knowledge credentials of credibility, legitimacy and saliency are (co-)produced in such settings. Hence, what participatory research does in the cases of the GAP project is not simply to provide new arenas, platforms or responsibilities for legitimising management through participation. It also allows for a reconfiguration of established actor roles, for example through the emergence of new types of actors such as the 'scientific fisherman' introduced above. Throughout Europe, we can today find an increasing trend where (primarily large-scale pelagic) fishing industries hire scientists to equip themselves with a proper scientific background, a phenomenon that has earlier been described as a 'communicative turnaround' evolving from a shared burden of proof in EU fisheries governance (Linke and Jentoft 2013).

Another analytic tool that appears useful to understand and get to grips with the intricacies of the collaborative practices attempted in the GAP case studies is the concept of 'boundary objects'. Originally the concept has been coined by Star and Griesemer (1989) to understand how collaboration is possible among actors from different backgrounds holding divergent views. Boundary objects according to them are 'both adaptable to different viewpoints and robust enough to maintain identity across them' (ibid: 387). The concepts of boundary objects and boundary work (the practice of negotiating boundaries) have recently been used in a fisheries context by Kari Stange (2017) to investigate knowledge exchange in stakeholder-led initiatives for producing management plans in EU fisheries management (see in particular Stange et al. 2016). We will return to the use and applicability of boundary work and boundary objects in the GAP context in Chap. 17.

2.2.2 Including Knowledge: Democratising Expertise

Coinciding with the 'participatory turn' described above, expert knowledge for policy use is increasingly affected by demands for justification beyond scientific means: through participation of stakeholders or so-called lay experts in procedures of knowledge production and advice (Irwin and Michael 2003; Lidskog 2008; Irwin and Horst 2016). This opening up of scientific authority and procedures has been referred to as a 'democratization of expertise' (Maasen and Weingart 2005). It implies not only that the scientific community is held more accountable for the societal use and utility of its knowledge production but also involves a shift from 'a legitimation through knowledge to a legitimation through participation' (ibid: 2; cf. COM 2001).

The turn towards stakeholder participation and the democratisation of expertise has been discussed intensively in the context of the provision of scientific advice for political decision-making (e.g. Jasanoff 1990; Carolan 2006; Lentsch and Weingart 2011; Pielke 2007). STS scholars like Sheila Jasanoff have emphasised the specific constraints arising for science in applied contexts, for which she invented the label 'regulatory science' (Jasanoff 1990). Strassheim and Kettunen (2014: 265) emphasise that the use of science advice has today 'become an integral and increasingly controversial part of policy making'. The controversial part relates to the difficulties of upholding an idea of basic, pure and objective science in applied contexts, where the provision of scientific expertise needs to fit specific, predefined policy requirements, as for example is the case with the institutional interface between scientific advice and policymaking in fisheries management described below. As noted already 30 years ago by Dorothy Nelkin, an apparent irony lies in the idea of scientific objectivity and its concurrent usefulness for policy: '… the greater the utility of science in political affairs, the less it can maintain its image of objectivity that has been the very source of its political value' (Nelkin 1987: 293; for similar discussions see Weinberg 1972; Collingridge and Reeve 1986; Yearley 2005: 160ff). The paradox noted by Nelkin has become a central research issue for STS scholarship. For example Bijker et al. (2009), in exploring the context of applied science, coined the notion of a 'paradox of scientific authority' relating to the basic question 'how can scientific advice be effective and influential in an age in which the status of science and/or scientists seems to be as low as it has ever been?' (Bijker et al. 2009: 1). And 'how can scientific advice still have some authority when developments in political culture have eroded the stature of so many classic institutions, and when STS research has demonstrated the constructed nature of scientific knowledge?' (ibid: 6). Departing from such observations, these authors raise a fundamental research question pertaining to the new governance perspective of our times: 'How can scientific advice still play a role in the democratic governance of technological cultures, where participation by citizens and by stakeholders increasingly complements the old institutional mechanisms of democracy? What is the new 'place for science advice' within such new arrangements for governance?' (ibid: 6).

The GAP project invites us to explore this 'new place for science advice' in the context of collaborative knowledge production between researchers and fisheries practitioners. As we discuss further in Chap. 17, the fisheries governance transition of which GAP is part of also represents a move towards a nested system design, where the top-down features of the TAC machine (like the CFP) are kept but can be extended with layers of localised units that allow capturing the diversity and complexity at the local level (Wilson 2009: 267). This is in fact the knowledge gap, which the collaborative research approach of GAP intends to fill, by activating the fine-grained, locally situated knowledges 'from the ground' for applications at higher scales. How governance systems and their partly international jurisdiction can be attentive to local levels and issues of scale are important parameters for whether institutional frameworks are conducive to bridge these knowledge gaps. As highlighted by Degnbol and Wilson (2008), this handling of complexity requires nested institutional structures that are linked across scales to enable possibilities for negotiation between actors at different levels. Important case study research, both from fisheries (e.g. Wilson et al. 2003) and other areas (e.g. Wynne 1992), have revealed the failures of ignoring local, situated knowledges and hence showed the limitations of exclusive scientific management approaches. The transition clarified with the GAP project and specifically with the 'scientific fisherman', represents a move beyond such knowledge divisions and allows to explore and authorize the situated knowledge of fishermen (see Chap. 17 for further discussion).

How then can such collaborations improve the knowledge base for decision-making in fisheries management? Through the case studies of participatory knowledge production in the GAP project, we aim to connect empirical observations to the wider questions raised in this chapter about democratising expertise, knowledge integration and participation in contemporary society. We are interested in how these processes work in practical applications, and how they may impact on the robustness of the resulting knowledge-base, intended for use in management and decision-making. In other words, we want to probe the processes as well as the outcomes of the participatory research exercises conducted in the 14 GAP case studies presented in this volume: What are the knowledge gaps that the GAP case studies are constructed to fill? Why are they not addressed by conventional designs? Can they be bridged through collaborative research? What characterises the relationship between scientists and stakeholders within the collaborative research projects? To what extent can collaborative research remedy the legitimacy deficits created by unresponsive management practices? Do the GAP case studies represent new modes of science–society relations, or do they reproduce a conventional and deferential relationship between science and lay clients? Are the case studies sites where scientists get access to new platforms for pursuing scientific research? Or are they arenas where fishermen get access to the resources of science for their own purposes? The overall GAP approach leads us to think that both of these alternatives can be fulfilled in practice – that is that collaborative research sometimes proves to be beneficial to both the scientists' as well as to the fishermen's interests. At the same time, it is relevant to assess how and at what degree these processes construct their specific subjects, like the 'scientific fishermen' (see above and Chap. 17).

On the one hand, these questions are of a more overarching nature, relevant for pursuing research on co-creating knowledge, governance transitions and science–society relations in general. However, on the other hand, the questions are pertinent to the overall GAP project and its case studies and will therefore be taken up and to some extent answered in Chap. 17, while a lot of more analytical research of course remains to be done in this domain.

2.2.3 *Institutional Reforms: From Top-Down Control Towards Recursive Interaction*

Just as the two previously mentioned themes require substantial institutional reform in order to take effect, they have concurrently also impacted on fisheries management discourses and the institutional reforms conducted over the last decades. Fisheries management structures and institutions have adapted to requirements of public participation and stakeholder involvement in policy, management and decision-making (Jentoft and McCay 1995; Kaplan and McCay 2004; St. Martin et al. 2007; Griffin 2013). A growing research agenda of 'fisheries social science' investigates these shifting governance perspectives by focusing on the consequences of ongoing transitions from top-down towards more participatory arrangements (e.g. Urquhart et al. 2014; Symes 2006; Symes and Hoefnagel 2010; Mackinson et al. 2011; Griffin 2013; Linke and Jentoft 2016). A part of this scholarly work focuses specifically on the social and institutional dimensions of knowledge interaction, the practices of knowledge inclusion and stakeholder's contributions to fisheries governance (e.g. Holm 2003; Holm and Soma 2016; Linke et al. 2011; Linke and Jentoft 2013, 2014; Griffin 2013, Mackinson and Middleton 2018).

In the EU, the turn towards increased participatory governance appears most pronounced with the establishment of EU Advisory Councils (ACs) as a product of the 2002 CFP reform (Penas-Lado 2016; Linke et al. 2011; Linke and Jentoft 2016;

Hatchard and Gray 2014). This governance shift takes place simultaneously with a stated wish for a transition from a single fish stock management approach towards an ecosystem perspective aiming to implement the Ecosystem-Based Approach to Fisheries (EBAF; cf. Garcia 2010). Connected to the EBAF, we find a new emphasis on maritime spatial planning (MSP), the tool which has been employed to organise the increasing interests for the use of the marine realm such as renewable energies or offshore oil and other mineral explorations as well as the expansion of aquaculture, interests which are expected to be augmented by the further establishment of the political agenda of 'Blue Growth' (EC 2017; see Arbo et al. 2018). While these new and emerging industries hold great potential for economic prosperity, they also bring about new conflicts for fisheries with regard to environmental and social challenges (Jentoft 2017). The joint trajectories of the CFP reform process to open up science–policy interactions for stakeholder involvement, the implementation of an EBAF, and the movement towards MSP under the Blue Growth paradigm expose new complexities for producing socially robust and relevant knowledge for policy- and decision-making (Ramirez-Monsalve et al. 2016a, b; Röckmann et al. 2015; Ballesteros et al. 2017; Mackinson and Middleton 2018). Shifting from the narrow management object of single fish stocks towards the more holistic management objectives of an EBAF requires a revision of the traditional, linear, annual management approach of fisheries, described earlier in this chapter as the 'TAC machine' (Holm and Nielsen 2004). The linear conception of science and policymaking in fisheries management, implying a clear boundary between the two domains, has today come under pressure with the shift to participation, and the democratisation of expertise, adding additional layers of complexity on the traditional science–policy interface. As mentioned above, these complexities are inserted from a broadening ecological perspective (instead of a single fish stock approach) as well as by a stronger commitment to the economic and social dimensions of sustainability under the new governance modes of fisheries. The shift therefore requires a linkage between the established science–policy procedures of the TAC machine, and new, more 'recursive interactions' (Weingart 1999) between science and other societal actors including their respective interests (Schwach et al. 2007; Ramirez-Monsalve et al. 2016a). As suggested above, this linkage can be envisioned with a so-called nested-system approach, as described by Doug Wilson. It would imply an arrangement of different spheres, organised like a Russian doll, in order to deal more appropriately with the layered dimensions of social, economic and ecological complexities (Wilson 2009: 276–79).

One example of an attempt to adapt fisheries management to these multifaceted transition requirements is currently pursued through the tool of Multiannual Plans (MAPs), inscribed in the recent CFP reform (Articles 9,10, cf. Ramirez-Monsalve et al. 2016b; Penas Lado 2016). MAPs are intended to include and achieve at least some objectives of ecosystem-based management, whereas a more fully developed EBAF framework faces more serious institutional challenges (Dickey-Collas 2014; Ramirez-Monsalve 2016b; Mackinson and Middleton 2018). However, one procedure facilitating the establishment of MAPs via recursive interactions is emerging with the participatory research practices between fishermen and scientists as conducted in the GAP project. Such collaborative exercises have multiple roots linked, on the one hand, to the idea of participatory governance in general as fetched out above and, on the other hand, linked to processes of including fisher's knowledge more thoroughly in policy and decision-making (cf. Hegland and Wilson 2009; Stange et al. 2016, Stange 2017; Röckmann et al. 2012). Another distinct root of participatory governance in fisheries lies in the concept of co-management. While we find a range of experiences and changing approaches to co-management in Europe (Linke and Bruckmeier 2015), the concept has perhaps most clearly been defined by Symes (2006: 113) as: 'systems in which responsibility for management is shared between the state and user groups, usually at the local level'. In bringing together the issues of stakeholder participation, co-management and the inclusion of fishermen's knowledge, the GAP approach both exemplifies and explores new types of questions that ought to be addressed to interactions between different actors under the reformed governance context.

One such a question is: How do the traditional formalised roles and functions of the three knowledge credentials, credibility, legitimacy and saliency, play out under the new recursive modes of interaction within a reformed management system. In a traditional (linear) view of science–society relations, they represent separate sources of authority (Bijker et al. 2009: 24ff; Cash et al. 2003; Wilson 2009). Under a new governance context, as explored in the GAP project, however, the boundaries between the three criteria become increasingly blurred and open to negotiation and interpretation among an increased number of actors. This makes a clear separation of their specific effects and functioning more difficult and requires trade-offs between them (Sarkki et al. 2014).

With respect to the nested system perspective mentioned above, we can imagine a layered approach of knowledge activation that includes the new challenges relating to the different scales which the overall GAP project approach exemplifies (for the span of the GAP cases, see Table 2.1 below): First, at the local level, scientists are brought in to explore and authorise local knowledge claims (ensuring credibility). Since this knowledge would need to be acted upon at some higher level (for legitimacy reasons), however, this may introduce new tensions. This can be referred to as a 'management wall' problem, that is that agreed credible and legitimised knowledge from lower levels does not lead to improved management actions because the higher-order system (e.g. TAC machine/CFP) is not capable of utilising such 'best available knowledge'. This might increase legitimacy problems, because the local knowledge claims are becoming more potent from being authorized by science: though participatory research. This can also imply a problem of saliency in reverse, since there is currently no obvious and predefined policy use for the activated local knowledge claims in the higher-order system of the TAC machine (see Chap. 17 for further elaborations on the 'management wall' problem).

The reformed CFP has been questioned with regard to the extent to which the ACs fulfil the purpose of such a layered knowledge activation process and for empowering stakeholders for participating in responsible ways in management and decision-making (Griffin 2013; Hatchard and Gray 2014; Linke and Jentoft 2013, 2014, 2016). The management wall appears to be made quite of concrete here. Emerging practices of participatory research on the other hand imply novel stages of stakeholder interactions and communication (Röckman et al. 2015), and can hence reveal the dynamics of knowledge production and interaction as they unfold. This calls for a new examination of how credibility, legitimacy and saliency of knowledge are co-produced, renegotiated and newly aligned under the recursive dynamics of participatory research.

As this suggests, the GAP approach features a model of interaction that fits nicely with key principles and perspectives driving the current reform trend in fisheries governance. The GAP project is a practical exemplification of the turn towards participation and knowledge inclusion described above. However, while this conclusion is an important starting point for analysing the GAP case study experiences, it does not guarantee a successful outcome. To make such a judgement, we must return to the questions with which we started and try to examine them from the perspective of nested systems: At which level can the knowledge gaps among science, policy and stakeholders be bridged appropriately through the collaborative research practices? And how can these practices, judged by the practical experience in the case studies, help in solving problems of unsustainable management and/or legitimacy deficits created by separating science, policy and stakeholders at higher levels? More specifically, do the case studies represent recursive knowledge practices, as the theoretical discourses reviewed above lead us to expect? And if so, at which level do they unfold and where in the system do they make (most) sense? Or do they merely end up reproducing a conventional assignment of roles among scientists, stakeholders and policymakers? While some of these questions are addressed again in Chap. 17, they urge us, particularly in those cases where the answers will begin with 'it depends…', to focus on the conditions that make a difference. How exactly are the case study projects related to reform processes? Are they directly linked to and/or informed by the reform? Or do they remain peripheral to such efforts?

2.3 The GAP Approach and Its Variability

In recent academic and policy discussions about fisheries management, we find a strong normative commitment to the issues discussed above, namely, that fishermen's knowledge has been ignored while it should be included, that participation has been too weak while it should be improved and that reforms, while pointing in the right direction, are too slow and too weak. The GAP project as such is centrally embedded in this normative discourse rather than strictly explorative. Its purpose was not primarily to analytically explore, understand and qualify the 'gap', but rather to demonstrate and challenge it in actual practice. While we as GAP participants share this view, the sentiment of this book is somewhat different. For the purpose of this book, the main focus is to suspend the normative commitment and attempt to be more analytical by relating to the three conceptual pillars and the resulting research questions fetched out above. What can the experiences from the GAP project tell us about participation, knowledge inclusion and institutional reform? The main empirical sources on which this book draws are the 14 GAP case studies (Table 2.1). In

Table 2.1 Overview of the GAP case studies

Chapter	Short Name	Title	Country
Chapter 3	Sustainability of brown crab fishery	Fishermen and scientists in the same boat: A story of collaboration in the UK south Devon crab fishery	United Kingdom
Chapter 4	Selectivity in Lake Vättern	Getting choosy about Whitefish in Lake Vättern: Using participatory approaches to improve fisheries selectivity	Sweden
Chapter 5	Mapping habitats and fishing	Understanding common collaboration in Galician small-scale fisheries: Validating a methodological toolbox through a process-oriented approach	Spain
Chapter 6	Management of herring	Information is the jam of the Western Baltic Herring sandwich: Bridging gaps between policy, stakeholders and science	Denmark
Chapter 7	Rare Wadden sea species	Aiming for by-catch: Collaborative monitoring of rare and migratory species in the Wadden Sea	Germany
Chapter 8	Fishing and habitats in the northern Adriatic Sea	The Italian Job: Navigating the (im)perfect storm of participatory fisheries research in the Northern Adriatic Sea	Italy
Chapter 9	Fishery monitoring for coastal cod	Trapped in the TAC Machine: Making a fisheries-based indicator system for coastal cod in Steigen, Norway	Norway
Chapter 10	Management of NW Mediterranean red shrimp	When fishermen take charge: The development of a management plan for the red shrimp fishery in Mediterranean Spain	Spain
Chapter 11	Multispecies and mixed fisheries in the North Sea	Does slow-burn collaboration deliver results? Towards collaborative development multi-annual multi-species management plans in North Sea mixed demersal fisheries	United Kingdom
Chapter 12	FADs in tuna fisheries	Action research in tropical tuna purse seine fisheries: Thoughts and perspectives	Spain/ France
Chapter 13	Baltic fisheries and Marine Spatial Planning	From planning for society to planning with society: Integration of coastal fisheries into the Maritime Spatial Planning	Estonia
Chapter 14	Discard sampling for flatfish fisheries	Implementing the landing obligation: An analysis of the gap between fishermen and policymakers in the Netherlands	Netherlands
Chapter 15	The Maltese Fisheries Management Zone	Taking the initiative on Maltese Trawl Industry Management: Industry and science collaboration on identifying nursery and spawning areas for Maltese trawl fisheries target species	Malta
Chapter 16	By-catch and discards of elasmobranchs	People, sharks and science: What can it take for industry-led research to make a difference to the management of elasmobranchs of conservation concern in UK waters?	United Kingdom

order to understand the outcomes and significance of them, we need to explore the process by which they were developed and look at the institutional context in which they are embedded in (this is done in Chap. 17).

All the GAP case studies are about collaboration, and although they express great variation in scope and maturity, they all aim to establish bridges across important divides, in particular those of knowledge between fisheries scientists and fishermen.

The process of initiating participatory research can have important bearings on how the details of the work materialises (Chuenpagdee and Jentoft 2007). With acute awareness of this, GAP was organised in two phases: GAP 1 identified shared needs for research and mobilised regional teams of researchers and fisheries stakeholders, and GAP 2 designed and carried out the research. This process has important consequences for the cross-cutting analysis of the knowledge dynamics of co-construction and delivery process among the case studies, because they were not strategically designed to study these processes as a single collective. Nevertheless, the emergent features and dynamics sit comfortably in the theoretical framework presented above with the three pillars and bring it to life with practical meanings, as portrayed most remarkably with the 'scientific fisherman' (see above and Chap. 17).

The case study selection process emphasised that the teams themselves identified and designed their project, a process which for many began in 2003 when the GAP idea was first conceived and scientists reached out to fishermen to form the embryos of collaboration through a co-design process. The 'selection' and formation of the case studies were initiated during the proposal writing stage and were sufficiently strong to withstand several knockbacks before funding finally came in 2008.

From the perspective that we study these process dynamics throughout the active life of the case studies, this implies some 'bias' that is important to acknowledge. For most of the cases, scientists took the initiative to instigate the joint research activities. Project development and review processes are hence deeply rooted in scientific culture and practice. At the same time, the case studies were challenging conventional approaches through their efforts to involve non-scientists in roles normally reserved for scientists. In the absence of institutionalised models for doing participatory research, the science partners were forced to take on chief roles for conducting the projects, being case study leaders and serving as writers, rapporteurs and communicators of the results. Thus, while the GAP project certainly is committed to the values of equal partnership and collaboration, such ideals are to some extent contradicted by the basic requirements on GAP as a research project as such.

However, this way of 'compromising' was a pragmatic solution to make the partnerships work when faced with barriers that would otherwise prevent stakeholder's participation in EU research projects in any meaningful way (Mackinson et al. 2011). Indeed, the GAP project could only be realised through science partners who were not, or at least not only, acting as gatekeepers of the conventional approach. Instead, they were committed to a different view, open to a more inclusive and responsive science ideal, which also provides opportunities for an educational dimension for scientists. In the same way, the fishermen and other stakeholder partners in GAP are not representative of those on the far sides of the gap, who have already concluded that scientists are biased against them. Hence, whereas one would perhaps wish for a project to explore the gaps in communication and understanding where they are at their deepest, this is not how the GAP project actually was set up and worked out. Instead, the GAP teams comprised partners that were all ready and open to collaborate, and the case studies hence featured situations where bridges were already in place.

As Table 2.1 indicates, the GAP case studies, all representing individual experiments and 'research in the wild' (Callon and Rabeharisoa 2003), vary tremendously in scale, complexity, ambition, resources, effectiveness, issues, financial values and the grounds being covered. This variation of the case studies is in itself a cause for reflection, because even the small sample of cases presented here indicates a massive amount of variability across European fisheries. This is an important message when engaging in fisheries issues, either through the lens of research or governance. We must therefore avoid generalisation and simplification and recognise the specific contexts that influence the performance of each individual case study. This variability will be taken up as an important element in our attempt to synthesise lessons from the GAP project in Chap. 17.

References

Arbo P, Knol M, Linke S, St. Martin K (2018) The transformation of the oceans and the future of marine social science. Maritime Stud 17(117). https://doi.org/10.1007/s40152-018-0117-5

Arnstein SR (1969) A ladder of citizen participation. J Am Inst Plann 35(4):216–224

Ballesteros M, Chapela R, Ramírez-Monsalve P, Raakjaer J, Hegland TJ, Nielsen KN, Laksá U, Degnbol P (2017) Do not shoot the messenger: ICES advice for an ecosystem approach to fisheries management in the European Union. ICES J Mar Sci 75(2):519–530. https://doi.org/10.1093/icesjms/fsx181

Bijker WE, Bal R, Hendriks R (2009) The paradox of scientific authority. MIT Press, Cambridge, MA

Braun K, Schultz S (2009) "… a certain amount of engineering involved": constructing the public in participatory governance arrangements. Public Underst Sci 19(4):403–419

Callon M, Rabeharisoa V (2003) Research 'in the wild' and the shaping of new social identities. Technol Soc 25:193–204

Callon M, Lascoumes P, Barthe Y (2009) Acting in an uncertain world: an essay on technical democracy. MIT Press, Cambridge, MA

Carolan MS (2006) Science, expertise, and the democratization of the decision-making process. Soc Nat Resour 19(7):661–668

Cash DW, Clark WC, Alcock F, Dickson NM, Eckley N, Guston DH, Jäger J et al (2003) Knowledge systems for sustainable development. Proc Natl Acad Sci U S A 100(14):8086–8091

CEC (2009) Commission of the European communities green paper. reform of the common fisheries policy. 163 final. CEC, Brussels

Chilvers J (2009) Deliberative and participatory approaches in environmental geography. In: Castree N, Demeritt D, Liverman D et al (eds) A companion to environmental geography. Wiley-Blackwell, Chichester, pp 400–417

Chilvers J, Kearnes M (2016) Science, democracy and emergent publics. In: Chilvers J, Kearnes M (eds) Remaking participation: science, environment and emergent publics. Routledge, London, pp 1–28

Chuenpagdee R, Jentoft S (2007) Step zero for fisheries co-management: what precedes implementation. Mar Policy 31:657–668
Clark WC, van Kerkhoff L, Lebel L, Gallopin GC (2016) Crafting usable knowledge for sustainable development. Proc Natl Acad Sci U S A 113(17):4570–4578
Collingridge D, Reeve D (1986) Science speaks to power: the role of experts in policymaking. St Martin's Press, New York
COM (2001) European governance: a white paper. Commission of the European Communities COM 428 final. Commission of the European Communities, Brussels
Daw T, Gray T (2005) Fisheries science and sustainability in international policy: a study of failure in the European Union's common fisheries policy. Mar Policy 29(3):189–197
Degnbol D, Wilson DC (2008) Spatial planning on the North Sea: a case of cross-scale linkages. Mar Policy 32:189–200
Dickey-Collas M (2014) Why the complex nature of integrated ecosystem assessments requires a flexible and adaptive approach. ICES J Mar Sci 71:1174–1182
Dubois M, Hadjimichael M, Raakjær J (2016) The rise of the scientific fisherman: Mobilising knowledge and negotiating user rights in the Devon inshore brown crab fishery, UK. Mar Policy 65:48–55
European Commission. (2017) Blue growth. Available at: https://ec.europa.eu/maritimeaffairs/policy/blue_growth_en
Finlayson AC (1994) Fishing for truth. A sociological analysis of northern cod stock assessments from 1977–1990. ISER, St. John's
Garcia SM (2010) Governance, science, and society: the ecosystem approach to fisheries. In: Quentin Grafton R, Hilborn RQ, Squires R, Tait D, Williams M (eds) Handbook of marine fisheries conservation and management. Oxford University Press, New York, pp 87–98
Gezelius S (2008) The arrival of modern fisheries management in the North Atlantic: a historical overview. In: Gezelius S, Raakjaer J (eds) Making fisheries management work. Springer, London, pp 27–40
Gray TS (2005) Participation in fisheries governance. Springer, Dordrecht
Griffin L (2013) Good governance, scale and power: a case study of North Sea fisheries. Routledge, New York
Hagendijk R, Irwin A (2006) Public deliberation and governance: engaging with science and technology in contemporary Europe. Minerva 44:167–184
Hatchard J, Gray T (2014) From RACs to advisory councils: lessons from North Sea dis- course for the 2014 reform of the European Common Fisheries Policy. Mar Policy 47:87–93
Hegland TJ (2012) Fishing for change in EU governance: excursions into the evolution of the Common Fisheries Policy. PhD thesis, Aalborg University
Hegland TJ, Wilson DC (2009) Participatory modelling in EU fisheries management: western Horse Mackerel and the Pelagic RAC. Maritime Stud 8(1):75–96
Hoefnagel E, Burnett A, Wilson DC (2006) The knowledge base of co-management. In: Motos L, Wilson DC (eds) The Knowledge base for fisheries management. Elsevier, Amsterdam, pp 85–108
Holm P (2003) Crossing the border: on the relationship between science and fishermen's knowledge in a resource management context. Maritime Stud 2(1):5–33
Holm P, Nielsen KN (2004) The TAC machine. In report of the working group on fishery systems. WGFS annual report. ICES, Copenhagen, pp 40–51
Holm P, Soma K (2016) Fishers' information in governance: a matter of trust. Curr Opin Environ Sustain 18:115–121
Hubbard J (2012) Changing regimes: governments, scientists and fishermen and the construction of fisheries policies in the North Atlantic 1850–2010. In: Starkey DJ, Heidbrink I (eds) From the 1850s to the early twenty-first century. A history of the North Atlantic fisheries. Verlag H.M. Hauscihld, Bremen, pp 145–147
Irwin A (2006) The politics of talk: coming to terms with the 'new' scientific governance. Soc Stud Sci 36(2):299–320

Irwin A, Horst M (2016) Engaging in a decentred world: overflows, ambiguities and the governance of climate change. In: Chilvers J, Kearnes M (eds) Remaking participation: science, environment and emergent publics. Routledge, London, pp 64–80
Irwin A, Michael M (2003) Science, social theory and public knowledge. Open University Press, Maidenhead
Jasanoff S (1990) The fifth branch: science advisors as policymakers. Harvard University Press, Cambridge, MA
Jentoft S (2017) Small-scale fisheries within maritime spatial planning: knowledge integration and power. J Environ Policy Plan 19(3):266–278
Jentoft S, McCay B (1995) User participation in fisheries management: lessons drawn from international experiences. Mar Policy 19(3):227–246
Kaplan IM, McCay B (2004) Cooperative research, co-management and the social dimension of fisheries science and management. Mar Policy 28:257–258
Khalilian S, Froese R, Proelss A, Requate T (2010) Designed for failure: a critique of the common fisheries policy of the European Union. Mar Policy 34(6):1178–1182
Lentsch J, Weingart P (2011) Introduction: the quest for quality as a challenge to scientific policy advice: an overdue debate? In: Lentsch J, Weingart P (eds) The politics of scientific advice. Cambridge University Press, Cambridge, MA, pp 3–18
Lidskog R (2008) Scientised citizens and democratised science. Re-assessing the expert-lay divide. J Risk Res 11(1):69–86
Linke S, Bruckmeier K (2015) Co-management in fisheries: experiences and changing approaches in Europe. Ocean Coast Manag 104:170–181
Linke S, Jentoft S (2013) A communicative turnaround: shifting the burden of proof in European sheries governance. Mar Policy 38:337–345
Linke S, Jentoft S (2014) Exploring the phronetic dimension of stakeholders' knowledge in EU sheries governance. Mar Policy 47:153–161
Linke S, Jentoft S (2016) Ideals, realities and paradoxes of stakeholder participation in EU sheries governance. Environ Sociol 2(2):144–154
Linke S, Dreyer M, Sellke P (2011) The regional advisory councils: what is their potential to incorporate stakeholder knowledge into fisheries governance? Ambio 40:133–143
Maasen S, Weingart P (eds) (2005) Democratization of expertise? Exploring novel forms of scientific advice in political decision-making. Springer, Dordrecht
Mackinson S, Middleton D (2018) Evolving the ecosystem approach in European fisheries: transferable lessons from New Zealand's experience in strengthening stakeholder involvement. Mar Policy 90:194–202
Mackinson S, Nøttestad L (1998) Combining local and scientific knowledge. Rev Fish Biol Fish 8(4):481–490
Mackinson S, Wilson DCK (2014) Building bridges among scientists and fishermen with participatory action research. In: Urquhart J, Acott T, Symes D, Zhao M (eds) Social issues in sustainable fisheries management. Springer, Dordrecht, pp 121–139
Mackinson S, Wilson DC, Galiay P, Deas B (2011) Engaging stakeholders in fisheries and marine research. Mar Policy 35:18–24
Mackinson S, Raicevich S, Kraan M, Magudia R, Borrow K (eds) (2015) Good practice guide: participatory Research in Fisheries Science. http://gap2.eu/outputs/pr-handbook/
Mangi S, Kupschus S, Mackinson S, Rodmell D, Lee A, Bourke E, Rossiter T, Masters J, Hetherington S, Catchpole T, Righton D (2018) Progress in designing and delivering effective fishing industry-science data collection in the UK. Fish Fish 19:622–642
Metzger J, Soneryd L, Linke S (2017) Enacting legitimate concerns: an agnostic approach to stakeholder participation in planning processes. Environ Plan A 49(11):2517–2535
Neis B (1992) Fishers' ecological knowledge and stock assessment in Newfoundland. Nfld Stud 8(2):155–178
Nelkin D (1987) Selling science: how the press covers science and technology. WH Freeman, New York

Nielsen KN (2008) Science|politics: boundary construction in mandated science – the case of ICES' advice on fisheries management. PhD Dissertation, University of Tromsø
Nielsen K, Holm P (2007) A brief catalogue of failures: framing evaluation and learning in fisheries resource management. Mar Policy 31:669–680
Penas Lado E (2016) The common fisheries policy: the quest for sustainability. Wiley Blackwell, Oxford
Phillipson J, Lowe P, Proctor A, Ruto E (2012) Stakeholder engagement and knowledge exchange in environmental research. J Environ Manag 95:56–65
Pielke RA Jr (2007) The honest broker: making sense of science in policy and politics. Cambridge University Press, Cambridge, MA
Pretty JN (1995) Participatory learning for sustainable agriculture. World Dev 23(8):1247–1263
Ramírez-Monsalve P, Raakjær J, Nielsen KN, Laksá U, Danielsen R, Degnbol D, Ballesteros M, Degnbol P (2016a) Institutional challenges for policy-making and fisheries advice to move to a full EAFM approach within the current governance structures for marine policies. Mar Policy 69:1–12
Ramírez-Monsalve P, Raakjær J, Nielsen KN, Santiago JL, Ballesteros M, Laksá U, Degnbol P (2016b) Ecosystem Approach to Fisheries Management (EAFM) in the EU – current science-policy-society interfaces and emerging requirements. Mar Policy 66:83–92
Röckmann C, Ulrich C, Dreyer M, Bell E, Borodzicz E, Haapasaari P, Hiis Hauge K, Howell D, Mäntyniemi S, Miller D, Tserpes G, Pastoors M (2012) The added value of participatory modelling in fisheries management – what has been learnt? Mar Policy 36:1072–1085
Röckmann C, van Leeuwen J, Goldsborough D, Kraan M, Piet G (2015) The interaction triangle as a tool for understanding stakeholder interactions in marine ecosystem based management. Mar Policy 52:155–162
Sarkki S, Niemela J, Tinch R, van den Hove S, Watt A, Young J (2014) Balancing credibility, relevance and legitimacy: a critical assessment of trade-offs in science–policy interfaces. Sci Public Policy 41:194–206
Schwach V, Bailly D, Christensen AS, Delaney AE, Degnbol P, van Densen WL, Holm P, McLay HA, Nolde Nielsen K, Pastoors MA, Reeves SA, Wilson DC (2007) Policy and knowledge in fisheries management: a policy brief. ICES J Mar Sci 64(4):789–803
St. Martin K, McCay B, Murray G, Johnson T, Oles B (2007) Communities, knowledge, and fisheries of the future Int. J Glob Environ Issues 7(2/3):221–239
Stange K (2017) Knowledge production at boundaries: an inquiry into collaborations to make management plans for European fisheries. PhD Dissertation. University of Wageningen
Stange K, van Leeuwen J, van Tatenhove J (2016) Boundary spaces, objects and activities in mixed-actor knowledge production: making fishery management plans in collaboration. Maritime Stud 15:14. https://doi.org/10.1186/s40152-016-0053-1
Star SL, Griesemer JR (1989) Institutional ecology, 'translations' and boundary objects: Amateurs and professionals in Berkeley's Museum of Vertebrate Zoology, 1907–39. Soc Stud Sci 19(3):387–420
Stephenson RL, Paul S, Pastoors MA, Kraan M, Holm P, Wiber M, Mackinson S, Dankel DJ, Brooks K, Benson A (2016) Integrating fishers' knowledge research in science and management. ICES J Mar Sci 6(1):1459–1465
Stirling A (2008) "Opening up" and "closing down" – power, participation, and pluralism in the social appraisal of technology. Sci Technol Hum Values 33:262–294
Strassheim H, Kettunen P (2014) When does evidence-based policy turn into policy-based evidence? configurations, contexts and mechanisms. Evid Policy 10(2):259–277
Symes D (2006) Fisheries governance: a coming of age for fisheries social science? Fish Res 81:113–117
Symes D, Hoefnagel E (2010) Fisheries policy, research and the social sciences in Europe: challenges for the 21st century. Mar Policy 34:268–275
Turnhout E, van Bommel S, Aarts N (2010) How participation creates citizens: participatory governance as performative practice. Ecol Soc 15(4):26

Urquhart J, Acott T, Symes D, Zhao M (2014) Social issues in sustainable fisheries man- agement. Springer, Dordrecht
Weinberg A (1972) Science and trans-science. Minerva 10:209–222
Weingart P (1999) Scientific expertise and political accountability: paradoxes of science in politics. Sci Public Policy 26(3):151–161
Wilson DC (2009) The paradoxes of transparency: science and the ecosystem approach to fisheries management in Europe. Amsterdam University Press, Amsterdam
Wilson DC, Raakjær Nielsen J, Degnbol P (2003) The fisheries co-management experience accomplishments, challenges and prospects. Springer, Dordrecht
Wynne B (1992) Misunderstood misunderstanding: social identities and public uptake of science. Public Underst Sci 1:281–304
Yearley S (2005) Making sense of science understanding the social study of science. SAGE, London

Chapter 3
Fishermen and Scientists in the Same Boat. A Story of Collaboration in the UK South Devon Crab Fishery

Emma Pearson, Ewan Hunter, Alan Steer, Kevin Arscott, and Paul J. B. Hart

Abstract The collaboration that forms the basis of the UK edible crab (*Cancer pagurus*) case study builds on a relationship between scientists and fishermen that was first established in 1996. The emphasis of the case study has been to develop awareness among fishermen of the need to be more involved in the management of the resource on which their livelihoods depend. To engage fishermen in management, we have worked together towards the development of an Individual-Based Model (IBM) of the south Devon crab fishery. The model replicates the dynamics of the fishery with crabs of varying size classes migrating into the exploited area and being removed from the area either as catch, natural mortality or emigration. The interplay between these factors can eventually be used to determine the level of fishing effort the fishery will sustain, once a good stock–recruit function is available. The ultimate aim is to enable fishermen to collect their own catch data and to use it together with the model to estimate a sustainable exploitation rate. This chapter describes how fishermen and scientists have worked in partnership during the case study. Initially, fishermen of the South Devon and Channel Shellfishermen's Association (SDCSA) were encouraged to participate by the Secretary of their association. Whilst the crabbers were initially passive, a core group became actively involved over the course of the project. Successes and shortfalls of the collaborative process are discussed along with key factors required to engage fishermen and scientists in the successful development of a bottom-up management approach.

Keywords Edible crab · Participatory research · Sampling at sea · Individual-based modelling · Local ecological knowledge

E. Pearson · P. J. B. Hart (✉)
Department of Neuroscience, Psychology and Behaviour, University of Leicester, Leicester, UK
e-mail: pbh@le.ac.uk

E. Hunter
Centre for Environment, Fisheries and Aquaculture Science (Cefas), Lowestoft, UK

A. Steer · K. Arscott
South Devon and Channel Shellfishermen's Association, Devon, UK

P. Holm et al. (eds.), *Collaborative Research in Fisheries*, MARE Publication Series 22, https://doi.org/10.1007/978-3-030-26784-1_3

3.1 Introduction

Current EU fisheries management is heavily dependent on top-down control, which has led to the disconnection and alienation of small-scale fishermen from policy and management decisions that directly impact their livelihoods. It is more often than not the case that inshore fishermen (whose activities are being regulated) are not directly involved in, or consulted, during the process of data collection, negotiations or the subsequent legislation of management measures. Small-scale inshore fishermen are greatly impacted by local management measures as they often do not have the opportunity to fish elsewhere. Here we demonstrate that through cooperation, scientists and fishermen have together developed tools that could enable fishermen to become actively involved in the management of the crab stocks they exploit.

Based in Devon, UK, the South Devon and Channel Shellfishermen's Association (SDCSA) is a group of inshore crab fishermen, processors and merchants. The fishermen have provided an ideal platform from which to launch a fishermen–scientist collaborative research project aimed at constructing a fishery-wide management tool. The SDCSA has already demonstrated their ability to establish novel fishermen-directed management measures. For instance, during the 1970s, they established, and have continued to operate, several seasonal open and closed trawling zones interspersed with potting only zones (Blyth et al. 2002). These zones are collectively called the Inshore Potting Agreement (IPA) and are located between Dartmouth in the east and Salcombe in the west (Fig. 3.1).

Originally, the IPA began as a successful, self-regulated and voluntary agreement between static and mobile fishing gear users to mitigate the loss of gear. In effect, the IPA created a fishery with a restricted spatial area, within which each vessel effectively fished its own 'territory'. These fixed 'territories' are useful from a scientific standpoint as they allow a time series of data to be collected for each fixed area. For decades, fishermen of the SDCSA have supported government scientists, such as the Inshore Fisheries and Conservation Authority (IFCA) (formally Sea

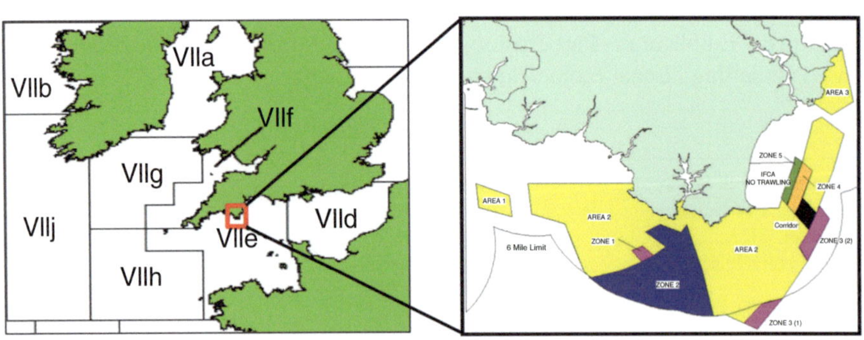

Fig. 3.1 Left: ICES area VIIe at the western end of the English Channel. Right: The IPA with areas permanently closed to trawling & scallop dredging marked in yellow and seasonally open and closed areas in green, orange, pink, black and blue

Fisheries Committee) and academics by supplying landings data upon request. However, fishermen have not received feedback on the data they had provided; thus, knowledge transfer has historically been in the direction of the scientist:

> "In the past we have had people come down and ask for data not really giving much of an explanation what it was for and didn't receive any feedback on the data. Which left fishermen very suspicious and unwilling to help." (4th Generation Fisherman).

Fishermen have shown an interest in contributing to research but have not had the opportunity to contribute their local ecological knowledge (LEK) to the process of formulating management measures. As a result, this study aimed to create a management tool for use by the crab fishermen in Devon to generate a sustainable fishery for the future. This would be a bottom-up management process with strong fisherman involvement.

The crab fishery in the English Channel forms about 20% of UK landings and the majority is caught off Devon and Cornwall (Elliott and Holden 2018). This statistic highlights the importance of the Devon crab fishery in terms of local socio-economics and the supply of crab to the market.

The crab fishery off the southwest of the UK, including the area in our study, was evaluated by the 2014 CEFAS stock assessment for the Western English Channel, which concluded that fishing effort was 'moderate to low', with an exploitation level close to that producing Maximum Sustainable Yield (MSY) (CEFAS 2014). Additionally, spawning stocks were rated as 'Good' and sufficient to sustain MSY.

The GAP2 method attempted to recruit all stakeholders in the creation of a sustainable fishery. Ideally, every actor in the fishery should be engaged at all stages in the creation of management measures. Neither fishermen nor scientists alone have the expertise, knowledge or influence to enact the fishery wide uptake of a management method. Therefore, collaboration is pertinent to the development of management tools for SDCSA fishermen to manage the stock they exploit.

3.2 Current Management

At present, the CEFAS stock assessments use fisheries-dependent data taken from Monthly Shellfish Activity Returns (MSARs). The MSAR forms, completed by fishermen, capture the number of days at sea and the weight and sex of crab landed. The spatial scale at which the data is recorded is defined by the size of the ICES fishing areas; the IPA fishery covers just 0.8% (470 km^2) of ICES area VIIe, which encompasses a total of 56,378 km^2 of sea (See Fig. 3.1). However, the large spatial scale used for these retrospectively applied stock assessments (conducted every 4 years) does not take into account local areas of intensive fishing, (such as the south Devon fishery), environmental variables or discard patterns.

In England, at the time of publication, fisheries are governed by three levels of legislation: European Commission (EC); national, for example Marine Management Organisation (MMO) and the Department for Environment Farming and Rural

Table 3.1 Management measures and their limitations applicable to the Devon and Severn IFCA area

Management measure	Limitations
MLS (more stringent than rest of UK)	150 mm for females (voluntarily in D & S)
Maximum vessel length (within 6nm)	15.24 m
Maximum pot limit	None
Use of edible crab as bait	Not allowed
Escape gaps in parlour pots/ creels	Yes
Towed gear restrictions	IPA and mid-channel blocks

Affairs (Defra); and regional through the local Inshore Fisheries and Conservation Authorities (IFCAs). The EC and national legislation demands that edible crabs in the UK waters are not landed below the Minimum Landing Size (MLS) across the carapace of 140 mm for females and not below 160 mm for males. Fishermen in the Devon and Severn IFCA have voluntarily set a 150 mm MLS for female crabs. A higher MLS ensures that a female has spawned several times before she is removed from the fishery (Warner 1977), so promoting the sustainability of the population. National legislation sets out management measures, which contribute to the current exploitation levels, such as the enforcement of the discarding of soft crabs, which are likely to have recently mated, and egg-bearing females, the future of the population. Additionally, escape gaps must be fitted to all parlour pots to allow juvenile crabs to escape before pots are hauled. These measures are summarised in Table 3.1.

Using MSARs data, CEFAS produce stock assessment reports that are used to design local and national management measures to align fishing effort with MSY. These reports are used by the Devon and Severn IFCA to produce local by-laws and national legislation, although at present, there are no limits on the catch weight landings or fishing effort. Unfortunately, the IFCA is under-resourced, with a budget of just £694,000 per year for 10 staff over an area of 3306 km^2,[1] and as such is not able to prosecute many transgressors. For example since 2011, there have been 11 fines issued in the form of a financial administrative penalty ranging between £250 and £3480 (average: £500) and four fines issued as simple cautions.[2]

All six transgressions, which have occurred during 2014, were still undergoing investigation when last determined in 2015 (Mat Mander, D&S IFCA, Personal Communication). These outcomes highlight the need for alternative management measures, which might be more successful than top-down enforcement in which fishermen have no input.

[1] IFCA Annual Plan 2018-19. https://www.devonandsevernifca.gov.uk/content/search?SearchText=Annual+Plan&SearchButton=Search (Last accessed 4th October 2018).

[2] https://www.devonandsevernifca.gov.uk (Last accessed 4th October 2018).

3.3 Available Science

There are a number of limitations of the current CEFAS stock assessment that lead to uncertainty. Crab stocks in English waters are currently assessed using the Length Cohort Analysis (LCA) method, which assumes that growth rates are constant across year classes. However, the discontinuous growth of crabs through moulting makes them notoriously difficult to age (Sheehy and Prior 2008). The LCA method also assumes that 'the fishery is operating over the entire stock' (CEFAS 2012), whereas the IPA covers just 0.8% of the stock assessment area (ICES Area VIIe). The disparity between area fished and area assessed creates problems for the design of well-directed management measures that are firmly rooted in local conditions. There is a need for a more restricted approach to assessment based on the actual area exploited. Furthermore, the LCA methodology does not take into account the life history traits of crabs, such as the continuous westward migration of females (Hunter et al. 2013). For example female crabs in the English Channel make a contra-natant migration from east to west down the English Channel, with no evidence of a reverse migration (Hunter et al. 2013). The LCA method does not consider migration as a variable that could affect biomass. Due to the one-way migration of female crabs, stock assessments should consider that the exploitable biomass in the IPA is strongly influenced by the biomass of crabs further to the east and the rate of migration of this biomass into the exploited area.

3.4 Description of the Case Study Process

3.4.1 Initial Relationship

Collaboration between the scientists and fishermen of this case study began in 2008, as part of the GAP1 Project, although the relationship between scientists at the University of Leicester and SDCSA crab fishermen has been ongoing since 1996. At that time crab fishermen were approached to take undergraduate students from the University of Leicester to sea, to experience a day in the life of an inshore fisherman. Due to the unique nature of the IPA, further studies were stimulated (Hart 1998; Kaiser et al. 2000; Blyth et al. 2002, 2004, 2006; Kaiser et al. 2007) to better understand how the then voluntary IPA agreement between mobile and static gear fishermen was maintained and what the conservation benefits were.

After research during the early 2000s, contact was maintained with the Secretary of the SDCSA and ideas were developed for further scientific research. The Secretary's motivation for further research was stimulated by his belief that the IPA was a great example of fishermen-directed management, which should be broadcast to a wider audience. In addition, economic benefits could be available to fisheries that can demonstrate sustainable exploitation such as a higher market value of the product (Johnston et al. 2001, Roheim et al. 2011). These ongoing connections meant that at the start of GAP2, the fishermen–scientist partnership was primed to begin collaboration.

3.4.2 GAP1

In 2008, at the start of GAP1, preliminary data on the spatiotemporal distribution of crab catches, landings and discards within the IPA were collected. This enabled an assessment of the feasibility of collecting sufficient data to allow for the development of a novel fishermen-directed stock assessment approach. The objective of GAP1 was to collect baseline data and develop the collaborative relationship between stakeholders to ensure the future success of GAP2.

3.4.3 GAP2

The GAP2 project started in 2011. Scientists used SDCSA monthly meetings as a platform to communicate their ideas and a provisional plan for the project to local fishermen. Fishermen were asked if they would be willing to take a scientist on-board their vessel once per month for a 12-month period and additionally contribute to seminars and discussions. At first, the 10–15 SDCSA members who regularly attended the monthly meetings did not show great enthusiasm for the proposed research but nonetheless agreed to take part. Their scepticism towards the project derived from previous encounters with scientists and management authorities as already outlined above.

3.4.4 Undertaking Project Work

In July 2011, a 12-month period of fieldwork for the project began. To ensure a representative sample of catches from the IPA, it was divided into eight areas (Fig. 3.2).

Information on the spatial distribution of fishermen's gear was obtained from Clark (2008). The secretary of the SDCSA provided fishermen's contact information. A list of fishermen in each area (approximately three to four) was prepared and contacted by phone, in alphabetical order. The aims and values of the project were explained to each fisherman individually. Fishermen were then asked if they would be willing to take part in the project by committing to taking a scientist to sea, each month, for 1 year. Once one fisherman had been found for an area (n=8), no other fishermen who fished in that area were contacted. A summary of the roles undertaken by fishermen and scientists for each task during the case study is outlined in Table 3.2.

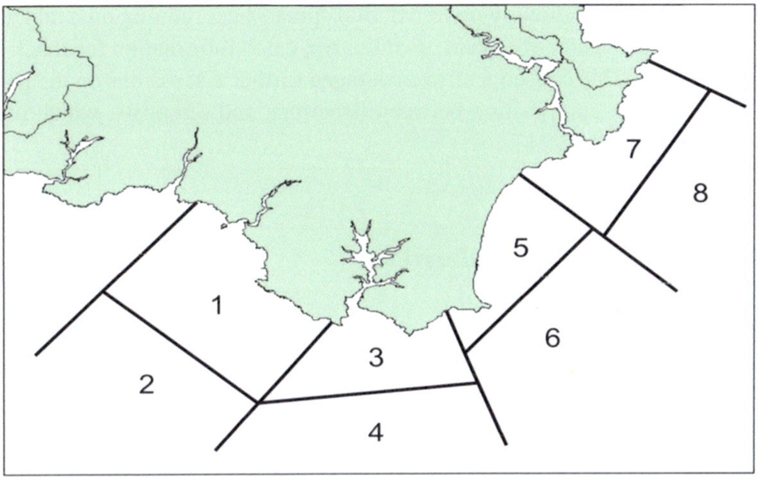

Fig. 3.2 A map of the eight areas used to representatively sample the catches of the IPA. The IPA was split into four zones west to east and two zones out from the shoreline, namely 0–3 nm and 3–6 nm

Table 3.2 A summary of the roles undertaken by fishermen and scientists during the various tasks of the case study

Case study tasks	Scientists' role	Fishermen's role
Data Collection	To record landings and discards in space and time. Analyse results and feedback findings.	Take scientist to sea and announce reason for each discard.
Semi-structured interview	Provide questions and structure for interviews. Analyse results and feedback findings.	Convey wealth of LEK regarding fishing and environmental factors to interviewer.
Substrate surveys	Create survey and distribute to fishermen. Analyse results and feedback findings.	Completing survey and share knowledge.
Modelling	Developing framework and code model.	Review model and provide feedback to ensure the model reflected the reality of the fishery.

3.5 Data Collection Process

Weather permitting, one trip per vessel per month was organised with each skipper. All the at sea work was done by the first author. Once on-board, as each pot was hauled and emptied the number and sex of individual crabs and whether they were to be landed or discarded was recorded. Discarded crabs were recorded as undersized (below MLS), soft-shelled or egg-bearing. In addition, all by-catch, whether marketable or not, such as various fish species, other crustaceans and various mollusc species were documented.

Data were recorded directly on to a digital spreadsheet running on a tablet. This allowed for the provision of instant, quantitative, catch information feedback to the skipper and crew. This consequently encouraged further discussions on the possible causalities of catch composition between fishermen and scientists, which assisted the participatory process.

3.5.1 Semi-Structured Interviews

During fishing trips, it became apparent that the vast store of fishermen' knowledge of the resource they exploited needed to be captured more coherently than through informal conversations whilst fishing. Therefore, semi-structured interviews were carried out with fishermen in their own homes at a convenient time of their choice to record their so-called Local Ecological Knowledge (LEK).

3.5.2 Surveys

In January 2014, a survey was mailed to all crab fishermen operating within the IPA (n = 46). The survey aimed to collect information regarding the seabed type within each fishermen's 'territory', leading to a comprehensive overview of the IPA's substrate type and location. Only 11% of fishermen replied to the survey. The poor response to the mailed survey highlighted the importance of face-to-face contact with fishermen in collecting fine-scale, fishery-wide data. The data on substrate type was to be used in the construction of the IBM of the fishery. As the fishermen's response was so poor and the information from those that did reply was so variable, data from EMODnet (http://www.emodnet.eu) was used instead.

3.5.3 Exchange Visit

In 2012, 13 members of the UK GAP2 hosted five members of the Norwegian GAP case study on a visit to Devon in October 2012. In return, five fishermen from south Devon visited the Norwegian cod (*Gadus morhua*) fishery in Steigen and the Lofoten Islands during April 2013. The project's scientists took part in both exchanges. The visit to Steigen gave the Devon fishermen the opportunity to learn about the fishing methods and management measures of the skrei cod fishery that takes place between the mainland of Norway and the Lofoten Islands in early spring. Other than the obvious benefits of learning about the Norwegian case study, this trip provided a platform for scientists and fishermen to socialise, integrate and learn together, which in turn built trust. The exchange visits were a great aid in highlighting the successes and shortfalls of the IPA and UK fisheries, in comparison with the

Norwegian cod fishery. The Norwegian fishermen were very interested in the auction in Brixham where they were impressed by the prices certain species achieved. In Norway, the price of the catch tends to be fixed by the fishermen's producer organisations there being no auction system. The Devon fishermen were made aware of the environmental issues created by a salmon farm visited and had the opportunity to join fishing trips on small gill net fishing boats exploiting cod.

3.6 Developing an Individual-Based Model of the Fishery

The details of the model and its results will be reported in a separate publication. The model was developed over a 4-year period and its initial form was presented to a small group of SDCSA members. They provided perspective on the initial structure and subsequent meetings helped to refine the model. Elements of the model are shown in Fig. 3.3. Output from the model in terms of crabs caught per day in

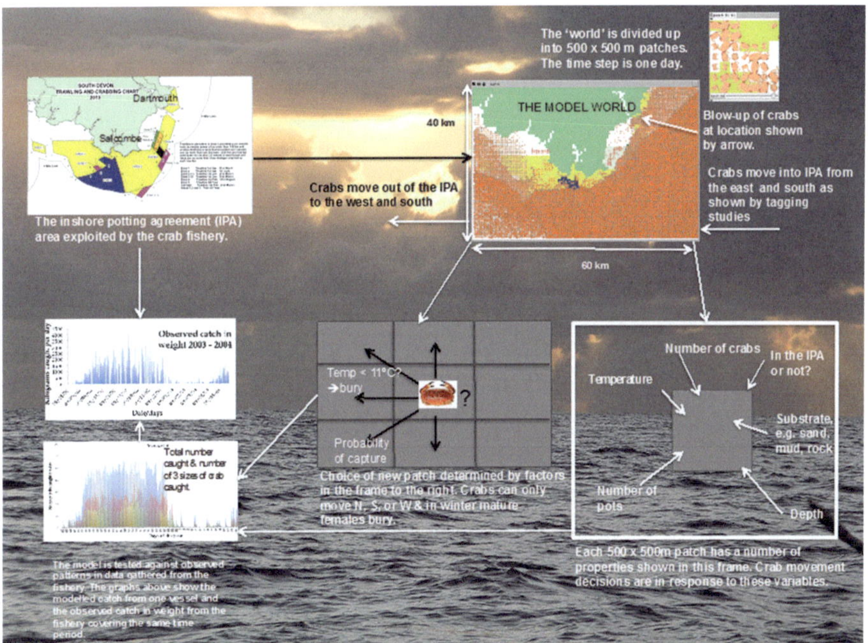

Fig. 3.3 A graphical representation of the model showing the real and modelled area exploited by the South Devon fishermen. The grey box on the right shows properties attached to each of the patches into which the sea area is divided each with side 500m. The grey box to its left shows the patches to which a crab could move together with two of the factors that affect its choice. If the crab buries, it stops moving

different parts of the fished area match the patterns shown by the data gathered from the field work and longer-term data extracted from a subset of fishermen's logbooks.

At present, there is no well-established stock–recruitment relationship for the crab stock. This is vital for determining the immigration rate of crabs into the fished area from the east and south. As a result, the numbers of crab caught in the model fishery cannot be properly calibrated against the actual number caught. Although the pattern of captures through time and space is well represented, further developments are needed to allow the model to be used for management. The most likely use will be to estimate the exploitation rate which produces a catch that can be sustained by the immigration rate of females. The existence of the model and its outputs might also help give fishermen greater credibility when discussing management issues with external management agencies or NGOs and in attaining some type of sustainability accreditation.

3.7 Results of a Collaborative Approach

As fishermen and scientists spent time collecting data at sea, the engagement of fishermen improved over the length of the project. The time at sea also facilitated one-to-one discussions on wide-ranging fishing-related topics. The majority of knowledge transfer and mutual learning took place at this time. Throughout the project, when fishermen's knowledge of a given phenomenon conflicted with that of current scientific understanding, the facts, as they are known, were offered by scientists and discussed with fishermen. (The correspondence of LEK with scientific findings was also analysed in more detail, see Pearson 2017.) Fishermen also educated scientists in the same way with their understanding of the fishery and crab behaviour. These informal, mutual-learning events provided a neutral arena for knowledge exchange, thus closing the knowledge gaps of fishermen and scientists and allowing for topics to be discussed in greater detail, leading to a deeper understanding from both parties.

Fishermen appreciate feedback; therefore, we aimed to provide this on a variety of occasions. Immediate feedback was given to fishermen regarding the day's fishing. Metrics were shown of total catch, the percentage of male and female crabs per trip and pot, total discards and by-catch, demonstrating the usefulness of good data collection to fishermen's everyday activities. This led to a greater understanding of why fine-scale data collection was important to be able to create and test a model of the fishery. The scientists being on-board aided these discussions and topics could be demonstrated or observed rather than simply being described in an interview.

Once all trips had taken place, the data from each trip were analysed and emailed to the skipper of each relevant vessel, alongside graphs detailing the seasonal variations in catch and catch composition. Landings and discard data were also presented to all fishermen who attended SDCSA meetings to provide a fishery-wide view of the temporal and spatial distribution of catches across the IPA. Throughout the project, when results were presented, fishermen were encouraged to give their views on results, such as how they could be explained by environmental processes and crab behaviour. Consequently, fishermen commented on how useful this was:

> Normally in other projects [data] just disappears, we don't know where it's gone, or what it is being used for. This concerns fishermen as they do not know if it might be used against them in future for stock management or assessments without them knowing. With GAP2 we will be left with some tangible information at the end of the project which we can use ourselves to improve our fishery in the future and data that we know is reliable as we were involved in the collection of it throughout the whole process. (Interview)

Fishermen further commented that it was interesting to see how the catches varied month by month across the IPA, possibly reflecting the movement of crabs throughout the IPA.

During fieldwork there was almost daily contact between fishermen and scientists. For the remainder of the GAP project we maintained regular contact with the crab fishermen by phone, email, social media and by attending monthly meetings of the SDCSA. It was evident that the continuity in communication between fishermen and scientists re-enforced the relationships that had been established during trips to sea:

> With your [participatory research] approach you came to us [fishermen] and explained exactly what you intended to do and what you would be using the information for and involved fishermen from the very beginning, which inherently builds trust and participation. Also, the fact that you came aboard the boats and collected the data yourself on multiple occasions over a long period of time really aided the understanding of your project by fishermen. Once the data had been collected from the fishermen you didn't just disappear but kept returning to our meetings informing us of progress made, explaining other technical issues to us and helping in our monthly meetings with other fishing related issues. This whole process led to multiple strands of cohesion, producing a very strong bond between fishermen and scientists. (Interview)

Within the tight-knit fishing community of south Devon, trust is paramount. When communicating with fishermen, it is vital to remain neutral and not to discuss the catches of other vessels with competitors. Trust was also maintained by promising anonymity of catch data and fishing grounds when showing data to people outside of the case study area. Losing trust with a single member of the fishing community could lead to repercussions with other members of the community. It is also important to show that even a scientist can get her/his hands dirty and be prepared to assist where required on the boat or the quayside. In other words, it is important for the scientist to become a part of the community but still retaining an objective view of events.

GAP scientists demonstrated their commitment to the project by completing monthly 500-mile roundtrips to attend monthly SDCSA meetings. These meetings often discussed management issues outside the remit of GAP where the scientists could provide an alternative viewpoint, explain technical issues or help with interactions with other bodies. For example the present efforts of the UK Government, following the Marine Strategy Framework Directive, to set up Marine Protected Areas (MPAs) has created problems for the crabbers. Scientists in the GAP partnership have written letters to Natural England, to make a case for the preservation of the IPA.

As time passed and trust developed, the fishermen became more forthcoming with ideas on how to collect environmental data, to capture their knowledge on the resource they exploit and to give ideas on how to automatically record their catch and discard data without the need for an on-board observer. The GAP project has led to other opportunities for IPA fishermen, such as taking part in the Prince's Trust's Fishing into the Future (FITF) initiative and has increased the confidence of fishermen to become involved in similar projects. Fishermen have also been encouraged to attend the annual GAP meetings, with one attending the 2013 meeting, two at the 2014 meeting and three in 2015.

3.8 Discussion

The disengagement of fishermen from the management of the fisheries they prosecute often leads to fishermen having misconceptions about the science behind stock assessments. This can also be compounded by poor explanation of a stock assessment by the scientists responsible. Common misconceptions are that fishermen believe data were not collected directly from the fishery that they exploit, and therefore, management measures are not relevant to their fishery (e.g. Degnbol 2003); that all crabs should be counted to be able to do an accurate stock assessment (as expressed by one crab fisherman to a Cefas representative) and that decades of previous exploitation without the fishery failing must mean that the fishery is sustainable. Most frustrating for fishermen is that they often feel that some employees of government bodies that carry out the stock assessments and enforce the resulting management measures do not regard them as credible sources of information.

A collaborative approach to setting up a management tool was deployed in three parts:

- Fishermen took scientists to sea with the main aim of collecting data but also to allow face-to-face interactions, which led to the development of trust and mutual understanding, as described.
- Through semi-structured interviews of local fishermen, scientists recorded fishermen's local ecological knowledge of the resource and of the environment.
- Data feedback and discussions took place regarding stock assessment methodology and the development of the IBM.

The key to the success of this project was the time spent on-board vessels building a rapport with fishermen. The level of interaction between fishermen and scientists within this project would not have been possible without these repeated sea trips. The simple act of being on the vessels together led to invaluable insights for both parties regarding the realities of fishing and of fishery science, which could not be achieved by any other means.

3.8.1 Shortfalls

Although a core of local fishermen have been involved in the UK GAP2 Project, this core constituted only about 20% of all IPA fishermen. The silent majority did not make an appearance at the monthly meetings and were not easily contactable. The SDCSA meetings were the primary platform used to engage fishermen with broader aspects of their business and with the GAP Project in particular. The meetings were open to all fishermen within the IPA, as one has to be a member of the SDCSA to fish in the IPA. The core of fishermen involved in the GAP project was largely similar to the core of IPA users who regularly attended SDCSA meetings. No attempt was made by scientists to engage fishermen who did not attend monthly meetings or take scientists to sea in the project. Upon reflection, this issue should have been tackled early in the project to increase engagement.

A further shortfall of this case study was that it was heavily scientist-led. As previously mentioned, the ideas and aims of the project were compiled by scientists and agreed by fishermen, whereas in a true collaboration, both parties would have mutually derived the aims of the project.

The physical distance between the University of Leicester and the research site in south Devon, (240 miles) in some instances, had a detrimental impact on the project. Fishermen were frequently contacted by telephone, text or email instead of face-to-face, which would have been more beneficial. The distance also restrained scientists from attending meetings other than the monthly SDCSA meeting, such as MCZ consultations and IFCA meetings.

Our long collaboration with the crabbers created a perception in us that most management agencies could not be trusted with information that might be used against the fishery. Therefore, it was a conscious decision not to involve the IFCA at the start of the project. In hindsight, the IFCA should have been part of the collaboration. This became evident when it was seen that the Principal Environmental Officer of the Severn and Devon IFCA attended most monthly meetings. Although there was always some antagonism between the IFCA representative and the fishermen, there was considerable mutual understanding and trust built up over several years. Had we understood this earlier, it would have removed our fears that association with the IFCA could have undermined the fishermen trust in the scientists.

While the ultimate aim of the project has remained constant, the milestones to achieve this aim have not been fixed. At the outset, there were no time-dependent expectations outlined to fishermen. This reduced the perception of failures, and as and when milestones were achieved they were perceived as successes, which built trust and a sense of achievement. Similarly, there were no initial goals regarding numbers of fishermen to engage. Fishermen were asked if they wanted to be involved and for those who decided to take part in the project, their motivation was purely intrinsic. This meant that fishermen were involved due to their belief in the values of the project rather than feeling that they 'ought' to be working with us. Therefore, future engagement was gained 'organically', usually by word-of-mouth from already engaged fishermen. This led to a strong and stable core of fishermen involved in the project. In this sense, the collaboration grew from the bottom-up.

3.8.2 Future Work

The next step would be to create legitimacy for the bottom-up approach that we have advocated. This needs to be done with the current management agencies such as the MMO, the Devon and Severn IFCA and DEFRA or else, the approach we have taken will not be absorbed into the established management system. As the GAP programme has no successor, it is very likely that the interest of crab fishermen in local management issues so painstakingly developed through this project will dissipate.

Additionally, for future catch levels to be set, fishermen will need to record their own data. The current method of stock assessment employed by CEFAS uses data reported by fishermen to calculate landing weights and fishing effort. Hence, in principle, the credibility of self-collected data should not be doubted by the management agencies. Additionally, the IBM will be subject to a scientific peer review process, independent from the fishermen, ensuring its credibility as a potential tool for use in managing the fishery.

3.9 Conclusion

The 'gap' between the Devon crab fishermen and scientists was reduced by our cooperation and by the end there was a strong working relationship in place between the two parties. With the aid of the IBM and a system for data collection established, there is the potential for fishermen to contribute much more to the management of their own resource. The deficiency in this vision is that there is now no one institution responsible for establishing the bottom-up system as the norm.

While this project was heavily scientist-led, it bridged the gap between fishermen and scientists through creating mutual exchange of data, opinions and ideas. Prior to this project, fishermen in the IPA had often been asked to supply data to various agencies but were not engaged in the final use of the data so supplied. The project has given some of the fishermen the skills and confidence to become involved in future projects and has highlighted the value of collaborative research to both scientists and fishermen.

To this end, the attitudes to research in cooperation with scientists of some of the fishermen working in the SDCSA have changed positively. Fishermen feel empowered as they are 'listened to' and now have a communication channel through which they can 'have their say' and be heard by local managers. By giving fishermen feedback from the data collection process, a positive reinforcing loop is created; the more information they receive, the more they want to know and the more they become involved in the research process.

The greatest future challenge is to translate the outcomes of this collaborative research into management measures most likely implemented by the existing agencies such as the Severn and Devon IFCA, MMO and the EU.

References

Blyth RE, Kaiser MJ, Edwards-Jones G, Hart PJB (2002) Voluntary management in an inshore fishery has conservation benefits. Environ Conserv 29(4):493–508

Blyth RE, Kaiser MJ, Edwards-Jones G, Hart PJB (2004) Implications of a zoned fishery management system for marine benthic communities. J Appl Ecol 41:951–961

Blyth-Skyrme RE, Kaiser MJ, Hiddink J, Edwards-Jones G, Hart PJB (2006) Conservation benefits of temperate marine protected areas: variation among fish species. Conserv Biol 20:811–820

CEFAS Stock Status (2012) Edible crab (*Cancer pagurus*) in the Western English Channel (2012). CEFAS, Lowestoft

CEFAS Stock Status (2014) Edible crab (*Cancer pagurus*) in the Western English Channel (2014). CEFAS, Lowestoft

Clark S (2008) South Devon potting effort survey 1998. Devon Sea Fisheries Committee Research Report 200802

Degnbol P (2003) Science and the user perspective: the gap co-management must address. In: Wilson DC, Nielsen JR, Degnbol P (eds) The fisheries co-management experience. Accomplishments, challenges and prospects. Kluwer Academic Publishers, Dordrecht, pp 31–49

Elliott M, Holden J (eds) (2018) UK sea fisheries statistics 2017. Marine Management Organisation, Newcastle upon Tyne

Hart PJB (1998) Enlarging the shadow of the future: avoiding conflict and conserving fish. In: Pitcher TJ, Hart PJB, Pauly D (eds) Reinventing fisheries management. Springer, London, pp 227–238

Hunter E, Eaton D, Stewart C, Lawler A, Smith MT (2013) Edible crabs "go west": migrations and incubation cycle of *Cancer pagurus* revealed by electronic tags. PLoS One 8(5):e63991. https://doi.org/10.1371/journal.pone.0063991

Johnston RJ, Wessells CR, Donath H, Asche F (2001) Measuring consumer preferences for ecolabeled seafood: an international comparison. J Agric Resour Econ 26(1):20–39

Kaiser MJ, Spence FE, Hart PJB (2000) Fishing gear restrictions and conservation of benthic habitat complexity. Conserv Biol 14(5):1512–1525

Kaiser MJ, Blyth-Skyrme RE, Hart PJB, Edwards-Jones G, Palmer D (2007) Evidence for greater reproductive output per unit area in areas protected from fishing. Can J Fish Aquat Sci 64:1284–1289

Pearson, E. (2017) A collaborative study to develop and facilitate a fisher directed stock assessment of *Cancer pagurus* in the Inshore Potting Area, South Devon. PhD Thesis, University of Leicester

Roheim CA, Asche F, Insignares Santos J (2011) The elusive price premium for ecolabelled products: evidence from seafood in the UK market. J Agric Econ 62(3):655–668

Sheehy MRJ, Prior AE (2008) Progress on an old question for stock assessment of the edible crab *Cancer pagurus*. Mar Ecol Prog Ser 353:191–202

Warner GF (1977) The biology of crabs. Van Nostrand Reinhold Company, New York

Chapter 4
Getting Choosy About Whitefish in Lake Vättern. Using Participatory Approaches to Improve Fisheries Selectivity

Alfred Sandström, Johnny Norrgård, Thomas Axenrot, Malin Setzer, and Tomas Jonsson

Abstract Our case study was channeled through a comanagement group and initiated by fishermen. The aim was to develop the selectivity of the fisheries in Lake Vättern through a collaboration of fishermen, regional managers, and scientists. The case study was planned collectively within the fisheries co-management group and through workshops with participating fishermen. Fishermen were given special permits to test various strategies adapted to their own experiences and the properties of their local fishing grounds. First, the population structure of the focal species, whitefish, was investigated in a joint study utilizing fishermen's experiential knowledge and genetic analyses of whitefish. The results showed that the population is divided into at least two unique stocks that ideally should be managed separately. Second, the results from studies on by-catch in different areas, seasons, and gears indicated a substantial potential to increase the selectivity in this fishery, particularly by targeting whitefish aggregations adjacent to spawning areas. Our study highlights the potential of the participatory approach when facilitating solutions to problems related to small-scale fisheries management. Nevertheless, we also identify some factors that might jeopardize the long-term success and dissemination of results from this case study. The recent discovery of high levels of dioxins and

A. Sandström (✉) · T. Axenrot
Department of Aquatic Resources, Institute of Freshwater Research,
Swedish University of Agricultural Sciences (SLU), Drottningholm, Sweden
e-mail: alfred.sandstrom@slu.se

J. Norrgård
Lake Vättern Society of Water Conservation, Jönköping, Sweden

M. Setzer
Lake Vättern Society of Water Conservation, Jönköping, Sweden

School of Bioscience, Ecological Modelling Group, University of Skövde, Skövde, Sweden

T. Jonsson
Department of Ecology, Swedish University of Agricultural Sciences (SLU),
Uppsala, Sweden

School of Bioscience, Ecological Modelling Group, University of Skövde, Skövde, Sweden

P. Holm et al. (eds.), *Collaborative Research in Fisheries*, MARE Publication Series 22, https://doi.org/10.1007/978-3-030-26784-1_4

dioxin-like PCBs in muscle tissue of whitefish might make it harder for fishermen to sell their catch. Moreover, since the comanagement group only has an advisory function, the Swedish national authority needs to take the initiative and first implement the suggested changes in management.

Keywords Participatory research · Comanagement · Selectivity · Lake fisheries · Whitefish

4.1 Introduction

The fisheries in Lake Vättern, Sweden, are of a small-scale and governed by regional and national authorities and thus unaffected by the Common Fisheries Policy (CFP). As in the case of many other large lakes in Europe, the recreational fisheries sector of Lake Vattern is of considerable importance and takes a markedly larger part of the catch of many species than the commercial fisheries (Cowx et al. 2010; Alenius and Hallden 2012). The fisheries in Lake Vättern have altered their alignment substantially during the last decade. The traditional fisheries with gill nets, which focused mainly on salmonid species such as whitefish (*Coregonus lavaretus*) and Arctic charr (*Salvelinus salvelinus*), have largely been replaced by a fishery on the introduced signal crayfish (*Pacifastacus leniusculus*). This has rapidly become the most economically important species, currently constituting 95% of the fishing industry´s income. Due to a combination of the switch to crayfish and the deteriorated stock status of the traditional focal species, particularly of Arctic charr, the commercial fish catches have been steadily declining since the mid-1970s.

As a means of remediation, new fisheries restrictions were implemented in Lake Vättern in 2005. Three large areas comprising 15% of the lake surface were closed for all fishing. Other restrictions included an increased minimum landing size limit for Arctic charr and brown trout (*Salmo trutta*), a substantial increase in the minimum mesh size of gill nets and additional protection of Arctic charr and brown trout spawning areas and spawning migration areas. These management efforts appear to have helped turn the tide with regard to the availability of most fish species in the lake (Sandström et al. 2014), but they have also made it very difficult for commercial fishermen to target whitefish, which formerly was very important to them in terms of economic benefits. Since Arctic charr and whitefish are exploited in a mixed gill net fishery, a conflict exists between the protection of a weak and sensitive species (the charr) and the potential of an increased fishery of an abundant and underexploited species (the whitefish) (Box 4.1).

Box 4.1: Lake Vättern

The lake has a surface size of 1893 km^2. It is a highly nutrient-poor lake, 4–6 μg P L^{-1}, with a Secchi disc depth (a water visibility measure) ranging from 10 to 18 m. The drainage area is relatively small (6360 km^2). The mean depth is 40 m and the maximum depth is 128 m (Fig. 4.1). There are approximately 20 commercial fishermen in Lake Vättern that are licensed to fish in common waters. The dominant gears are gillnets and crayfish traps. There is also an extensive recreational fisheries sector, mainly targeting Arctic charr, trout, stocked salmon, perch and Northern pike. For these species, the estimated catches in the recreational fisheries are higher than in the commercial fisheries. The fish communities in Lake Vättern are monitored through two major programs: hydroacoustic surveys combined with midwater trawling and benthic multimesh gillnets (Fig. 4.1. Axenrot 2014; Sandström et al. 2014).

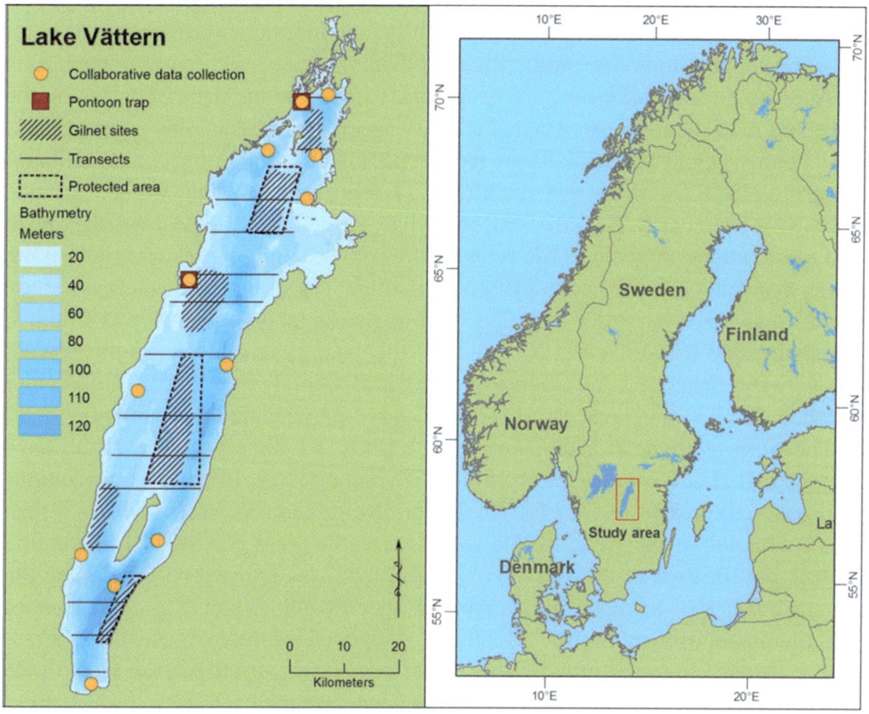

Fig. 4.1 Map of Lake Vättern and the different sampling areas of the monitoring program using gillnets and hydroacoustics combined with midwater trawling and the collaborative sampling of this case study

4.1.1 A Case Study Planned and Run by a Fisheries Comanagement Group

Parallel to the rapid change of the fishery and the crisis of many fish stocks, a new form of governance was implemented in the early 2000s in Lake Vättern. As a part of a national Swedish initiative to facilitate new fisheries management regimes, a fisheries comanagement group was founded in 2004–2005, with Lake Vättern Water Conservation Society as the lead partner. The comanagement group has since emerged as an important arena for management advice, conflict resolution, general discussions, and information exchange between different groups. The fisheries, however, are still mainly regulated by the national authority (SWAM). Thus, the group has no formal mandate for regulating fisheries and, similar to the Advisory Councils (ACs) in the EU marine fisheries, only has an advisory function (Fiskeriverket 2007). Nonetheless, their advice is in most cases taken seriously and implemented by the national authority (van Mastrigt 2013).

This case study grew out of discussions in the comanagement group about potential collaborative projects involving different stakeholders as a means to solve urgent management problems. The most critical problems have been alternative remediation measures to protect and restore the Arctic charr stock and the issue of balancing the contribution of the introduced signal crayfish species to the economy and its potential threat to the environment. There has been a long tradition of scientists collaborating with local fishermen in Lake Vättern (Tiselius 1723; Widegren 1863; Ekman 1916; Svärdson 1957). Nevertheless, there has, to the best of our knowledge, been fairly little attention given to the roles of the collaborating parties and the driving forces that influence the process of collaboration. The ambition of our case study was to draw on the earlier experience from the collaboration that was initiated already in 2005 (Fiskeriverket 2007) and further strengthen and deepen the interaction between fishermen and scientists, using it as a catalyst for improving a wider collaboration between local resource users and researchers in management.

The work required to undertake the case study was first discussed in the Lake Vättern fisheries comanagement group. All participating fisheries organizations were asked to present an idea for a collaborative research project. The group than held a referendum and voted for the various research ideas presented. The winning project, put forward by the commercial fishermen, was focused on developing the whitefish fisheries, learning more about this species and aiming to increase catches and simultaneously minimize by-catches of undersized fish of sensitive species such as Arctic charr and brown trout. Other suggestions by fishermen were (a) to study the effects of the introduced signal crayfish on the lake food-webs and (b) to study the population dynamics and occurring conflicts in the exploitation of Arctic charr. The project related to developing whitefish fisheries gained support from many participants, since the problem of by-catch was recognized as a serious issue and that whitefish is of limited interest to exploit by other groups of fishermen. Enhanced fishing of whitefish could also give commercial fishermen an additional alternative to signal crayfish that today dominate the economy of the fisheries.

Potentially, this could render a more diverse commercial fisheries sector, less sensitive to changes in the catches of one single species. Furthermore, higher fisheries mortality of whitefish was believed to lead to decreasing competition for common resources both within the whitefish stock as well as with other competing species such as Arctic char and brown trout. This could be hypothesized to lead to a faster growth rate of both whitefish as well as competing species.

After deciding on the topic, the comanagement group agreed to the following objectives for the case study:

- To strengthen Lake Vättern Fisheries Comanagement Group by cooperating in a common project with clear objectives and thus developing the collaboration and knowledge exchange among fisheries stakeholders, managers, and scientists.
- To develop an economically profitable and sustainable fishery of whitefish with minimum by-catch of sensitive species such as Arctic charr and brown trout.

All the case study work was channeled through the co-management group, which was involved in planning, organizing and communicating results from the study. Results obtained during the project period were also presented and analyzed by this group.

4.1.2 Developing the Design of Collaborative Data Collection

This case study utilized a variety of collaborative approaches. The process of collaboration (and some of the inspiration to structure this text) was based partly on the organizational principles put forward by Johnson and van Densen (2007). The collaborative approaches were also partly inspired by the good practice guide for participatory research that was achieved during GAP1 (Mackinson et al. 2008) and by the approaches summarized in Berkes et al. (2001) and Huntington (2000).

Previous experience of emerging comanagement groups suggests that they could empower stakeholders and contribute to knowledge exchange between fisheries stakeholders, managers and researchers (Dietz et al. 2003). However, quantitative empirical studies on the contribution of co-management to the success of fisheries management are often limited by the complexity of the issue, making them rather difficult to interpret (Berkes 2009; Linke and Bruckmeier 2015). We hypothesize that a co-management group could be beneficial and contribute favorably when implementing a collaborative case study as suggested by Piriz (2004). Preferably co-management can help create what Carlsson and Berkes (2005) describe as "a continuous problem-solving process, rather than a fixed state and joint learning within problem-solving networks." Since the GAP project was initiated in 2008 and the co-management group formed in 2004–2005, it was possible to build trust and address conflicts at an early stage, long before the actual start of the case study. Using a comanagement group as a planning platform was considered to be positive to the overall work and has been a convenient forum to reach all relevant stakeholders on a regular basis.

The planning and design of the work of the case study was also conducted in a series of workshops with all the participating fishermen. The first workshop was spent mainly on communicating and discussing the objectives of the case study and agreeing on a basic framework for the studies undertaken in collaboration between fishermen and scientists. Besides containing discussions and presentations, the workshop also had external experts participating (gear selectivity experts, social scientists, etc.) who were invited to inspire and bring new ideas for the case study to the table. To be able to quantitatively test how to improve the selectivity in the fishery, a general approach was discussed and agreed upon in the workshops with fishermen, scientists, and regional managers. This meant that participating fishermen followed a common sampling design, but that each fisherman was allowed to make smaller adjustments of the sampling design if it was needed to adapt it to the conditions of their fishing areas. As an example, fishermen agreed to test the effect of fishing depth on selectivity, but the depth range tested was not exactly the same for all fishermen, since depth may vary between fishing areas. In total, around twenty fishermen participated in the case study (thirteen commercial, two subsistence, and three recreational fishermen). Since fishermen in Lake Vättern operate close to their homes and harbors, the fishing areas were scattered throughout the lake, covering the majority of the open basin but only to a limited extent in Northern archipelago areas (Fig. 4.1). Fishermen collected data together with GAP2 scientists in collaborative field trips and by themselves during their normal fishing activities.

Besides workshops with all participants, focused conversations (Gray 2005) took place in less formal environments, close to harbors, in boats and in fishermen´s kitchens. These meetings with individual fishermen helped tailor the design of the studies to each person's individual needs. Fishermen discussed ideas about the study design with scientists and then adapted it to their own, individual fishing areas. This freedom proved to be a positive incentive for fishermen to remain involved in the case study. Both workshops and focused conversations with individuals were used to connect with fishermen and give and receive feedback on the progress of the case study.

The outcome of the workshops and many meetings with the comanagement group was a decision to conduct a number of studies to meet the objectives of the case study. The studies decided upon included:

- A study on the population structure of whitefish using a combination of genetics and fishermen's experiential knowledge
- Several trials to test approaches to mitigate by-catch problems of undersized fish in both the commercial gillnet fishery and the recreational trolling fishery
- A test of a new innovation in fishing gear as an alternative fishing method to mitigate by-catches
- A test of the possibility to locate spawning aggregations of whitefish using sophisticated scientific hydro-acoustic equipment mounted on a traditional fishing boat

4.2 Collaborative Data Collection and Analyses

4.2.1 Whitefish Population Structure

It became clear during the formulation of the research objectives and during the first workshop of the case study that fishermen recognized many different types of whitefish stocks and that this diversity of whitefish was an important issue for the fishermen, since it could have significance for the management of the stocks. Whitefish is also a species known for its marked resource polymorphism and has often been used as a model species to further the understanding of speciation processes in fishes (Vonlanthen et al. 2012; Hudson et al. 2011). Studying the population structure of the species was considered an important first prerequisite to solving problems with selectivity and assessing if management advice should be given for more than one specific stock. We thus designed a study in collaboration with fishermen to investigate ecomorphological variations and population structures of whitefish. The results from this work are described in more detail in Sandström et al. (unpublished).

The fishermen differentiate between several unique subpopulations of whitefish, each with its own characteristics. Even though there are a large variety of different names and forms of whitefish described both by fishermen and existing literature, three main categories can be identified (see images below). Fishermen separate stocks mainly on the basis of their morphology, time and place of spawning, and diet. To further assess information provided by fishermen, we collected samples of whitefish from a fish-monitoring program with survey stations distributed randomly across the entire lake (Fig. 4.1), thus covering all available habitats and the spawning sites of various forms of whitefish recognized by fishermen. Sampled whitefish specimens were analyzed for age, growth, diet, morphology and genetic structure using microsatellite DNA. The analysis showed small but statistically significant patterns and revealed that there are at least two unique populations of whitefish in Lake Vättern. Also, there was a clear indication of an isolation by distance pattern in one of the identified populations but not in the other, indicating differences in spatial distribution and in the choice of spawning sites. Fishermen expressed concern that the scientific analysis did not reveal genetic differences between all the three stocks identified by them. Given that fish spawn in very different places and that their time of spawning differs for up to three months, fishermen argued that these stocks must have limited gene flow between them. Other aspects of the results were confirmed by fishermen, particularly the differences between stocks in diet and spatial distribution patterns. Our results suggest that the two main whitefish types should ideally be managed separately, although difficulties exist in terms of robustness of classification based only on a visual inspection and hence may make such a management strategy difficult to develop in practice unless the stocks are targeted at their spawning sites (Fig. 4.2).

Fig. 4.2 Images of the different whitefish populations identified by fishermen. Upper – "River whitefish", middle – "Deep spawning whitefish" and lower image – "Lesser sparsely rakered whitefish." Note the elongated snout of the upper "river whitefish." It is this characteristic that is used by fishermen to identify that population. (Photos by A. Asp, F. Engdahl, and A. Sandström)

4.2.2 *Improving the Selectivity of the Gillnet Fishery*

The results from this work are described in more detail in Setzer et al. (2017). The main aim of this case study was to optimize the balance between the catch of the focal species whitefish and the by-catch of other species. To study this issue, fishermen were given special permits from regional authorities to conduct experimental fishing trials as a part of their normal fishing activity. These trials explored, above all, the importance of fishing sites, gears, fishing areas, and fishing season. Many fishermen wanted to test a specific strategy that could help improve selectivity. Some strategies aimed to target whitefish when they aggregate to spawn and others when they aggregate to forage on the eggs of other spawning fishes (mainly vendace and smelt). The survey data were either collected through cosampling or self-sampling. Fishermen mainly collected data together with scientists in collaborative, joint survey trips, out of which 37% were truly collaborative: the crew was composed of both scientists and fishermen. In the rest of the trips, data were collected only by fishermen but in accordance with the agreed survey design. In total, gill-nets were set and lifted at 564 different stations that were set during 184 different survey trips at 30 experimental sites. In order to assess selectivity in these trials, we compared catches with a gill-net monitoring program. We also constructed a gillnet selectivity model (Jonsson et al. 2013), a development of the SELECT approach

(Millar and Fryer 1999), to compare the retention probabilities of various species with actual catches in the trials (Fig. 4.3).

The results from the gillnet trials suggest that it is possible to markedly reduce the by-catch of undersized Arctic charr and brown trout if the fishing effort is targeted to areas and seasons where whitefish aggregate and the documented by-catch is low. The average catch of all trips during the joint selectivity project was 26 kg per 1000 m net – well above the target viability level of 6.2 kg set in collaboration with fishermen based on previous catch records. Of all undersized fish caught in the trials, 75% were alive and could be released back into the lake, albeit with an uncertain fate. Besides for releasing undersized fish, fishermen also occasionally released fish that were over the minimum legal size or of spawning color so that they could spawn. The overall by-catch in the collaborative trials was significantly lower than the by-catch of undersized fish by nets with similar mesh sizes (43 mm) during gillnet monitoring in the summer when stocks have more of an overlapping spatial distribution. The amount of whitefish in relation to undersized Arctic charr and brown trout per effort was almost 45 times higher in the trials as compared to when gillnets were distributed randomly as part of monitoring in the summer. In 61% of the fishing trips, there was no by-catch at all. The most promising sites, with a low quota between by-catch and whitefish catches, were all adjacent to whitefish spawning areas. The best time periods were autumn and winter – close to spawning periods when whitefish aggregate. Fishermen, when asked about optimal time of fishing, said that it is not possible to prepare the valuable roe of whitefish if they are caught too close to spawning, thus favoring fishing close to the spawning period as opposed to during spawning. As predicted in selectivity models, the by-catch increased with decreasing mesh-size in all species. On the other hand, given that the potential catch of whitefish peaked at 40–43-mm mesh size, we recommend that the allowed mesh size could be as large as 43 mm, given that the specific fishing site has a documented low rate of by-catch. Some fishermen also stressed that larger whitefish are easier to sell and that parts of the catch caught in 40-mm mesh size nets was too small. In

Fig. 4.3 Left: Undersized and/or spawning colored Arctic char and trout being removed carefully from gillnets and released back in the lake by Johnny Norrgård. Right: GAP2 scientist Thomas Axenrot and fisherman Daniel Ståhl are pulling gillnets in L. Vättern in February as ground truth for hydroacoustic studies. (Photos by J. Alcalde and A. Asp)

areas with moderate by-catches, we recommend a higher mesh-size of at least 46 mm to avoid by-catch. This trial engaged fishermen more than in other activities within the case study, mainly because it addressed a difficult problem that they felt was important to solve. In addition to contributing their experiential knowledge related to fishing sites and seasons, fishermen were able to contribute significantly in all aspects related to the technological adaptations of gillnets (mesh-sizes, thread diameter, net height, how the nets are sewn, etc.).

4.2.3 *Test of Pontoon Traps*

The results from this work are described in more detail in Setzer et al. (2017). Three fishermen and scientists from the GAP2 project collaborated to test the potential of using a recent Swedish invention for whitefish fisheries, the pontoon trap, as a means to enhance the selectivity of the fisheries. To the best of our knowledge, this gear has never been used before in a lake. The trap is a passive fishing gear that is elevated to the surface using compressed air inflated into cushions attached to the top and bottom of the trap chamber (Hemmingsson et al. 2008). To further evaluate the selectivity of this gear, we mounted a selection bar at the side of the chamber (Fig. 4.4). The selection bars were covered with a Dyneema ® net every second time the trap was emptied. Thus, half of the time fishing took place with a selection bar and half the time without a selection bar.

At one site, whitefish catch was considerable (630 kg), whereas at the other, it was dismal (3 kg). Eight other fish species were also part of the catch. Perch, northern pike, and bream were landed and sold. Arctic charr, eel, burbot, roach, and ruffe were released alive. There was a significant difference in the size of whitefish as well as in the number of small-bodied species when comparing periods with a selection bar with periods without a selection bar. To summarize, pontoon traps could be a valuable addition to the arsenal of lake fishermen. Even though we used one of the smallest types available, mainly built to catch perch, we were able to get a significant catch of whitefish. The catch was highest during the spawning period when

Fig. 4.4 The pontoon trap chamber (left) and the selection bars (right). (Photo by A. Asp)

whitefish migrate closer to the shoreline. All by-catches could be released alive and many fish were apparently also able to escape through the selection bars. Fishermen were satisfied with the catch rate and the easy-to-handle construction used when emptying the trap. They were, however, skeptical about the robustness of the trap during harsh weather conditions and extreme catches and were also worried about the trap being too expensive relative to its catch potential.

4.2.4 Release Mortality of Arctic Charr in the Trolling Fishery

Besides the commercial fisheries, there is also an important recreational fishery in Lake Vättern. Recreational fishers faced similar by-catch problems to those in the Arctic charr commercial fishery. A significant proportion of the catch in the recreational fishery is often released back into the water (up to 68% according to Alenius and Hallden 2013). The main reasons for returning fish to the water are that fish are smaller than the legally permitted size, that the daily catch quota has already been exhausted, and that there are ethical reasons for adopting a catch-and-release approach. To understand the magnitude of these problems and to find measures to address them, we conducted a study with a group of recreational fishermen (previously published in Swedish by Norrgård et al. 2015) to investigate the release mortality of Arctic charr in Lake Vättern. We studied the potential release mortality by following the fate of released fish. A subsample (38 out of 62 in total) of the released individuals was followed by equipping them with acoustic transmitters in order to follow their fate over a long time period after their release. The results showed that roughly one-third of the released fish die, mostly soon after being released. The most critical period for released fish was the first hour after their release. Of the released fish, 10% died immediately after their release due to injuries related to the fishing method, 19% died due to predation from birds while swimming close to the surface, and 2% died after diving down from the surface to deeper waters. The study was conducted in late summer when fishing intensity is the highest. During this part of the year, water temperature is normally the highest and Arctic charr aggregate at deeper waters, potentially making them more sensitive to being caught and released. We conclude that release mortality can significantly reduce the recovery rate of the Arctic charr stock and that this phenomenon deserves more attention among fishermen as well as managers. Further attention should be allocated to developing measures to mitigate this problem and develop guidelines for handling of fishes that are released, preferably in collaboration with local recreational fishermen.

4.2.5 *Using Acoustics to Monitor Deep-Spawning Aggregations of Whitefish in Collaboration with Fishermen*

The results from this work are described in more detail in Sandström et al. (2016). As described earlier, one of the best alternatives to promote an efficient fishery of whitefish with limited by-catches is to target whitefish adjacent to their spawning areas in the very deepest rifts in the midwinter months (December-February). To investigate whether it would be possible to identify spawning aggregations and estimate their density and biomass, a pilot study was conducted using hydroacoustics in cooperation with a local fisherman. The hydroacoustic equipment could be fixed on the fishing boat if a few minor adjustments are made. This assessment showed that using hydroacoustics in collaboration with fishermen could provide information that would enable managers to identify specific areas where fishermen could get permits for targeting deep-spawning whitefish in the winter when by-catches of other species normally are lower. The biomass of whitefish in the investigated area was estimated to be 238 tons. This figure should be seen as an absolute maximum, given that some whitefish could have been confused with other species, mainly burbot. If the biomass of deep-spawning whitefish were homogenously distributed all over the deep rift (which constitutes 13% of the lake surface area), the maximum estimated biomass of the spawning stock would be 1 763 tons. Even though these estimates are relatively uncertain, they confirm that whitefish stock is considerable and that it is currently being exploited at a low rate. Winter catches of whitefish, before the GAP2 project started, were below 1 ton, whereas they have now increased to 3–6 tons annually. To summarize, this study confirms that there is a discrete stock that is aggregating and spawning in the deep rifts of Lake Vättern from mid-December to early February. Using hydroacoustics in collaboration with fishermen could be a promising method to find suitable fishing areas and further assess this unique phenomenon (Fig. 4.5).

Fig. 4.5 A transducer was fixed to a tow body mounted on a fishing boat (left); GAP2 scientist Thomas Axenrot (right) is temporarily working as a "captain" on the fishing boat; the computer used to log acoustic data on whitefish can be seen to his right. (Photos by A. Asp)

4.3 Discussion

An important determinant of long-term success of participatory research is the incentives that stakeholders have to participate. In our case, fishermen had numerous incentives to partake in the collaboration. One main factor that motivated fishermen to participate was the fact that they shaped the research topic in line with the problems they wanted to solve. Another factor was that the project enabled fishermen to target fishing areas that were previously closed. Fishermen were also, hopefully, motivated by the actual collaboration and learning experience. During the project, there were several interviews with participating fishermen and members of the comanagement group (Garavito-Bermúdez et al. 2014; Stöhr et al. 2014; Lundholm and Stöhr 2014; Jacobsen et al. 2011). Stöhr et al. (2014) recognized the crisis in the fishery experienced by fishermen in the early 2000s as another important driving force that helped to develop trust between stakeholders and hence act as a complementary incentive to move toward a new governance regime. Another important aspect for this case study was the time-span of the endeavor. The project was initiated already in 2008, and since then, we have been able to work continuously with fishermen, enabling us to learn together and develop trust on a personal level. This has been beneficial in terms of outcomes and hopefully also for the collaboration in the long-term.

Another interview survey was conducted by van Mastrigt (2013) as part of her assessment of whether the current collaborative governance regime has improved the sustainability of Lake Vättern's fisheries. She concluded that most fishermen were happy with the collaboration with scientists and that the main incentive that made them participate was the possibility of having an improved fishery in the long run. Second to that was their genuine interest in learning more about their resources and the ecology of the lake.

Most small-scale fisheries share certain important universal prerequisites required for collaborative research. Since fishermen mainly use small boats, the fishing enterprises are also small, constituting mainly of a few persons operating close to home or the harbor. Consequently, they are not necessarily as organized as the large-scale, marine fishers. In our case, however, there was a relatively high degree of organization among fisheries stakeholders, further supported by the highly organized comanagement group. But our experience from other lakes and coastal areas is that this may not always be the case.

Compared to large-scale fisheries, the majority of the exploited fish populations in Lake Vättern can be considered as data-poor stocks and the availability of knowledge on the fisheries and stocks they rely on are often scattered and less available (Berkes et al. 2001). Collaborations with fishermen thus might be of larger relative importance in small-scale fisheries as a cost-efficient way to improve assessments of the fisheries and the status of important stocks. Given the limited size of fishing areas, fishermen often have a very deep knowledge of their respective fishing grounds. It could also be hypothesized that lake fishermen have a more holistic view of the ecosystem they rely on than large-scale fishers due to the closed nature of these ecosystems (Garavito-Bermúdez et al. 2014).

One important issue highlighted in many previous studies has been the importance of finding a balance between participating fishermen's freedom to plan and design their own activities and ensuring that the study design is optimal for conducting a powerful statistical analysis. On the one hand, giving fishermen the freedom to adopt certain ideas to their specific fishing conditions has been a positive incentive for them to engage in the project. Fishermen are used to working in a flexible manner and adapting to rapid changes in weather and the behavior of their target fishes. On the other hand, without a common sampling strategy, the data collected are more difficult to analyze and, in the worst-case scenario, impossible to utilize in testing the hypotheses of the collaborative project. This dilemma can be partly solved by using some of the more advanced and flexible modeling techniques that are now available (GAMs, mixed modeling, etc.). To a certain extent, some of these statistical problems can also be overcome by being able to collect very large datasets when multiple fishermen are participating, thus enhancing the power of analyses when the number of observations is increased. In our case, the fact that a common framework for structuring the design of the data collection was decided upon also helped.

Perhaps the largest setback during the project was the discovery that the level of certain types of dioxins and dioxin-like PCBs in the muscle tissues of whitefish was close to and, in some cases, even above the EUs threshold-levels for human consumption. Sweden, Finland, and Latvia have been able to negotiate with the commission to get a derogation for fishermen to sell certain species with levels of organic toxins that occasionally exceed EU thresholds for organic toxins (EU commission regulation No. 1259/2011). This derogation, however, is not applicable to whitefish, as at the time of negotiations, it was not known that whitefish had high levels of toxins. This means that the future of the whitefish fishery is uncertain, at least in the short-run. Whereas the levels of PCBs are decreasing in Arctic charr from Lake Vättern at a rate of 3% and that of dioxins at a rate of 4% per year (Danielsson and Bignert 2009), there is unfortunately no equivalent time-series data available on whitefish. It is, however, likely that levels of toxins are decreasing in whitefish, as well as in other fish species in Lake Vättern. The levels of dioxins and dioxin-like PCBs, the major problem in Lake Vättern and neighboring Lake Vänern, in whitefish are strongly correlated to the fat content of the fish (Karlsson et al. 2018). At the moment, the whitefish stock in L. Vättern is also extremely old for an exploited fish population. Mean ages of catches have increased following the decreased exploitation rate and are currently often over 10 years. If fishing pressure would increase, leading to a decreased mean age of the population, levels of dioxins and dioxin-like PCBs could be hypothesized to decrease, and thus, they would most likely seldom exceed the critical levels set by the EU for fish sold on the market. It must be emphasized that this issue has currently not been resolved and that it is unclear what long-term impacts this might have on the whitefish fishery.

The ability of comanagement groups to foster better organizational structures and collective flexibility have been viewed as vital to the prospect of such fisheries to rapidly adapt to the introduction of new management tools or regulation systems (Gray 2005). However, if, as in our case, the comanagement group lacks the direct mandate to regulate the fishery, there is a risk that participating stakeholders become disappointed and lose their interest in collaboration. Besides the feeling of disappointment as a result of not having a complete mandate over fishing regulations, stakeholders feel that the process of reaching a desired result has been too slow. This was highlighted by Stöhr et al. (2014) in their analysis of the processes determining the success of comanagement groups (where Lake Vättern was included as one of the case-studies).

4.4 Summary and Conclusions

The results from our case study are the product of four years of intense collaborative research. We believe that our experiences have generic qualities relevant to others interested in adopting a participatory approach when working with fisheries science in small-scale fisheries and inland fisheries in particular. Our results show that it is possible to solve urgent management problems of relevance to stakeholders by having fishermen and scientists collaborate. Fishermen identified a form of whitefish that rarely before has been reported in Scandinavian waters and assisted researchers in finding the exact spawning sites of these fish. They also came up with several ideas of importance to the technical development of the fisheries. Thus, fishermen helped to increase the value of the research. This testifies to the added value of combining different sources of knowledge and experience.

The main objective of this case study was to test whether there was potential to simultaneously improve the selectivity and efficiency of this fishery. We believe that our results show that there is a significant potential to reach relatively ambitious selectivity targets and still have a reasonably profitable fisheries, given that fishermen can sell their catch for a decent price. The product of the case study was compiled as a set of recommendations (Setzer et al. 2017) for the comanagement group and also forwarded to responsible authorities. The recommendations cover (i) future collaborations among fishermen, scientists, and regional managers, (ii) fisheries management regulations, (iii) research on L. Vättern fishes and fisheries, and (iv) implications for monitoring of fish stocks in L. Vättern.

Acknowledgments First and foremost, we are very thankful for the kind reception given to us by all the participating fishermen. This helped sustain the project even in the harshest winter conditions. Without their help, this research would not have been possible. We would like to especially acknowledge the support and engagement of all the participants in Lake Vättern Fisheries comanagement group. We are also very thankful for the assistance provided by the Master's student Audrey van Mastrigt and the PhD student Mikael Lundin (now Dr. Lundin!). We express our gratitude to Anders Asp, Michael Bergström, Malin Hällbom, Fredrik Engdahl, Tanja Martins, Magnus Andersson, Eva Bergstrand, Anders Kinnerbäck, and Daniel Rydberg for valuable assistance with the field work and for helping out with laboratory analyses. Eeva Jansson, Tore Prestegaard, Stefan Palm, Johan Dannewitz, and Sara Bergek are acknowledged for their support with the analyses of whitefish genetics and population structure. We are also thankful for the support given by the GAP2 consortium. Part of this work was financed by SWAM, the Swedish Agency for Marine and Water Management. The study on release mortality of Arctic charr was partly funded by habitat restoration funds of the County Administration Board of Jönköping, Karlstad University, and the European Fisheries Fund. The work pertaining to developing a gillnet selectivity model was funded by grants from Formas (The Swedish Research Council for Environment, Agricultural Sciences and Spatial Planning) and the European Science Foundation (ESF). Finally, we are thankful for the support given by our contacts at SWAM, Fredrik Ljunghager, and Jens Persson.

References

Alenius B, Halldén A (2012) Fritidsfisket i Vättern 2010 – Sammanställning av enkätsvar och fältobservationer. Vätternvårdsförbundets rapport, 114. ("The recreational fisheries in Lake Vättern 2010 – summary of a questionnaire and a field study")

Alenius B, Hallden A (2013) Fritidsfisket i Vättern 2010. Sammanställning av enkätsvar och fältobservationer. Rapport från Vätternvårdsförbundet. Nr 114. 84 pp. (In Swedish)

Axenrot T (2014) Vätterns pelagiska fiskbestånd. In: Årsskrift 2013. Vätternvårdsförbundet rapport 112, 76–89. ("Pelagic fishes in Lake Vättern")

Berkes F (2009) Evolution of co-management: role of knowledge generation, bridging organizations and social learning. J Environ Manag 90(5):1692–1702. https://doi.org/10.1016/j.jenvman.2008.12.001

Berkes F, Mahon R, McConney P, Pollnac R, Pomeroy R (2001) Managing small-scale fisheries. Alternative directions and methods. International Development Research Centre, Ottawa

Carlsson L, Berkes F (2005) Co-management: concepts and methodological implications. J Environ Manag 75:65–76. https://doi.org/10.1016/j.jenvman.2004.11.008

Cowx IG, Arlinghaus R, Cooke SJ (2010) Harmonising recreational fisheries and conservation objectives for aquatic biodiversity in inland waters. J Fish Biol 76:2194–2215. https://doi.org/10.1111/j.1095-8649.2010.02686.x

Danielsson S, Bignert A (2009) Miljögifter i röding från Vättern, 1965–2006. In: Miljögifter i fisk och kräftor (Ed. Måns Lindell). Vättenvårdsförbundet rapport, 101. ("Environmental toxins in Arctic charr from Lake Vättern 1965–2006")

Dietz T, Ostrom E, Stern P (2003) The struggle to govern the commons. Science 302(5652):1907–1912. https://doi.org/10.1126/science.1091015

Ekman S (1916) Om Vätterns näbbsik. Svensk Fiskeritidskrift 25(4), 101–107. ("On the beak whitefish of L. Vättern")

Fiskeriverket (2007) Regional och lokal samförvaltning av fiske. Rapportering av regeringsuppdrag. 147 pp. ("Regional and local fisheries co-management. Report to the Swedish government.")

Garavito-Bermúdez D, Lundholm C, Crona B (2014) Systems thinking and resource management: linking a conceptual framework on systems thinking with experiential knowledge. Environ Educ Res 22(1):89–110. https://doi.org/10.1080/13504622.2014.936307.

Gray T (2005) Participation in fisheries governance. Springer Verlag, Dordrecht

Hemmingsson M, Fjälling A, Lunneryd S-G (2008) The pontoon trap: description and function of a seal-safe trap-net. Fish Res 93:357–359. https://doi.org/10.1016/j.fishres.2008.06.013

Hudson AG, Vonlanthen P, Seehausen O (2011) Rapid parallel adaptive radiations from a single hybridogenic ancestral population. Proc R Soc Lond B 278:58–66. https://doi.org/10.1098/rspb.2010.0925

Huntington HP (2000) Using traditional ecological knowledge in science: methods and applications. Ecol Appl 10:1270–1274. https://doi.org/10.2307/2641282

Jacobsen RB, Wilson DCK, Ramirez-Monsalve P (2011) Empowerment and regulation – dilemmas in participatory fisheries science. Fish Fish 13(3):291–302. https://doi.org/10.1111/j.1467-2979.2011.00434.x

Johnson TR, van Densen WLT (2007) Benefits and organization of cooperative research for fisheries management. ICES J Mar Sci 64:834–840. https://doi.org/10.1093/icesjms/fsm014

Jonsson T, Setzer M, Pope JG, Sandström A (2013) Addressing catch mechanisms in gillnets improves modelling of selectivity and estimates of mortality rates: a case study using survey data on an endangered stock of Arctic charr. Can J Fish Aquat Sci 70(10):1477–1487. https://doi.org/10.1139/cjfas-2012-0472.

Karlsson M, Andersson G, Bohman P, Hållén J, Sandström A, Viktor T (2018) Dioxiner i fet fisk - Hot och utvecklingsmöjligheter för svenskt småskaligt kust- och insjöfisk. IVL Rapport B 2301. 32 pp

Linke S, Bruckmeier K (2015) Co-management in fisheries – experiences and changing approaches in Europe. Ocean Coast Manag 104:170–181. https://doi.org/10.1016/j.ocecoaman.2014.11.017

Lundholm C, Stöhr C (2014) Stakeholder dialogues and shared understanding: the case of co-managing fisheries in Sweden. Special issue on communication for and about sustainability. Sustainability 6(7):4525–4536. https://doi.org/10.3390/su6074525

Mackinson S, Neville S, Raicevich S, Worsøe Clausen L (eds) (2008) Good practice guide to participatory research between fisheries stakeholders and scientists. GAP project deliverable 1, 23 pp

Millar RB, Fryer RJ (1999) Estimating the size-selection curves of towed gears, trap, nets and hooks. Rev Fish Biol Fish 9:89–116. https://doi.org/10.1023/A1008838220001.

Norrgård J, Sandström A, Alenius B (2015) Överlevnad hos återutsatt röding vid trollingfiske. Vätternvårdsförbundet rapport, 117, 20 pp. ("Survival of released Arctic charr in trolling fisheries")

Piriz L (2004) Hauling home the co-management of coastal fisheries: a study on institutional barriers to fishermen's involvement in the management of coastal fisheries on the West coast of Sweden. Dissertation, Göteborg University, Sweden

Sandström A, Andersson M, Edsman L, Degerman E, Hammar J, Ragnarsson-Stabo H (2014) Fiskets fångster och trender för Vätterns kommersiella fiskarter. In: Årsskrift 2013. Vätternvårdsförbundet rapport, 112, 76–89. ("Fishery catches and trends in the status for commercial fishes in Lake Vättern")

Sandström A, Asp A, Axenrot T, Ståhl J (2016) Hydroakustisk undersökning vintertid av djuplekande sik i Vättern. Vättern Fakta 2:2016. 13 pp. ("Mobile hydro-acoustics to monitor deep-spawning aggregations of whitefish in Lake Vättern.")

Sandström A, Jansson E, Dannewitz J, Bergek S, Palm S, Prestegaard T, Holm P, Norrgård J (Unpublished) Cryptic or not so cryptic after all – a collaboration between fishermen and scientists confirm divergence in polymorphic populations of lake whitefish

Setzer M, Sandström A, Norrgård J, Ragnarsson-Stabo† H 2017. Utveckling av sikfisket i Vättern – ett samverkansprojekt med fiskare och forskare. Vätternvårdsförbundet rapport 125, 47 sidor. ("Developing whitefish fisheries in L. Vättern – a collaboratory project involving fishermen and scientists.")

Stöhr C, Lundholm C, Crona B, Chabay I (2014) Stakeholder participation and sustainable fisheries: an integrative framework for assessing adaptive comanagement processes. Ecol Soc 19(3):1–14. https://doi.org/10.5751/ES-06638-190314

Svärdson G (1957) The coregonid problem VI. The palearctic species. Report from the Institute of Freshwater Research Drottningholm 38:267–356

Tiselius D (1723) Uthförlig beskrifning öfver den stora Swea och Giötha Siön Wätter. Uppsala 1723. ("Extensive description of the great Swea and Giötha Lake Wätter")

Van Mastrigt A (2013) Fishery co-management, a sustainable way to develop fisheries? Masters thesis, University of Groningen, 63 pages

Vonlanthen P, Bittner D, Hudson AG, Young KA, Müller R, Lundsgaard-Hansen B, Roy D, Di Piazza S, Largiader CR, Seehausen O (2012) Eutrophication causes speciation reversal in whitefish adaptive radiations. Nature 482:357–362. https://doi.org/10.1038/nature10824

Widegren H (1863) Bidrag till kännedom om Sveriges salmonider. Översigt af Vetenskaps-Akademiens Förhandlingar (1863), 1–78. ("Contribution to the knowledge on Sweden's salmonids")

Chapter 5
Understanding Fishermen-Scientist Collaboration in Galician Small-Scale Fisheries (NW Spain): Validating a Methodological Toolbox Through a Process-Oriented Approach

Duarte Vidal, Pablo Pita, Juan Freire, and Ramón Muiño

Abstract In the early 1990s, Galician Regional Government introduced a community-based fishery management system for some sedentary-invertebrate resources. This meant the introduction of territorial use rights, which provided greater autonomy to decision-making, regulation, control, management and resources assessment. Nevertheless, after three decades, shallow achievements were built upon the collaboration between science and fisheries. The GAP Galician Case Study was born as an attempt to strengthen long-term collaboration among fishermen, scientist, and policy-makers. Taking into account the background and lessons learnt in similar processes, the research team tried together with relevant fishing actors, to collect meaningful information, to initially conduct a building process of a marine reserve, establishing new management formulas through the exploration of use rights for local users. Social and political constraints resulted in an adjustment of the original approach, focusing the CS on the collection of information on marine ecosystems and fisheries. Recovering local knowledge through the design and implementation of participatory tools, it was possible to address accurate spatial information about marine ecosystems and fishing activity, always challenging to achieve in small-scale fisheries. Beside technical outcomes, the Galician CS was an

D. Vidal (✉)
Department of Fisheries Socioeconomic, Centro Tecnolóxico do Mar – Fundación CETMAR, Vigo, Spain

P. Pita
Department of Applied Economics, Faculty of Economics and Business Administration, University of Santiago de Compostela, Santiago de Compostela, Spain

J. Freire
Tecnológico de Monterrey, Business School, Mexico City, Mexico

R. Muiño
Department of Biology, Faculty of Science, University of Coruña, Coruña, Spain

P. Holm et al. (eds.), *Collaborative Research in Fisheries*, MARE Publication Series 22, https://doi.org/10.1007/978-3-030-26784-1_5

experiment to put some light into process-oriented approaches that enable collaboration among fishermen, scientists and policy-makers.

Keywords Galician fisheries · Territorial user rights · Experience-based knowledge · Métier · Small-scale fisheries · Comanagement

5.1 Introduction

Galicia is a region which has been historically linked to the sea. According to the Statute of Autonomy of Galicia[1], the Autonomous Community of Galicia has exclusive competencies into fisheries management over the estuaries and inland waters[2] (Fig. 5.1).

Since 1990, Galician artisanal fisheries have been governed by a dual management system, a top-down command-and-control and a comanagement system. Beyond the centralized and hegemonic command-and-control management, the comanagement system is based on a bottom-up approach, where regional managers provided an assignment of territorial user fishing rights for coastal communities (TURFs).

In spite of some conflictive beginnings, this initiative succeeded in facilitating the recovery of shellfish grounds, allowed for the improvement of organization, while it enhanced responsibility and compliance in the management and control of marine resources by the fishermen and coastal communities.

This social change was accompanied by the creation of a key figure in the whole process, namely the technical assistance (TA)[3], meaning persons hired as technicians by regional government to act as intermediary between policy-makers and fishermen.

Traditionally, policy-makers, scientists, and the fishing sector have relationships based on mistrust, which further causes a lack of communication, transparency, and asymmetry in management (Reed 2008). Therefore, the main role for the TA is to try to connect different types of knowledge and be able to take them to a specific aim.

[1] Ley Orgánica 1/1981, de 6 de abril, de Estatuto de Autonomía para Galicia «BOE» núm. 101, de 28 de abril de 1981 Última modificación: 17 de julio de 2010 Referencia: BOE-A-1981-9564

[2] Strip of sea adjacent to the territory and internal waters of the coastal State, where the State exercises full sovereignty both over surface waters, supra-adjacent air space, seabed and subsoil. The maximum limit of the territorial sea is 12 miles (UNCLOS, art. 2, 3 y 4). Article 149.1.19ª of the Constitution grants the State exclusive competence in matters of marine fishing, "notwithstanding the competences granted to the Autonomous Communities within the regulation of the sector" (Suarez de Vivero and Martinez 2007)

[3] Technical assistances, subsidized since 1993 by the government, have the role of scientific-technical advisors who act as liaisons between fishermen' organizations and the administration. Hence, they have a prominent role in the management of some resources, particularly those under community exploitation models based on TURFs.

Fig. 5.1 Legal inshore waters limits in the Autonomous Region of Galicia

GAP was born as an attempt to demonstrate the role and value of stakeholder-driven science within the context of fisheries governance, and to establish an active long-term collaboration between scientists and fishermen. This could improve the assessment of biological and social factors linked to the coastal fleet.

The Marine Resources and Fisheries Group (RMyP) of the University of A Coruña was involved in several research projects for the design and implementation of marine reserves to inshore Galician waters. This was an opportunity to address the concept of TURFs, trying to extend it beyond the actual shellfish grounds and sedentary-invertebrate marine resources. The new formula of Marine Reserves of Fishing Interest applied in Galicia can be understood as a "Trojan horse" for the exploration of TURFs and to provide management plan recommendations for both low- and high-mobility resources.

A set of conflicts came into light during the first stage of the design of a new marine reserve proposal, at the coastal community of Aguiño. What was agreed between local fishermen and scientists from RMyP was to work toward generating accurate information for the study area, through the involvement of a greater number of fishermen and guilds that potentially use that fishing area, both frequently and temporarily. It sought to reduce the conflict and to provide new arguments for policy-makers to support a future proposal for a marine reserve within the study area. Nevertheless, neighboring guilds whose fishers would fish in the proposed area refused to continue the ongoing process. It was thus prioritized to obtain data from those guilds accepted to be involved, and was hoped that once some data was obtained, this would be an incentive from other guilds to take part.

The Galician CS is an example to strengthen fishermen-scientist collaboration through the use of a shared knowledge methodology, previously tested in several projects. Data gathered from a "bottom-up" approach support social learning, resulting in increased social and cultural recognition of fishermen and lowered barriers that separate the two cognitive realities. This consequently results in common objectives emerging and building of alliance (social networks) through social interaction processes (McCay and Jentoft 1996; Berkes and Folke 2001; Wiber et al. 2004; Chuenpagdee et al. 2006; Bull et al. 2008; Berkes 2009; Prell 2009; Reed et al. 2009, 2010; Hage et al. 2010).

The Galician CS is trying to respond to issues on how collaboration, coresponsibility, and understanding were created by partnerships; the blocks, which constrained the ongoing actions; lessons learnt from the applied methodology; could technical assistance play a new role? However, the Galician CS will not be able to explain how the collaboration settled to explore the introduction of new TURFs, beyond the previous experiences on Marine Reserves of Fishing Interest.

5.2 The Dual Galician Fisheries Management System

The Galician CS describes the design and implementation process of a participatory methodology aiming to strengthen a long-term collaboration among managers, scientist, and fishermen.

The theoretical framework of this research lies on how different types of knowledge can be complementary in the pursuit of a common goal. Collaboration is first enabled by accepting new forms of knowledge. The inclusion of Experience Based Knowledge (EBK) (Davis and Ruddle 2010; Ruddle 2000) allows scientific information to be complemented along the process. Scientific literature suggest that a combination of both types of knowledge not only can be useful for the empowerment of local communities (Silvano and Valbo-Jorgensen 2008; Huntington 2000, Reed et al. 2009), but can also facilitate a more effective, easy, and accurate monitoring and fisheries management (Raymond et al. 2010; Tippett et al. 2007). It can also allow for more updated knowledge of fishing activity and its impact on resources (Berkes et al. 2000).

The study mainly focuses on coastal fleet, since they generate significant employment (García-Negro et al. 2009; Villasante 2009) and income for fishing communities.

Currently, the coastal fleet constitutes 90.8% of the total Galician fishing vessels. The coastal fleet is comprised of vessels between <12 and 24 meter length (Table 5.1). Coastal vessels work on a "day-to-day" basis, although it does not mean all of them share similarities in social and economic structure. In fact, great variations exist between different vessel segments in terms of adoption and investment in technology, on-board organization and distribution, fishing strategies, and fleet dis-

Table 5.1 Galician fishing vessels according to main fishing grounds and length (meters). 2015

	<12		12–18		18–24		>24		Total	
Fishing grounds	N°	%	N°	%	N°	%	N°	%	N°	%
NW Atlantic	**3829**	**88,70**	**261**	**6,05**	**97**	**2,25**	**130**	**3,01**	**4317**	**95,81**
SSF	3812	96,02	158	3,98					3970	91,96
Trawler					4	5,06	75	94,94	79	1,83
Seiner	14	9,33	68	45,33	63	42,00	5	3,33	150	3,47
Guillnets	1	2,70	20	54,05	14	37,84	2	5,41	37	0,86
Bottom longline	2	7,69	14	53,85	9	34,62	1	3,85	26	0,60
Surface longline			1	1,82	7	12,73	47	85,45	55	1,27
EU fishery					**2**	**2,63**	**74**	**97,37**	**76**	**292,31**
Trawler							30	100,00	30	39,47
Bottom longline					2	4,35	44	95,65	46	60,53
International fishery			**1**	**0,88**	**3**	**2,65**	**109**	**96,46**	**113**	**376,67**
Trawler							34	100,00	34	30,09
Seiner							3	100,00	3	2,65
Surface longline			1	1,32	3	3,95	72	94,74	76	67,26
Total	**3829**		**262**		**102**		**313**		**4506**	

Source: www.pescadegalicia.com

tribution (García-Negro 2005). Thus, vessels below 12-meter length are basically considered within the joint small-scale fisheries (SSF) (Table 5.1), which are characterized by a selective use of gear (multigear and multitarget), limited environmental harm, less sophisticated technology (Freire and García-Allut 2000; García-Allut and Freire 2002), lower investment costs, and higher values per unit of catches (Chuenpagdee et al. 2006).

If we set aside both the community EU fishery and the high seas, the Regional Galician Government is set in a context of a dual fisheries management structure: A centralized command-and-control; a top-down system regulated by effort mechanism, temporary cessation, subsidies, and regional/national (authority) surveillance and control (Table 5.2).

This type of management is mainly focused on mobile target species (*Maja brachydactyla, Necora puber, Octopus vulgaris, Merluccius merluccius, Sardina pilchardus, Trachurus trachurus, Sepia officinalis, Maja brachydactyla, Lophius piscatorius, Raja undulata, Scomber scombrus, Scomber japonicus, Dicentrarchus labrax, Pollachius pollachius*, among others). The centralized management system is characterized by low compliance, high levels of intersectoral poaching, and a consistent decline of most of the fish and crustaceans catches (Freire and García-Allut 2000).

Moreover, another management system emerged in the 1990s: a community-based co-operative system for sedentary marine resources where access to natural grounds is granted by the regional body to licensed fishermen belonging to "guilds," through an annual exploitation plan (Molares and Freire 2003). Based on the allocation and assignment of TURFs, local communities gain importance not just as

Table 5.2 Current management systems in Galician fisheries

Characteristics	Command-and-control	Community based (from 1990s)
Property rights	Regional government	Communitarian (territory)
	No access restrictions	Access limits
Decision-making		
Flows	Top-down	Bottom-up
Institutions	Autonomous government	Fishermen organizations: Guilds
Knowledge	Scientific	Scientific + traditional (EBK)
Regulations	Catches: Daily quotas per vessel/ fisher/crew	Catches: TACs (using daily quotas per fishermen/vessel and total annual effort)
	Effort: Gears, closed seasons, minimum sizes (sexes, reproductive females)	Effort: Minimum sizes, seasons, rotations
Target species	Mobile marine resources	Sedentary-invertebrates resources
Surveillance control		
Who	Government	Government + fishermen
Punishment	Legal	Legal + social

authorized users but as "proprietors" who possess collective-choice rights and can thus be involved in the management of barnacles, scallops, razor clams, cockles, clams, or sea urchins. It also means that as a proprietor, they can authorize how resources may be utilized and who may access resources (Schlager and Ostrom 1992).

The exploitation annual plan, frequently supported by the TA, defines the different units of the comanagement system and has the following components (Molares and Freire 2003; Macho et al. 2013):

- Organizational components: number of authorized fishers, working days, gear permitted, fishing grounds allowed (no-take zones or temporary access areas), monitoring, and control (designated sites to control catches).
- Biological components: target species, stock assessment (sampling and self-sampling methods).
- Economic components: economic and production objectives, catch quotas (per day), first sale and market production, and a financial program (including yearly incomes and expenses).

A dual management system coexists not only with regards to different fisheries but also at the same fleet level. It would be normal for a fisherman to choose to carry out a fishing activity within a fishery regulated under TURFs and the next day decide to carry out a fishing activity managed under a command-and-control system. It is particularly relevant how both systems intersect and especially within the SSF. Legal procedures in SSF do permit the use of five different gears at the same fishing annual cycle. For artisanal fishermen, it is quite common to assume a different "behavior" (not just in terms of fishing), depending on the management system followed.

Therefore, most of the artisanal fishermen are immersed in both management systems. In spite of the existence of TURFs since 1990, there are still significant open access fisheries (common octopus, spider crab, fish rock, flat fish, etc.), which are regulated by central management. Here, the role of fishermen is basically advisory (Pomeroy and Berkes 1997), neither are participating, nor are responsible actors into decision-making.

5.3 Description of the CS Process

The Galician CS aimed to go deeper into the findings of the first phase of the project, which was one of the first trials to create a long-term collaborative process between scientists and fishermen from the Guild of Carreira-Aguiño.

As stated above, the main target of the CS was to develop a methodology for the construction of an accurate database of fishing activity into an area where a Marine Reserve of Fishing Interest was being designed by a working group of local fishermen and scientists (Fernández-Vidal and Muiño 2014). Following a public presentation of the state of the marine reserve design process by members of the Working Group, some disagreeing voices from the local fishing community suggested the need to establish strong and stable links with the neighboring Guilds, not just in order to get their support, but also to involve them during the decision-making of concrete measures. The political context and its underlying power relations need to be placed on a table (Coglan and Shani 2006) into participatory research.

Taking into account the background and the high level of confidence acquired by RMyP within the fishing community, it was relatively easy to continue the job. The Guild of Carreira-Aguiño led the communication process with the neighboring guilds, with the facilitation of the Galician Fishermen Federation (GFF), an entity with legal standing and a decade of experience as interlocutor between fishermen and policy-makers.

However, socio-political issues arising between different players and a limited capacity to resolve them came up and only those guilds who were like-minded got involved into the project. Thus, the CS extended the participation to neighboring Guilds of Cambados and Ribeira (Fig. 5.2).

The initial focus to design a marine reserve as a management tool was replaced by a more focused view on obtaining data, on the uses and distribution of fishing activity into the study area. The new approach was established based on a participatory methodology to strength transparency and understanding between partners. This was applied through three steps:

- Defining the role of stakeholders and scientists: in order to establish confident collaborative partnerships, it was necessary to clarify the different roles within the CS; scientists needed to rebuild their role as experts and fishermen to review their view against scientists.

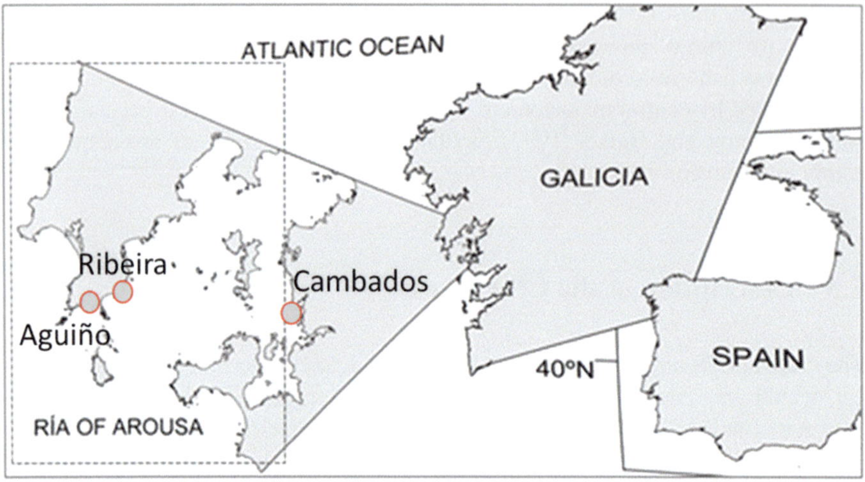

Fig. 5.2 Study area GAP2 Galician CS

- Designing and implementing of tools for data collection: using social research techniques, three kind of data were obtained:
  - Information on substrates.
  - Information on fishing grounds and target species.
  - Information on métiers included in the study area.
- Monitoring of fishing activity: the activity of the main métiers obtained in previous stage was monitored through logbooks and loggers.

5.3.1 *Define the Role of Stakeholders and Scientists*

Usually, when a fishermen's organization approach to a scientific to resolve different issues, what they are really demanding is not a long-term preservation solution but rather support or funding to solve (in the short term) one or more issues that lead to the catches lost or decrease (Freire 2005).

The Galician CS was born assuming a change related to the role of science, from an expert knowledge and with a "high profile" in management to an actor with a "low profile." The role of the facilitator was pointed out – as a figure to support productive and affective interactions among the stakeholders; being responsible for condensing, analyzing, and minoring back the outcomes of each meeting (van Buuren and Loorbach 2009) without taking part in decision-making.

That "low profile" was driven along the whole process. The way science approached fishermen was different; it were the guilds, supported by TA, who decided who, how, and when to participate assuming responsibilities and leading the next steps of the research.

The GFF was in charge to get in contact with the guilds as it is frequently the role of GFF to organize meetings with policy-makers and guild members in marine and fishing affairs. TA was acting as an intermediary between fishermen and scientist and their main role was to reduce asymmetries and to avoid miscommunication. Their strategic positioning as a member of the guild gives them recognition from the local community, and at the same time their expert knowledge allows them to engage in an expert's dialogue with scientist and policy-makers.

Several workshops were held between scientists (facilitators), fishermen and TAs; the following resolutions were adopted:

1. *Fleet segment to be included.* The most representative type of fleet was chosen in each of the three guilds. The fishing fleet of the Guild of Carreira-Aguiño, Ribeira, and Cambados is mostly focused on the exploitation of low-mobility resources, like barnacles, razor clams, sea urchins, scallops, cockles, and clams, and all of them are managed by TURFs. Nevertheless, the Guilds of Carreira-Aguiño and Ribeira (only a few boats in the Guild of Carreira-Aguiño) are characterized by high number of fishing vessels which target resources that are highly mobile and are managed under a command-and-control system. The Guild of Ribeira is mainly focused on octopus (fishing pots vessels), with a number of vessels focusing on this catch throughout the fishing cycle, while in the Guild of Cambados, there are purse seiners targeting small pelagic.
2. *Selection of the study area.* The study area was delimited according particular fleet segments use of marine spaces. The selected area covers 456 km^2.
3. *Selection of fishermen.* Fishermen involvement was voluntary without any participation enforcement mechanism. The head of each guild contacted fishermen from the local guild to participate in the research according to principles of urgency, leadership (Mitchell and Shortell 2000; Selman 2004) and a "personal" tie (relationship). Other fishermen were approached by the project through a "snowball technique" (Luyet et al. 2012). Following the first stages of the project, fishermen not previously selected showed an interest and were thus included to participate.
4. *Data treatment and confidentiality.* Usually fishermen do not trust scientists because they tend to perceive the latter as government staff with whom there are communication barriers. One fisherman said the following, highlighting the social prejudices about "science'" that many fishermen have:

   > "We are tired of taking biologists on board. Whenever they come we lose a day because we end up showing them around and not fishing (laughs)." (Excerpt from an interview with a fisherman).

Management of knowledge in an "open-access" and "open-competition" environment is of vital importance to artisanal fishermen, precisely because shared knowledge will improve the fishing of future generations. However, fishermen are sometimes reluctant to share information outside their family circles, particularly when other members of the community (direct competition) are present. In order to deal with mistrust, a series of basic principles for action were proposed. These formed the basis for the study and for the planned interaction between scientists and fishermen:

- Voluntary participation of fishermen.
- Confidentiality in the use and treatment of information.
- Continuous communication among fishermen, TAs, GFF, and the research team.
- Validation of results within the fishing sector.
- Sharing of results with the fishing sector.

5.3.2 *Design and Implementation of Tools for Gathering Data*

To achieve the objectives, the scientific team has been using ethnographic methods to discover and record EBK. EBK was collected using two different techniques:

First, mapping exercises (semistructured interviews) were based on a set script that defined the overall direction of interviews and the subjects to be addressed. Fishermen were provided with a series of printed maps showing the study area for which they were supposed to gather information. Transparent acetate sheets were placed over the maps. The interviewee kept the rhythm on the interview, while the formulation of questions was left to the discretion of the interviewer. In order to make the task easier, the area was divided into five zones. Fishermen had ensured that maps had information about existing substrates, gears used, main species, and fishing grounds (Fig. 5.3).

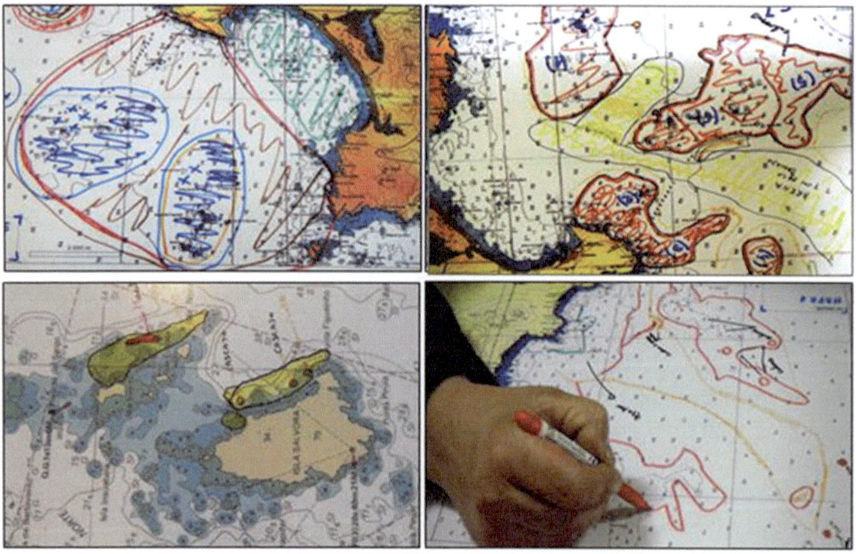

Fig. 5.3 Mapping result (EBK) obtained from fishermen by the scientific team

All information obtained was subsequently processed. Workshops were conducted to validate the results obtained (mapping substrates and fishing grounds) with those fishermen who provided data. Mistakes regarding results, strategies, techniques, and processes were updated (Shepard 2008).

In the second step, a structured and semidirected personal interview recorded technical information about vessels, especially information about target species and gears, and their relationship with the yearly fishing cycle during 2011. The purpose was to obtain information in the most efficient way, leaving some room and flexibility to interviewees in terms of how they answered questions. All information gathered from structured interviews was integrated with those data obtained from the mapping exercises. It helped to identify the different métiers working on the case study. Métier selection was based on gear, target species, fishing grounds, and seasonality (annual fishing cycle).

5.3.3 Monitoring Fishing Activity

A one and a half year voluntary fishery-monitoring program was designed keeping in mind the selected métier (see Results section). Low-cost global positioning system (GPS) data loggers and fishing logbooks were used to determine effort distribution and to assign catches per unit effort (CPUE) to specific fishing grounds within the study area. The logbook design was agreed jointly among fishermen in several workshops designed for this purpose. An operational protocol was also designed through the creation of a specific timeline of tasks and functions. Each institution had its functions clearly identified:

- Technical Assistances:
  - Dissemination of information within their organization.
  - Provide fishermen involved with GPS loggers and logbooks.
  - Download GPS logger data with the designated software.
  - Collect filled-in logbooks weekly.
  - Download GPS data weekly.
  - Communicate any developments in the activity (i.e., addition and loss of participants, obstacles, contributions, etc.).
- Galician Fishermen Federation:
  - Corporative broadcasting and communication of the study.
  - Calling meetings and workshops.
  - Logistics support and supervision of TAs.
- Research team:
  - Presentation and communication of the study.
  - Coordination of activities.
  - Training of TA and fishermen.

- Collection, validation, analysis, and interpretation of the information.
- Presentation of the results.

The starting point for the vessels monitoring stage was a series of training workshops (aimed at training in the use of loggers, logbook design, and protocols). These workshops were conducted for the TA and fishermen involved in the CS. The ultimate aim of the case study was to share results with the users. This is something that does not happen often. Sharing results with fishermen can help them for better agreements and common objectives and strategies. It will also help them in their interactions with external actors who have political power or make decisions regarding fishing territories.

5.4 Results

5.4.1 Building Partnerships

A total number of 47 stakeholders from different institutions took part in the GAP2 Galician CS. They were divided into six categories (Table 5.3).

Seven scientists were involved in the project and two PhD students. It is worth pointing out that policy-makers were not involved in the study. The reasons for their absence will be explained in the discussion section.

Twenty-seven workshops, unequally distributed across the different phases (Table 5.4), were held through the course of the case study. The most number of workshops was held during the planning session (11) where, in addition to presenting the study, the area was defined, organizations and fishermen to be involved were selected, and the roles to be played were assigned to each entity.

A large number of people participated in these sessions. The first batch of fishermen to participate was decided upon. In general, fishermen were enthusiastic although there was some reluctance on their part because they were concerned about confidentiality vis-a-vis the information they shared. Training sessions (8) were also frequent as they were a priority on the development of the different stages. The whole training sessions were scheduled by the scientific team. The validation and return data sessions took place within the approach of engaging fishermen in the review and generation of new knowledge.

Table 5.3 Stakeholders involved into Galician CS

Category	Fishermen	Technical assistant	Nonprofit organization[a]	Scientists	Policy-makers	PhD students
Number (n)	31	3	4	7	0	2

[a]Galician Fishermen Federation

Table 5.4 Communication activities scheduled into Galician CS

Subject	Sessions	Workshop activities
Planning	4	Presentation of the CS: Role assignment
	3	Selection of the CS: Fishermen and study area
	4	Arrangements and logistics
Training	2	Logbook design
	2	Gathering and processing data
	2	Use of logbooks and loggers
	2	Review
Validation	2	Checking performance of loggers
	3	Mapping substrates
	2	Mapping fishing grounds
Return	1	Substrate mapping report

5.4.2 Ecosystem Mapping and "Métier" Identification

5.4.2.1 Cartography of Substrates

Generally, fishermen distinguished two broad substrate types: hard and soft substrates. They were also able to identify several bottom subtypes for each of these general substrates (Fig. 5.4). The highest variability of subtypes was found within the soft substrates, associated basically to shellfish grounds. It is worth noting that each of the three subtypes of rocky substrates corresponds to a particular habitat. Intersection areas were identified as priority habitats constituting the boundaries between sandy and rocky bottoms.

5.4.2.2 Cartography of Fishing Grounds

Fishermen provided information about the main fishing grounds of 40 commercial species. Most of these species were fish (63%) but also included mollusks (23%), crustaceans (13%), and echinoderms (3%). On several occasions, fishermen provided information about nursery areas, as well as information pertaining to the higher and lower productive areas (in terms of abundance and quality). A scale of importance, based on a series of criteria, was established. The final product was a map of the main fishing grounds for each target species (Fig. 5.5).

5.4.2.3 Identification of Métiers

Fifteen métiers were identified in Galician CS (Table 5.5). Each of these métiers was characterized according to gear, target species, and seasonality (annual fishing cycle), using the set of information gathered in the previous stages.

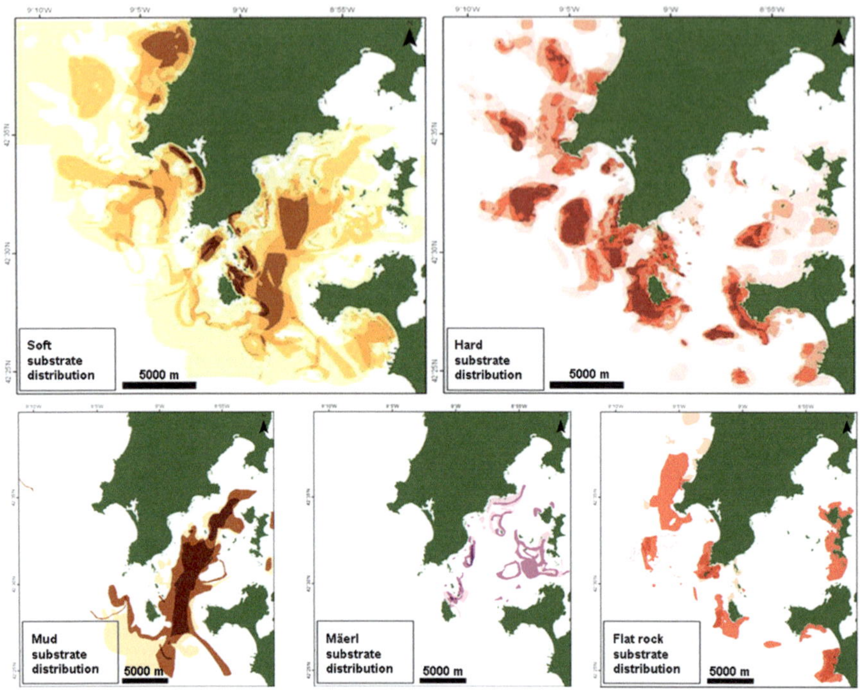

Fig. 5.4 Distribution of different substrate types and subtypes gathered through EBK

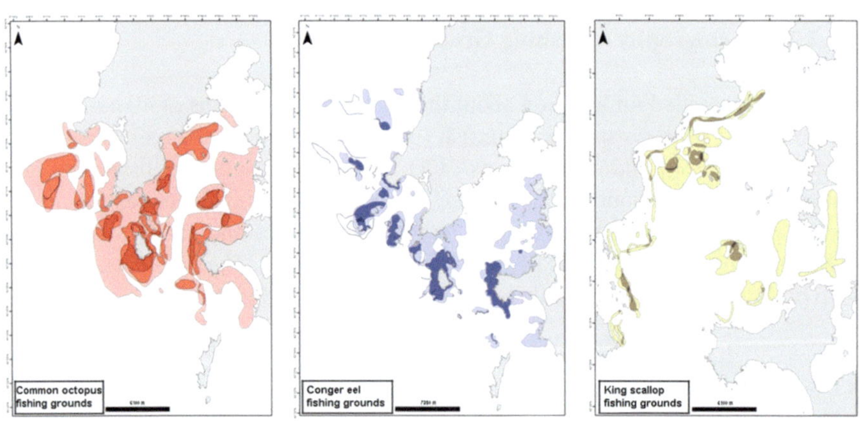

Fig. 5.5 Distribution of fishing grounds related to target species gathered through EBK

Table 5.5 Preliminary métiers obtained

	Gear		Target species		
Métier	Local name	English name	Scientific name	Seasonality (1)	Management system
1	Boliche (2)	Trawl	*Loligo vulgaris*	Jul-Oct	Command and control
2	Buceo	Diving	*Paracentrotus lividus*	Jan-mar	Community based / TURFs
3	Buceo	Diving	*Ensis arcuatus*	Apr-Sep	Community based / TURFs
Ensis siliqua					
4	Cerco	P. Seiner	*Scomber scombrus*	Jan-Dec	Command and control (TAC)
			Trachurus trachurus		
			Sardina pilchardus		
5	Cerco	P. Seiner	*Ammodytidae* spp.	Apr-Oct	Command and control
6	Miño	Trammel net	*Maja brachydactyla*	Dec-may	Command and control
			Labrus bergylta	Jan-Dec	Command and control
			Sepia officinalis		
7	Nasa	Trap	*Trisopterus luscus*	Jan-Dec	Command and control
Conger conger					
8	Nasa	Trap	*Palaemon serratus*	Jul-Jan	Command and control
			Necora puber		
9	Nasa	Trap	*Octopus vulgaris*	Jan-Dec	Command and control
10	Palangrillo	Bottom longline	*Dicentrarchus labrax*	Jan-Dec	Command and control
Pollachius pollachius					
11	Palangrillo	Bottom longline	*Conger conger*	Jan-Dec	Command and control
12	Raspeta	Hand scraper	*Pollicipes pollicipes*	Jan-Dec (3)	Community based / TURFs
13	Rastro ameixa	Clam rake	*Venerupis pullastra*	Jan-Dec	Community based / TURFs
			Venerupis rhomboides	Jan-Dec	Community based / TURFs
14	Rastro vieira	Scallop dredge	*Pecten maximus*	Nov-mar	Community based / TURFs
Aequipecten opercularis					
15	Trasmallo	Trammel net	*Labrus bergylta*	Jan-Dec	Command and control
			Maja brachydactyla	Nov-may	Command and control

Just five out of all the métiers identified are managed by a community-based system linked to TURFs; the rest of the métiers are regulated by a command-and-control system where regional authorities disposed legal requirements to establish fishing rules and assure compliance. The degree of overlapping between the two

Table 5.6 Main métiers identified in the monitoring vessel stage

Métier	Gear (license)	Target specie	Track	Logbook	%[a]
2	Diving	*Paracentrotus lividus*	166	103	54.8
3	Diving	*Ensis arquatus*	37	36	75.7
4	Purse seine	*Scomber scomber*	232	223	28.5
6	Trammel net	*Maja brachydactyla*	107	61	10.3
9	Trap	*Octopus vulgaris*	79	47	41.8
11	Bottom longline	*Conger conger*	6	17	0.0
12	Hand scraper	*Pollicipes pollicipes*	32	35	50.0
13	Clam rake	*Venerupis pullastra*	101	41	25.7
		Total	**760**	**563**	

[a]Overlaps on the number of tracks and logbooks on the same fishing day and boat

different management systems into artisanal vessels is very common; if we take an example of an artisanal boat with a clam rake and a trap licenses, it is often a case where the skipper gather clams earlier week and octopus at the end of the week. Contrary to those fisheries that two different management systems, the purse seiners vessels (semi-industrial fleet), which focus their business into a monospecific harvest strategy, are immersed just in a centralized management, based on total allowable catches (TACs).

Not all the métiers identified during the monitoring stage were further included, due to low-rates of representativeness of number of vessels, volume of catches, and seasonality.

In addition, as participation was on a voluntary basis, it was not possible to cover a minimum number of stakeholders in all the métiers. Thus, only those with high representability level were finally selected. Therefore, eight métiers were selected for the next stage of monitoring fishing activity (Table 5.6).

5.4.3 *Monitoring Fishing Activity*

5.4.3.1 Monitoring Vessel

The number of tracks, logbooks, and overlaps of different métiers varied significantly. In some cases, the number of data loggers (tracks) gathered was higher than data collected by logbooks. In others, data loggers and logbooks did not correspond to the same fishing day, which meant different overlapping degrees (Table 5.6).

All métiers showed distinct differences regarding the percentage that track data and logbook data show an overlap. Métier 3 had the highest percentages of agreement (<75%) followed by métiers 2 and 12 (≤50%) (Table 5.6). It is important to point out that those métiers are management by TURFs, which means that there is close and accurate monitoring done by TAs. The role of the TA here was appropriate to understand the degree of fishermen compliance along the fishing-monitoring stage.

Although this was not initially, a priority it is worth noting the number of tracks obtained by the purse seine fleet as it showed a high occurrence of the purse-seine fleet within the study area.

5.5 Discussion

5.5.1 About the Process

The main goal of the current research was to explore a different collaboration approach to strengthen the management of low- and high-mobility resources. Galician marine reserves and the link to TURFs was the theoretical framework on which we built new management tools and recommendations. As the design of a new marine reserve proposal reached a point of no progress within the study area, research objectives had to be renegotiated and rescheduled. This modified the strategy, and different roles needed to be assigned.

In the short-term, the analysis focused on the stakeholders' contributions to embrace the complexity of fishery management (Symes and Hoefnagel 2010). Historically, the high technical language of scientific knowledge has contributed to a decrease in the collaboration between scientists and fishermen. Also, the role assumed by scientists, as an expert, did not accommodate for the fisherman's knowledge (EBK), leading to an increasing isolation and rejection from one to another (Jentoft and McCay 1995; Wilson 2003; Jacobsen et al. 2012). The Galician CS contributes by presenting a new way of configuring knowledge exchange into stakeholder collaboration.

The biggest number of participants of the Galician CS came from local fishing communities: 80.8% of the participants were members of institutions and organizations from the fishing sector, and 66.0% of the group being active fishermen.

Academic institutions also contributed with a significant number of participants in the different phases of the project. The University of A Coruña linked from RMyP was the scientific institution mostly involved in the study. Scientists came from different disciplines and contributed with their expertise at different stages of the case study. In the early stages, social scientists were instrumental to the framing and development of the project, while biologists and PhD students helped to conceptualize and to consolidate the research. The role of scientist's team as a facilitator (low-profile) helped fishermen to feel like the owners of the information produced. Additionally, the TAs were essential for the creation of boundary spaces (Stange et al. 2016) as they took part in the organization and research coordination.

On the other hand, the role of the policy-maker was reduced to a secondary level. Though they could help by providing data from official sources, there was the danger that their presence could result in an asymmetric social relationship (Alegret 2002). Their absence provided an entirely living process, far from control and supervision. It was important to put effort into the collection of fishing data, so that stakeholders could have a technical advantage (in the form of scientific arguments) to put pressure in further meetings with policy-makers. The following steps would be that NGOs and other groups from civil society should be included, as well as policy-makers.

The large number of workshops conducted could suggest that scientists did develop an active and fluent communication with fishermen. Usually, joint workshops result in disconnected presentations, with mutual accusations and low rates of agreement. The Galician CS workshops, on the contrary, were no longer seen as being top-down but rather as a process of knowledge creation, debate, and decision-making (horizontal and equitable). Fishermen and their organizations were integrated into the project from the outset as both the judge and the jury.

Decisions were consensual and shared with fishermen' organizations. This had been one of their main demands as earlier there was a significant distance between scientists and fishermen (Freire 2005). The experience of TAs acting as mediators (Macho et al. 2013) helped scientists and fishermen to interact and ensured that information flowed both ways.

Each stage had its purpose; the aim of the training sessions was to engage fishermen and TA in the design and implementation of efficient processes, the validation sessions created boundary spaces to resolve different issues regarding data collection and data analysis, driving the stakeholder interest, while ongoing and future actions were evaluated during the validation sessions in order to promote equity, learning, trust, and respect between partners (Luyet 2012).

5.5.2 About Ecosystem Mapping and Métier Identification

Official statistics do possess certain limitations as standardized monitoring methods are applied to different types of fishing production systems despite the fact that artisanal and coastal fishing are enormously heterogeneous, diverse, and complex (Cambiè et al. 2012). Thus, scientists would usually promote scientific information and management solutions not grounded in the local reality and thus rejected fishermen, which means that other methods are required that would allow us to understand the fishing production system at the local level (Salas et al. 2004; Salas and Gaertner 2004). The Galician CS provided a new set of information about the fishing grounds, target distribution, seasonality, gear distribution, and effort in a way produced and thus agreed jointly.

Additionally, proposals adopted by policy-makers (in the command-and-control management) aiming to decrease fishing effort are frequently promoted. There is limited knowledge around the effort actually applied and how to measure it. Spatial information is also usually absent in fisheries management policies (Pita et al. 2008). Galician CS is incorporating spatial references vis-a-vis fishing, which are facilitating the identification and characterization of different métier units.

Information about substrates shows a clear separation between fleets fishing in sandy and mixed bottom areas (associated with shellfish gathering) and other fleets fishing in rocky bottom and intersection areas. This is important to know as these intersection areas (named beiradas = borders), which mark the boundaries between rocky and sandy bottoms, can only be delimited accurately through EBK. This is because nautical maps do not have the accuracy, nor the information required of other fishing factors such as seasonality. Beiradas are particularly important as they are highly productive areas and especially relevant as some target species used it as nurseries habitats (feeding grounds).

In addition, the integration of information provided allowed us to conclude that the presence of a particular substrate (which "priori" might be suitable for the occurrence of a particular species) does not always correspond with the presence of a fishing ground. While scientists would probably include such areas in their management models, the reality is that fishermen do not use the whole sea uniformly; thus, the integration of this type of information collected through EBK would allow the development of ad hoc management models for each particular area, gear, and/or resource.

Information obtained with respect to métiers is very different from official data. Currently, open access fisheries, however, are not managed through métier units. Only those managed under TURFs are somewhat similar to métiers. This means that 10 out of the 15 métiers identified do not take into account their métier unit for management, which can cause a considerable imbalance in their assessment, monitoring, and planning in the short and medium terms. There are métiers oriented to both single-species and multispecies fisheries. In spite of commonalities between the two, they are composed of different fishing units. Thus, métiers could be sharing the same target species, but they could be using different fishing gears and/or fishing in different fishing grounds. Therefore, understanding all the métier units within the same marine space is central to multifactorial management, which is different from traditional management based almost exclusively on the concept of gears.

5.5.3 *Monitoring Fishing Activity*

Monitoring vessels are meant to oversee the distribution and use of marine space by fleets. Also, they help estimate fishing effort and link it to specific areas. Daily and stable monitoring helps track catches and value different fishing grounds.

Social participation models involve mapping complex processes that are influenced by historical, political, and cultural contexts (Stenseke 2009; Abelson et al. 2007; Sabatier et al. 2005). Activities with a greater technical (science) component temporarily displace fishermen from the spotlight, resulting in decreased fisher participation. It was, however, in such technical phases such as monitoring of fishing that often extends beyond 18 months that fishermen involvement was most relevant. Because of lack of continuous fisher participation, the number of tracks and logbooks was highly variable during this phase. In addition, the lack of overlap between tracks and logbooks resulted in poor fisher involvement with this activity overall. Fishermen had the following to say the monitoring stage:

> "When we come back to shore after spending over eight hours at the sea, we just want to get home (…)."

It would be wrong to say that the selection of fishermen in the early stages (the GFF initially selected the fishermen with the help of fishermen leaders) triggered low participation in the monitoring stage. In fact, the initial selection of fishermen was done according to personal affinities of fishermen leaders and did not result in what Jentoft has called fragmented representation, since the participation of fishermen both in the communication and ecosystem mapping activities was high. However, basing the selection of informants on personal (and/or political) affinities may generate conflicts related to partiality in later stages of the study (Jentoft and McCay 1995). Perhaps a wider call for participation could have been solicited from the beginning, making it possible to identify potential participants according to different interests.

In any case, the dynamics of the CS fostered the inclusion of new fishermen who had not been part in previous stages, suggesting an open and flexible partnership. The inclusion of these fishermen was useful in the stabilization and increase of participation rates, especially along the monitoring stage. The involvement of a new group of fishermen who had strong and dependent relationships with TURFs and hence were aware of collaborative management generated a new level of collaboration within the project. It does suggest community-based system provides confident incentives to involve fishermen in participatory research than command-and-control, where cooperation is always a real challenge. The reason can be understood through an understanding of empowerment and ownership, where fishing rights would act as social contract, providing privileges to the local fishermen who take them with a moral responsibility (Lam and Pauly 2010).

5.6 Conclusion

The Galician CS is an attempt to describe a process-oriented approach concerning what many authors call as "transition management" (Loorbach 2010; Wittmayer and Schäpke 2014). To summarize and to assess the process, the 5 dilemmas for participatory fisheries science are used (Jacobsen et al. 2012):

- The fishermen-scientists relationship: As already mentioned, no policy-maker was involved along the three project steps and the scientists did not discuss the results with policy-makers (it is a step that has been proposed for the future) in order to ensure this good relationship.
- The scientific background, training-skills in social science, and the relevance of TA as key actors were essential to keep a close relationship among actors. Their critical role must be understood keeping in mind the interpersonal relationships with fishermen guilds. The social relevance of TA lies in their role as intermediary, a role adopted to generate changes at the community-level, providing leadership in view of representing common interest (Pohl et al. 2010). Precisely, one of

the most important issues to strength understanding was the language. The incorporation of local and contextual fishermen knowledge entailed the transformation of power relations (Wilson 2003; Greenwood and Levin 2007), into an equal balance and democracy arena.

- A protocol of confidentiality was established regarding data gathered at the first steps of the research project. The research team was permitted to use data in scientific forums, but the ownership would belong to fishermen guilds. It meant that fishermen and their organizations could make a particular use of data as they considered appropriate. The reason was to return the data to providers, as a tangible political capital to use as against possible threats or to promote future sustainable fishing actions. It strengthened confidence into the relationship among science and fisheries, through an honest code of conduct between both parties.
- Fishing is an activity pressed by time. Uncertainty and hard work conditions on board are common. Also, fishermen are pushed by scientists (and policy-makers) to get involved in research projects and thus there is limited time left to attend all the demands. Into command-and-control system, participation of fishermen is always forced as they did not receive any meaningful incentive. On the other hand, community-based management does bring fluent collaboration among fishermen and TA, as it provides clear incentives in terms of rights and moral responsibilities for commoners (Lam and Pauly 2010).
- Communication was a key factor along the process. From a neutral position, the research team understood the communication as a process without expert communicators. Dialogue flowed between both parts, and the skills of the scientist taking the role of facilitator assisted to promote joint reflection toward a common understanding, as part of learning by doing. Based on respect, openness and deliberation research team managed to establish a common language (Pohl et al. 2010).

Galician CS as a participatory experiment was able to create spaces for learning between all actors, based on a long-term vision. Nevertheless, the results obtained so far showed that a stable and long-lasting collaboration relationship has not been fully achieved. The following steps were not reached, and there are several reasons for the lack of continuity. The economic crisis in Spain and particularly in Galicia, had a strong impact on fisheries science; with many research groups like the one that conducted this proposal being disjointed, there were few possibilities to survive. The absence of clear leadership, though proved important for many parts of this project, caused a setback in the promotion of this kind of research project at a higher level institutionally. In addition, the regional government is still not very much in favor of reviewing the allocation of new fishing rights within inshore waters. Thus, into the last decade a few initiatives have attempted the promotion of collaborative approaches into the design of new user-rights that have not been effectively accompanied. It is necessary to re-open the path of promoting discussion spaces within the public regional, national and european agenda.

References

Abelson J, Forest PG, Eyles J, Casebeer A, Martin E, Mackean G (2007) Examining the role of context in the implementation of a deliberative public participation experiment: results from a Canadian comparative study. Soc Sci Med 64(10):2115–2128. https://doi.org/10.1016/j.socscimed.2007.01.013

Alegret JL (2002) Gobernabilidad, legitimidad y discurso científico: el papel de las ciencias sociales en la gestión de la pesca de bajura. Zainak Cuadernos de Antropología-Etnografía 21:13–25

Berkes F (2009) Evolution of co-management: role of knowledge generation, bridging organizations and social learning. J Environ Manag 90(5):1692–1702. https://doi.org/10.1016/j.jenvman.2008.12.001

Berkes F, Folke C (2001) Back to the future: ecosystem dynamics and local knowledge. In: Gunderson L, Holling CS (eds) Panarchy: understanding transformations in human and natural systems. Island Press, Washington, DC. in press

Berkes F, Colding J, Folke C (2000) Rediscovery of traditional ecological knowledge as adaptive management. Ecol Appl 10(5):1251–1262. https://doi.org/10.1890/1051-0761(2000)010[1251:ROTEKA]2.0.CO;2

Bull R, Petts J, Evans J (2008) Social learning from public engagement: dreaming the impossible? J Environ Plan Manag 51(5):701–716. https://doi.org/10.1080/09640560802208814

Cambiè G, Ouréns R, Vidal DF, Carabel S, Freire J (2012) Economic performance of coastal fisheries in Galicia (NW Spain): case study of the Cíes Islands. Aquat Living Resour 25(02):195–204. https://doi.org/10.1051/alr/2012010

Chuenpagdee R, Liguori L, Palomares MLD, Pauly D (2006) Bottom-up, global estimates of small-scale marine fisheries catches. Fisheries Centre research reports 14(8). Fisheries Centre, University of British Columbia, Vancouver, 105 pp.

Coglan D, Shani AB (2006) Roles, politics and ethics in action research design. Syst Pract Action Res 18(6):533–546. https://doi.org/10.1007/s11213-005-9465-3

Davis A, Ruddle K (2010) Constructing confidence: rational skepticism and systematic enquiry in local ecological knowledge research. Ecol Appl 20(3):880–894. https://doi.org/10.1890/09-0422.1

Fernández-Vidal D, Muiño R (2014) Fact or fiction? Assessing governance and co-management of marine reserves of fishing interest in Cedeira and lira (NW Spain). Mar Policy 47:15–22. https://doi.org/10.1016/j.marpol.2014.01.016

Freire J (2005, February 04) El papel de la investigación científica como apoyo al sector pesquero artesanal. Retrieved from https://nomada.blogs.com/jfreire/2005/02/ciencia_y_pesca.html

Freire J, García-Allut A (2000) Socioeconomic and biological causes of management failures in European artisanal fisheries: the case of Galicia (NW Spain). Mar Policy 24:375–284. https://doi.org/10.1016/S0308-597X(00)00013-0

García-Allut A, Freire J (2002) Procesos de producción pesquera e incertidumbre: La comercialización de los productos pesqueros en la pesca artesanal de Galicia. In: Andar o Mar, II Xornadas Internacionais de Cultura Tradicional. Association Canle de Lira, Carnota, pp 151–178

García-Negro MC (2005) Importancia económica del sector pesquero gallego: análisis de las tablas input output pesca-conserva gallegas. Dissertation, Universidade de Santiago de Compostela

García-Negro MC, Villasante S, Carballo Penela A, Rodríguez Rodríguez G (2009) Estimating the economic effect of the prestige oil spill on the death coast (NW Spain) fisheries. Mar Policy 33:8–23. https://doi.org/10.1016/j.marpol.2008.03.011

Greenwood DJ, Levin M (2007) Introduction to action research. Social research for social change, 2nd edn. Sage Publications, Inc., Thousand Oaks

Hage M, Leroy P, Petersen AC (2010) Stakeholder participation in environmental knowledge production. Futures 42(3):254–264. https://doi.org/10.1016/j.futures.2009.11.011

Huntington HP (2000) Using traditional ecological knowledge in science: methods and applications. Ecol Appl 10(5):1270–1274. https://doi.org/10.1890/1051-0761(2000)010[1270:UTEKIS]2.0.CO;2

Jacobsen RB, Wilson DC, Ramirez-Monsalve P (2012) Empowerment and regulation–dilemmas in participatory fisheries science. Fish Fish 13(3):291–302. https://doi.org/10.1111/j.1467-2979.2011.00434.x
Jentoft S, McCay B (1995) User participation in fisheries management: lessons drawn from international experiences. Mar Policy 19(3):227–246. https://doi.org/10.1016/0308-597X(94)00010-P
Lam ME, Pauly D (2010) Who is right to fish? Evolving a social contract for ethical fisheries. Ecol Soc 15(3):16. [online] URL: http://www.ecologyandsociety.org/vol15/iss3/art16/
Loorbach D (2010) Transition management for sustainable development: a prescriptive, complexity-based governance framework. Governance 23(1):161–183. https://doi.org/10.1111/j.1468-0491.2009.01471.x
Luyet V, Schlaepfer R, Parlange MB, Buttler A (2012) A framework to implement stakeholder participation in environmental projects. J Environ Manag 111:213–219. https://doi.org/10.1016/j.jenvman.2012.06.026
Macho G, Naya I, Freire J, Villasante S, Molares J (2013) The key role of the barefoot fisheries advisors in the co-managed TURF system of Galicia (NW Spain). Ambio 42:1057–1069. https://doi.org/10.1007/s13280-013-0460-0
McCay BJ, Jentoft S (1996) From the bottom up: participatory issues in fisheries management. Soc Nat Resour 9(3):237–250. https://doi.org/10.1080/08941929609380969
Mitchell SM, Shortell SM (2000) The governance and management of effective community health partnerships: a typology for research, policy, and practice. Milbank Q 78(2):241–289. https://doi.org/10.1111/1468-0009.00170
Molares J, Freire J (2003) Development and perspectives for community-based management of the goose barnacle (Pollicipes pollicipes) fisheries in Galicia (NW Spain). Fish Res 65(1–3):485–492. https://doi.org/10.1016/j.fishres.2003.09.034
Pita P, Freire J, García-Allut A (2008) How to assign a catch value to fishing grounds when fisheries statistics are not spatially explicit. Sci Mar 72:693–699. https://doi.org/10.3989/scimar.2008.72n4693
Pohl C, Rist S, Zimmermann A, Fry P, Gurung GS, Schneider F, Speranza CI, Kiteme B, Boillat S, Serrano E, Hirsch Hadorn G, Wiesmann U (2010) Researchers' roles in knowledge coproduction: experience from sustainability research in Kenya, Switzerland, Bolivia and Nepal. Sci Public Policy 37(4):267–281. https://doi.org/10.3152/030234210X496628
Pomeroy RS, Berkes F (1997) Two to tango: the role of government in fisheries co-management. Mar Policy 21(5):465–480. https://doi.org/10.1016/S0308-597X(97)00017-1
Prell C, Hubacek K, Reed M (2009) Stakeholder analysis and social network analysis in natural resource management. Soc Nat Resour 22(6):501–518. https://doi.org/10.1080/08941920802199202
Raymond CM, Fazey I, Reed MS, Stringer LC, Robinson GM, Evely AC (2010) Integrating local and scientific knowledge for environmental management. J Environ Manag 91(8):1766–1777. https://doi.org/10.1016/j.jenvman.2010.03.023
Reed MS (2008) Stakeholder participation for environmental management: a literature review. Biol Conserv 141(10):2417–2431. https://doi.org/10.1016/j.biocon.2008.07.014
Reed MS, Graves A, Dandy N, Posthumus H, Hubacek K, Morris J, Stringer LC (2009) Who's in and why? A typology of stakeholder analysis methods for natural resource management. J Environ Manag 90(5):1933–1949. https://doi.org/10.1016/j.jenvman.2009.01.001
Reed MS, Evely AC, Cundill G, Fazey I, Glass J, Laing A, Newig J, Parrish B, Prell C, Raymond C, Stringer LC (2010) What is social learning? Ecol Soc 15(4). Retrieved from http://www.jstor.org/stable/26268235
Ruddle K (2000) Systems of knowledge: dialogue, relationships and process. Environ Dev Sustain 2(3–4):277–304. https://doi.org/10.1023/A:1011470209408
Sabatier P, Focht W, Lubell M, Trachtenberg Z, Vedlitz A, Matlock M (2005) Collaborative approaches to watershed management. In: Swimming upstream: collaborative approaches to watershed management. MIT Press, Cambridge, MA

Salas S, Gaertner D (2004) The behavioural dynamics of fishers: management implications. Fish Fish 5(2):153–167. https://doi.org/10.1111/j.1467-2979.2004.00146.x

Salas S, Sumaila UR, Pitcher T (2004) Short-term decisions of small-scale fishers selecting alternative target species: a choice model. Can J Fish Aquat Sci 61(3):374–383. https://doi.org/10.1139/f04-007

Schlager E, Ostrom E (1992) Property-rights regimes and natural resources: a conceptual analysis. Land Econ:249–262. https://doi.org/10.2307/3146375

Selman P (2004) Community participation in the planning and management of cultural landscapes. J Environ Plan Manag 47(3):365–392. https://doi.org/10.1080/0964056042000216519

Shepard L (2008) La evaluación en el aula. Instituto Nacional para la Evaluación de la Educación, México (Translate from the English version, Classroom Assessment) In: Robert Brennan ACE/Praeger Westport (eds.) Educational measurement, 4th edn. Greengood Publishing Group, Colorado, pp 623–46

Silvano RA, Valbo-Jørgensen J (2008) Beyond fishermen's tales: contributions of fishers' local ecological knowledge to fish ecology and fisheries management. Environ Dev Sustain 10(5):657–675. https://doi.org/10.1007/s10668-008-9149-0

Stange K, van Leeuwen J, van Tatenhove J (2016) Boundary spaces, objects and activities in mixed-actor knowledge production: making fishery management plans in collaboration. Maritime Stud 15(1):14. https://doi.org/10.1186/s40152-016-0053-1

Stenseke M (2009) Local participation in cultural landscape maintenance: lessons from Sweden. Land Use Policy 26(2):214–223. https://doi.org/10.1016/j.landusepol.2008.01.005

Suarez de Vivero J. L, Martinez I (2007) Propuesta de regionalización de las aguas jurisdiccionales españolas (subdivisiones). La política marítima y la planificación espacial: aplicación metodológica al arco atlántico-mediterráneo (Golfo de Cádiz y Mar de Alborán) Proyecto MEC (SEJ22007–66487/GEOG). Universidad de Sevilla

Symes D, Hoefnagel E (2010) Fisheries policy, research and the social sciences in Europe: challenges for the 21st century. Mar Policy 34(2):268–275. https://doi.org/10.1016/j.marpol.2009.07.006

Tippett J, Handley JF, Ravetz J (2007) Meeting the challenges of sustainable development—a conceptual appraisal of a new methodology for participatory ecological planning. Prog Plan 67(1):9–98. https://doi.org/10.1016/j.progress.2006.12.004

Van Buuren A, Loorbach D (2009) Policy innovation in isolation? Conditions for policy renewal by transition arenas and pilot projects. Public Manag Rev 11(3):375–392. https://doi.org/10.1080/14719030902798289

Villasante S (2009) Sobre la sostenibilidad de las pesquerías artesanales de Galicia: Conservando hoy para las futuras generaciones. Retrieved from Instituto de Estudios Marinos para la Nutrición y el Bienestar (INESMA)

Wiber M, Berkes F, Charles A, Kearney J (2004) Participatory research supporting community-based fishery management. Mar Policy 28(6):459–468. https://doi.org/10.1016/j.marpol.2003.10.020

Wilson DC (2003) Examining the two cultures theory of fisheries knowledge: the case of bluefish management. Soc Nat Resour 16:491–508. https://doi.org/10.1080/08941920309150

Wittmayer JM, Schäpke N (2014) Action, research and participation: roles of researchers in sustainability transitions. Sustain Sci 9(4):483–496. https://doi.org/10.1007/s11625-014-0258-4

Chapter 6
Information Is the Jam of the Western Baltic Herring Sandwich: Bridging Gaps Between Policy, Stakeholders and Science

Lotte Worsøe Clausen, Verena Ohms, Christian Olesen, Reine Johansson, and Peter Hopkins

Abstract The management of the herring fishery in the Skagerrak-Kattegat and Western Baltic is challenging because of the complex dynamics of several herring populations with different migration patterns and population dynamics and a complex fleet structure fishing for herring. To address these complexities and policy issues related to management areas and EU borders, management has become increasingly complex and non-transparent. The Case Study (CS) explores how things could be simplified and rationalised by developing a Multi-Annual Management Plan (MAMP – previously referred to as Long-Term Management Plan (LTMP)) that could provide predictability and stability. Industry stakeholders from two Advisory Councils (the Pelagic and Baltic Sea ACs), representatives from management, from national governments and the EU, and scientists collaborated to

Peter Hopkins expresses personal views and is not representing the official position of the Commission.

L. W. Clausen (✉)
DTU Aqua, National Institute of Aquatic Resources,
Technical University of Denmark (DTU), Lyngby, Denmark

International Council for the Exploration of the Sea (ICES), Copenhagen, Denmark
e-mail: Lotte.worsoe.clausen@ices.dk

V. Ohms
Pelagic Advisory Council (PelAC), Zoetermeer, The Netherlands

C. Olesen
Danish Pelagic Producers Organisation (DPPO), Copenhagen, Denmark

R. Johansson
Baltic Sea Advisory Council (BSAC), Copenhagen, Denmark

P. Hopkins
Directorate-General for Maritime Affairs and Fisheries, European Commission,
Brussel, Belgium

P. Holm et al. (eds.), *Collaborative Research in Fisheries*, MARE Publication
Series 22, https://doi.org/10.1007/978-3-030-26784-1_6

improve stock management by developing a robust MAMP. A common understanding of relevant scientific and political issues is a prerequisite for the successful formulation and implementation of a MAMP. The CS focused on getting all relevant stakeholders involved and successfully engaged the European fishing industry through the ACs and, to some extent, the management bodies involved within the EU. The two ACs have only limited overlap in their membership. Until this CS came about, there was very little collaboration in providing advice on Western Baltic Spring Spawning (WBSS) herring management. With the opportunity to collectively formulate management objectives for the fishery, the two ACs, science partners and management representatives narrowed the gaps between stakeholders. In this chapter, we review the achievements of the project and analyse the effectiveness of the collaborative process and how it affected the individual endeavours of the scientific and stakeholder groups. The chapter describes the process of creating an arena where a management framework and procedures could be discussed by all involved so as to establish a communication platform for setting objectives, management clauses and evaluation criteria for the requested MAMP. We also reflect on the concept of the 'GAP Method' and how such science-management-stakeholder collaboration can contribute to overall fisheries governance.

Keywords Herring · Migration · Management areas · Complex policy · MAMP

6.1 Introduction

Western Baltic spring spawning herring (WBSS) is a relatively small stock, but is managed by means of a highly complex governance scheme. It spawns in the western Baltic Sea where it is exploited by several EU fishing fleets. It then migrates into the Kattegat, Skagerrak and eastern North Sea areas, where it mixes with North Sea autumn spawning herring (NSAS), following an extremely variable age and season-dependent pattern. There it is exploited by EU and non-EU fleets. Every year, a TAC for WBSS is set for the two management areas and quota allocations between fleets and areas are negotiated. This poses demanding scientific challenges and requires complicated political processes of resource allocation among fishing fleets. Such a complex management system is unstable and unpredictable for all involved, in particular for the fishermen who depend on WBSS for their income.

The WBSS herring case study (CS) sets out to explore the potential for simplifying and rationalising an unsatisfactory complex fisheries management situation by developing a Multi-Annual Management Plan (MAMP), which could provide predictability and stability for all. Stakeholders of two Advisory Councils (the Pelagic and Baltic Sea ACs), representatives from management from national governments and the EU, as well as scientists engaged in collaboration worked together to improve stock management through the joint development of a robust MAMP.

6.2 Background

The management of the herring fisheries in the Skagerrak-Kattegat and Western Baltic is challenging because of the complex dynamics of several herring populations in the area, each with different migration patterns and population dynamics. These herring populations are managed as two main stock complexes (for readability from here referred to as the North Sea Autumn Spawners (NSAS) stock and the WBSS stock). The two stocks are utilised by four different fleets targeting a mixture of stocks across their distribution areas.

Given the complexity in biology and the fleets utilising the resource, the management is correspondingly complicated. The herring fishery is regulated according to area specific Total Allowable Catches (TACs) for human consumption fisheries and for bycatches of herring taken in the small meshed industrial fishery for reduction purposes (herring catches in the Div. IIIa and SD 22–24, see Fig. 6.2). But the assessment and fisheries advice are stock-based (e.g. for WBSS), to which estimates of potential WBSS catches from the neighbouring area of the eastern North Sea are added and from which estimates of potential NSAS catches within the area are subtracted.

Hence, the biology is complex and the fishery involves a mixture of interests. Moreover, the need to give advice across management areas and stocks makes the yearly provision of advice complicated. On top of that, both EU and non-EU fleets operate in the Skagerrak which means that any advice must be part of the yearly negotiation process between the EU and Norway on shared fishery resources. These negotiations are non-transparent and the resulting TACs often lead to frustration for both fishermen and scientists because the reasoning behind the results is frequently unknown to the stakeholders.

6.3 Disentangling Complex Management Applying the 'GAP Method'

The WBSS is managed through a complex governance scheme despite its relatively small stock size and relatively low economic value. Various attempts have been made to simplify the advisory process. Nevertheless, every year all parties involved end up frustrated and confused. Here is a typical quote from an AC secretary: 'This is the third year in a row now that at first I think I understand the fishery of this stock (mainly thanks to GAP), only to find out that I don't understand anything about that stock/the fishery when reading the ICES advice. All these different fleets and TACs are so confusing! How is anybody supposed to understand anything?' The stakeholder from the commission shares this view: 'To manage herring in the Skagerrak we have to juggle with scientific issues related to the migration and variable mixing of stocks as well as sensitive political considerations arising from the fact that the North Sea component is shared with Norway, and with the conflicting interests between EU Member States regarding the division of WBSS catches between the Baltic and the Skagerrak'. And an EU industry representative stated: 'It is frustrating that the TAC year after year is set through horse trading. How are the vessel owners supposed to make sound long-term investments when the basis of the economic return is so unpredictable?' A transparent and comprehensible process is obviously needed by all those involved in WBSS management.

The assessment and TAC calculations for WBSS are complicated which this adds to the frustration when the actual TAC does not appear to match with the perceived stocks observed at sea. In the past, attempts have been made, for example, by the International Council for the Exploration of the Sea (ICES) to make the process clearer and get managers, scientists and stakeholders to meet. But this has rarely led to anything other than reports and advice which again were confusing and did not include the views of all participants (e.g. Workshop on procedures to establish the appropriate level of the mixed herring TAC (Spring Western Baltic (WBSS) and Autumn Spawning North Sea (NSAS) stocks) in Skagerrak and Kattegat (Division IIIa ICES 2010). These meetings and the closed process at the end of the advisory process at the EC-Norway negotiations left both scientists and stakeholders with a sense of not having any influence on the final outcome.

We suggest that the complexity of the management scheme is the underlying reason for this. The scientific assessment cannot provide transparent advice because the harvest control rules used to set fleet-wise catch options have not been discussed and agreed by all partners. They cannot be discussed and agreed upon by all parties, because their underlying scientific basis and consequences for the stocks have not been properly reviewed by ICES.

The fragmentation and absence of common platforms for the involved parties to discuss the management objectives means that the decision-making process has been split up. The WBSS management scenario thus appeared to be a perfect case study for 'the GAP approach' (see Chap. 1) to set up a common platform. Lack of trust and cooperation were standing in the way of the formulation of appropriate management measures. Accordingly, the CS was set up to develop a MAMP for WBSS based on the development of mutually agreed management objectives and TAC setting procedures that could be accepted by all involved parties.

6.4 Description of the CS Process

The project was actually born even before the GAP1 project. The scientist leading the WBSS CS had been having discussions with some of the stakeholders (mainly representatives from the human consumption fishery) in her earlier work dealing with the yearly correction of misreported catches. These discussions were done by phone, and the true general behaviour of the fleet was described. This gave the scientists the chance to correct the catches feeding into the assessment. During these conversations, the idea of a project to simplify the very complicated advice came up and was consequently successfully carried out in GAP1. A couple of meetings were

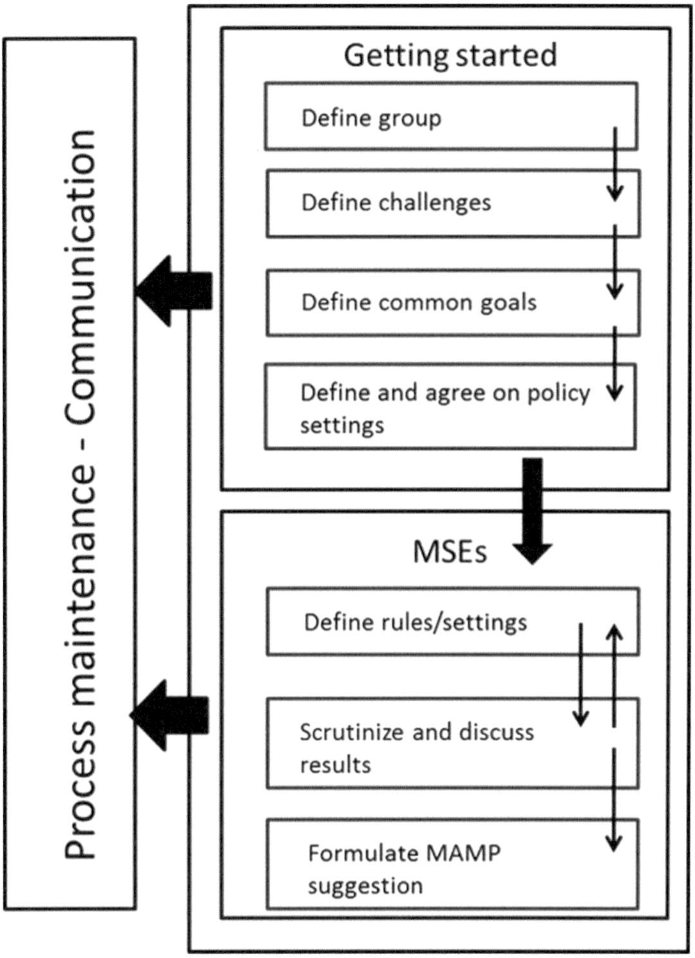

Fig. 6.1 Schematic structure and workplan for the CS (MSE: Management Strategy Evaluation)

held as part of GAP1 that gave the scientists and a few other stakeholder partners the chance to discuss and together develop their research ideas and needs on a specific topic based on participatory research. The biology of the WBSS herring was discussed in these meetings as well as the need for a simple and stable management system. This led to initial ideas of what the management objectives in a MAMP could be. Everyone agreed that it was really important to involve the Pelagic AC, the Baltic AC, Norway and representatives from management and science in formulating a MAMP.

These ideas were then carried further in the CS in GAP2. Figure 6.1 illustrates the schematic structure and workflow in the CS. It was decided that the workflow and communication on project progress between the partners should be given priority and be an objective in itself. Furthermore, it was agreed that principles of management should be developed through discussions based on priorities for the fishery, science, other stakeholders and management, because it was agreed that a precondition for a successful outcome was for all parties to feel they owned the end result. Thus, the main aim was to develop the 'Process maintenance – Communication' part of the CS structure between all participants and convey information on science, policy and fisheries. The various meetings under the CS were predominantly arranged by the coordination-group, but given the well-established relationship between the participants in the CS, there were also many emails, phone calls and small meetings. The lead scientist frequently sent out information-status emails to all, keeping track of the more 'extracurricular' developments in the CS.

The CS has successfully made use of the Message Box (Baron 2010) and brainstorming exercises (e.g. so-called re-tailored snow-carding, see Duke and Geurts 2004) in meetings to grasp and visualise the perceptions regarding the main points in MAMP for the complex herring stock in the Western Baltic and adjacent waters. However, the main tool has been the good, positive, inclusive and multi-directional dialogue of the CS. This dialogue has taken place at meetings, as well as more informally through emails, phone calls, web-services and in the other occasions when participants met (such as other meetings). It has undoubtedly been an advantage that most of the participants in the CS knew each other prior to the onset of the CS. This made dialogue smoother and less formal. Although not stated in the initial CS description, the coordination of the CS was quickly taken on by the lead scientist and the two secretaries from Pelagic AC (PELAC) and the Baltic Sea AC (BSAC) through a 'steering committee'. This ensured that all participants were represented and facilitated communication between participants since the ACs had contact with the relevant EU stakeholders.

The first main objective of the scientists and stakeholders was to establish contact with relevant participants beyond the group in GAP1. It was agreed upon to draw on respective networks to find potential participants for the first meeting of the CS. The agenda for this first meeting (or seminar) was drawn up by the scientist and the PELAC secretary in order to make it comprehensive enough to match all ideas and expectations for the CS. While planning the seminar, several meetings between the scientist and the stakeholders were held which made it easy to formulate an agenda, objectives and participants. It soon emerged that it was important to widen the group of those involved in the CS in order to promote a 'space' for discussion of the appropriate management framework and procedures between all relevant parties. The CS was to some degree successful in reaching all relevant participants for this first seminar (but see below). From then on, the collaboration continued throughout the CS.

From the start the main challenge for the CS was bringing representatives from all stakeholders and managers to the table. This was so because the Norwegians did not have the same incentive or backup to join the CS due to the political nature of the CS. Initially, this created much frustration and mistrust in the project, but through discussion and by drawing on experience from EC managers and stakeholders it could be addressed. It soon emerged that there are several layers in the policy process. Given the role of Norway in the more closed processes, it is not straightfor-

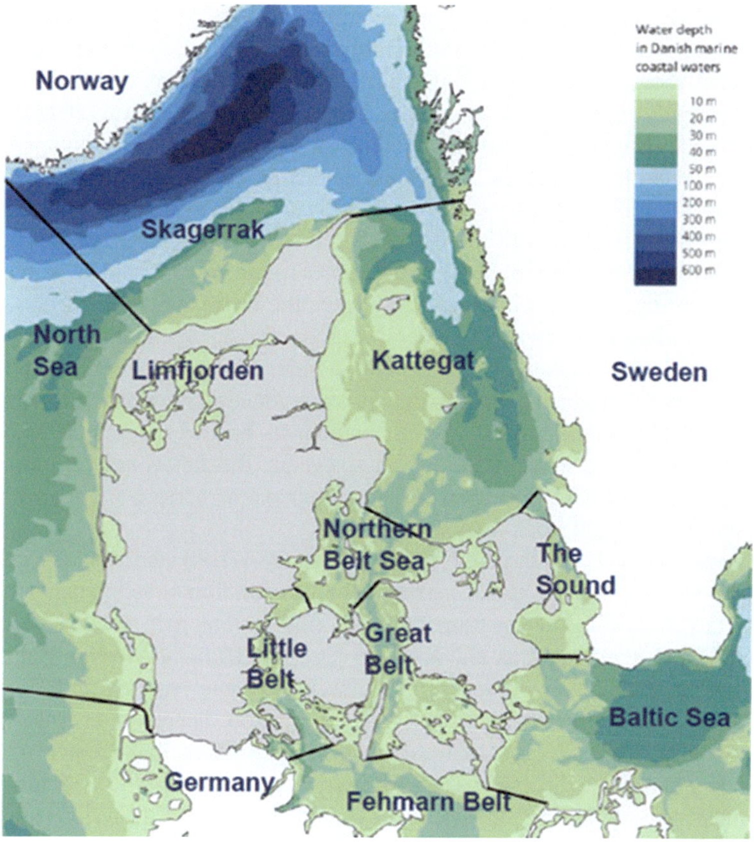

Fig. 6.2 The transition area between the Baltic and North Sea. The Skagerrak and the Kattegat constitute Division IIIa, whereas the Belts and part of the Baltic constitute Sub-Divisions 22–24

ward for them to participate in any research project which may potentially have a policy effect after completion (such as the proposal for a MAMP). One EC representative said the following: "When I participate in things like this, everybody knows that I'm here as me and cannot be seen as taking the entire EC into confidence, whereas a Norwegian (manager) would very easily be taken as representing Norway and thus making statements on the entire country's behalf".

The Skagerrak-Kattegat-Western Baltic is the transition area between the brackish Baltic Sea and the North Sea (Fig. 6.2). The hydrographical and geographical characteristics of this transitional area of Skagerrak-Kattegat and Danish waters framed by islands and fjords and the archipelago and bays on the Swedish side have given rise to a number of more or less separate herring populations in the Skagerrak-Kattegat-Western Baltic. Historically, no less than 25 such populations have been recorded (Poulsen 1975).

In recent times, these populations are known as the WBSS herring stock that spawn and use the area as nursery and migrate through it. The WBSS is composed of a rather large complex of herring populations from the Skagerrak, Kattegat and Inner Danish waters and the Western Baltic (e.g. Rügen, Schlei and Flensburg). Very recently, part of the Central Baltic herring stock was seen in the Western Baltic area, apparently using it as a summer feeding area (Gröhsler et al. 2013). In the Skagerrak and Kattegat areas mixed feeding aggregations generally comprise herring from the North Sea and the area spanning the transition zone between the North Sea and the Baltic Sea proper (here collectively referred to as the 'Western Baltic Sea'). Given the mixing with the NSAS, the ICES Herring Assessment Working Group (HAWG) makes use of biological samples routinely collected to estimate the stock composition of the annual catches. Since the beginning of the 1990s, there have been routine analyses of stock compositions of the herring populations in the North Sea and adjacent areas in commercial samples for stock assessment and management purposes. Recent developments in stock identification methodology have made it possible to monitor local stock components beyond the general spawning components of spring-autumn-and-winter spawners. But this is not yet part of the routine treatment of herring catches. All spring-spawning herring are treated as one stock, despite the local stock diversity.

However, it is not only the biological nature of the WBSS stock that complicates management. The Western Baltic herring fishery is a multinational fishery that seasonally targets herring in the transition area (the eastern parts of the North Sea (Eastern IVa,b), the Skagerrak and Kattegat (Division IIIa) and Western Baltic (SD 22–24)). The main fleets come from Denmark, Sweden, Norway and Germany, while Poland has a minor fishing activity in the area too. There are three fleet types which have a TAC allocated across areas in a complex management scheme (Fig. 6.3). Fleet C is a directed fishery for herring with trawlers (with 32 mm minimum mesh size) and purse seiners. Fleet D contains all trawl fisheries with mesh size less than 32 mm and small Swedish purse seiners without individual herring quotas. For most of the landings taken by this fleet, herring is landed as bycatch: Danish and Swedish bycatches of herring from the sprat; Norway pout and blue-whiting fisheries are included in fleet D. Finally, Fleet F is a directed fishery for herring and some as bycatch in a directed sprat fishery in Subdivisions 22–24. The sprat fishery in SD 22–24 has a certain amount of herring bycatch which is closely monitored and counted against the sprat quota (up to 8% herring allowed).

Hence, the three fleets target different parts of the WBSS stock, which has given rise to diverging interests with respect to management objectives for the stock.

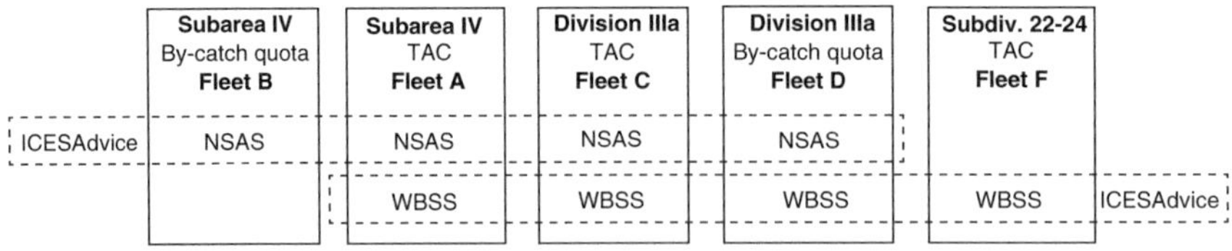

Fig. 6.3 Allocation of WBSS and NSAS across fleets and areas (ICES 2015). ICES provides an overall TAC for WBSS and NSAS, respectively, and then guide upon request the managers on the potential split between fleets and areas based on herring migration and fishery pattern

Although Fleet D is getting smaller, the small-meshed fishery is still carried out in the area and is responsible for approximately 8% of the total herring landings from Skagerrak-Kattegat.

Having defined the challenges in terms of biology and the fishery, common ground was found to move forward and 'define and agree on policy settings' (Fig. 6.1). The first step was to define management objectives that all could agree on (the first seminar). Next it was necessary to discuss and agree on the more management-rule-related factors (second large meeting).

After doing this, the second part of the CS (MSE box in Fig. 6.1) was embarked upon. This was an iterative process where the actual formulation of the MAMP settings, to be tested in the final Management Strategy Evaluation (MSE), was developed through a series of smaller meetings between stakeholders and scientists and larger on-line meetings with the entire group.

The continuous dialogue between stakeholders and scientists in the CS has enabled information-flow on the state of the art of stock biology and the fishery. During the meetings, the management partners have been able to give their input on how to integrate shared management objectives with operational policy.

6.5 Results of the Collaboration in the CS

The CS has worked towards reaching a common ground for managing WBSS. This has involved exploration of data on stock dynamics as well as setting terms for a MAMP. This has meant some degree of participatory modelling and considerable effort with respect to communication and information sharing. The results will be described under the two pillars of the CS: Biology and Management.

6.5.1 Biology

6.5.1.1 Reconstruction of Historical Catches

Area misreporting (reporting catch as belonging to an area other than where it was taken from) has probably taken place in IIIa and the adjacent North Sea, where catches of WBSS or NSAS fished in the North Sea have been reported as fished in IIIa. This misreporting has been due to the size differences of herring in the two areas, where the optimal sized herring could be caught in the North Sea more efficiently for a longer period of the year and thus result in a better overall outcome for the fishermen. Herring caught in IIIa were less attractive to land for the major part of the fishing season. Area misreporting is understood to have taken place for at least the Danish catches from 1997 to 2008. Area misreporting ceased from 2008 onwards as from that time on management has allowed optional transfers (flexibility in terms of where to take the IIIa TAC), allowing fishermen to catch part of the TAC for IIIa in the North Sea.

In order to get correct time-series data of catches for assessment purposes, Danish reported landings each year from 2002 to 2009 were corrected for misreporting based on information from the industry, weekly evaluation of the fishing trips and, since 2004, by means of VMS data. All Norwegian herring catches in IIIa in the period 1995–2001 are understood to have been taken in the North Sea. This was corrected in the assessment time-series. It is unclear to what extent Swedish catches reported in IIIa in the period 1991–2008 have been correctly reported in terms of area. As in the case of Denmark, it is suspected that some of Sweden's North Sea catches have been reported as IIIa catches. So more accurate time-series data of catches from the period prior to 2008, based on collaboration in the CS, has been made available for the assessment. It is unlikely that after 2008, there has been misreporting in the Danish and Swedish fishery due mainly to new regulations prohibiting vessels from fishing in two management areas in one trip. Also, given new flexibility regarding where to take the IIIa TAC from has reduced the incentive for area misreporting.

Stakeholders have been involved in the quality checking of data using historical knowledge, logbooks and 'grey data' in order to increase the precision of the available data. To begin with this was subject to variability in terms of accuracy and at times was also historically quite patchy. While aligning catch data of WBSS it became apparent that the sampling for herring stock identity did not accurately represent what was happening in the actual fishery. The stakeholders pointed out this mismatch. When sampling commercial catches to assess the biological composition of the proportions of the two herring stocks, it is crucial that the sampling scheme and coverage mirror the actual distribution of the fishery. By comparing the sampling coverage to the reported catches using the ICES rectangle over the period 2002–2011, it was evident that catches were concentrated in the north-western part of area IIIa whereas the sampling intensity was highest in the north-eastern area (Fig. 6.4).

As a result of the discussions in the CS, the sampling coverage has improved over recent years, currently covering the entire distribution area and following the spatial and temporal distribution of the catches. There is still however room for improvement. Recent sampling very poorly covers area IVaE as industry representatives claim that there are very few catches in this area.

6.5.1.2 WBSS Stock Complexity

Collaboration between scientists and stakeholders required alignment in the way both perceived the WBSS stock complexity. The occurrence of several local populations of herring along the coastline of Skagerrak has on several occasions been communicated by local fishermen to scientists in the context of a mismatch between the assessment results of WBSS and what was observed at sea. Since recruitment to the WBSS stock is driven by more by the southern part of the stock (the 'Rügen

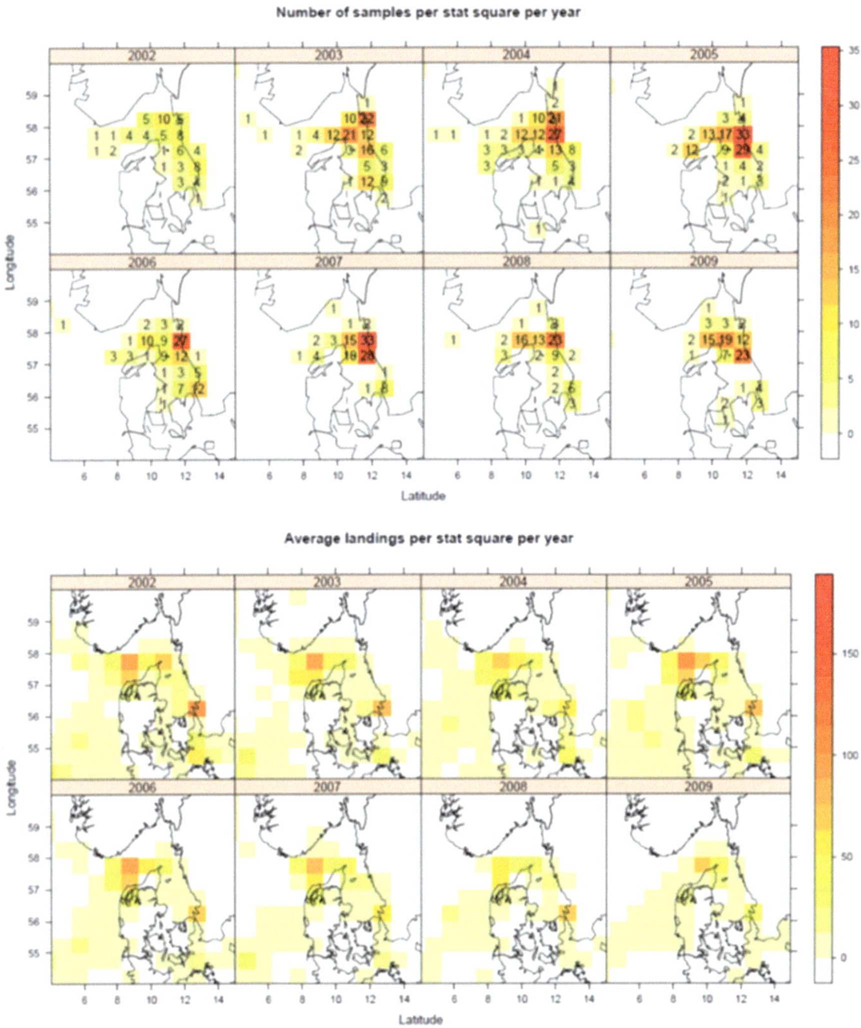

Fig. 6.4 Number of samples by ICES rectangle (upper panel) and average landings in tonnes per year by ICES rectangle (lover panel) over the period 2002–2009

herring'), the scientists took upon the task of exploring the mixture of populations within the WBSS stock, based on observations by stakeholders. Existing baseline data on herring genetics were supplemented with samples taken from local populations with the guidance from local fishermen. These were then put together with an analysis of otolith shape in order to assess the applicability of this parameter as a population indicator.

Although the exact quantification of the various populations to the overall WBSS stock is yet to be analysed, the way in which fishermen and scientists view stock complexity is now much more similar after continuous dialogue between them

Table 6.1 Identified populations in the WBSS with their main spawning time annotated in parenthesis, s = spring, w = winter. The WBSS are grouped by colour indicating differentiation power. X marks recorded occurrence of the population in the area

Stock name	Herring population	Western Baltic	Belts and Sound	Kattegat	Skagerrak/Eastern North Sea
Western Baltic Spring Spawners (WBSS)	Rügen (s,w)	X	X	X	X
	Schlei fjord (s)	X			
	Kiel fjord (s)	X			
	Kattegat (s)		X	X	X
	Lillebælt (s,w)		X	X	X
	Limfjord (s)			X	X
	Ringkøbing fjord (s)			X	X
	Skagerrak (s)			X	X

during the GAP CS. The otolith shape proved to be a promising tool for identifying population diversity within the WBSS stock. This will ultimately lead to a 'complexity' parameter as input to the assessment with the aim to make the output more accurate in terms of population diversity and contribution to the overall WBSS stock (Table 6.1).

6.5.2 *Management*

Throughout the CS, fishery stakeholders across the PELAC and BSAC were very actively involved in trying to synthesise their views and aims regarding management of herring in IIIa, keeping in mind the scientific (biological) aspects of the fishery. Initially, the two groups of fishery stakeholders had different opinions on various parts of a MAMP: the share of the total WBSS TAC for each of the two major management areas (Division IIIa and SD 22–24) and whether all fleets catching WBSS herring should be included in a MAMP (this would mean also including North Sea fleets, because a minor part of the WBSS migrates into the eastern parts of the North Sea). The permitted inter-annual variation clause on TAC variation was also a point of disagreement. During the first seminar, these differences were resolved through group work where all possible management objectives were listed and subsequently prioritised using the so-called snow-carding approach (Duke and Geurts 2004). Figure 6.5 shows the final choice of management objectives identified by the CS. The over-arching objectives of a MAMP were that the plan should be Specific, Measureable, Achievable, Realistic and Time-bound (SMART). It should be simple to grasp for all stakeholders and, most importantly, give a high and stable yield based on a sustainable fishing mortality.

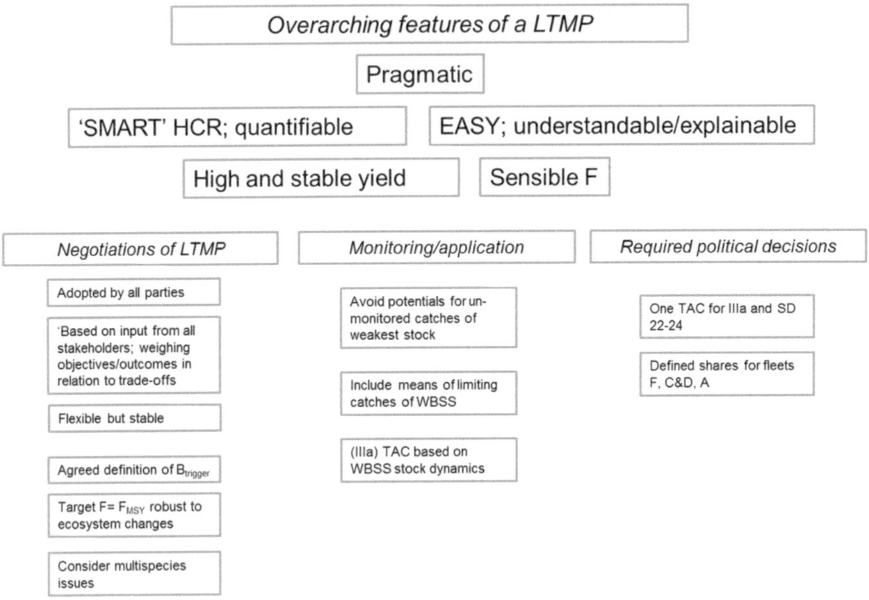

Fig. 6.5 A schematic overview of the agreed management objectives for the WBSS

The more detailed objectives of MAMP dealt with area-based management and how this can be done in a more transparent and stable manner than before. Many times, before the CS was developed, the separation between the TAC areas (IIIa and SD 22–24) had been the cause of disagreement between stakeholders. Hence, after defining the overarching management objectives, the CS set out to settle the more 'politically sensitive' features of a MAMP.

6.5.3 *Political Issues*

The split of WBSS TAC between the Baltic and the Skagerrak is a political decision. The Member States (MS) have had very different opinions on the matter. Germany's interest is in the Baltic fishery, Sweden's interest is in the Skagerrak fishery. Additionally, the TAC for Norway in the Skagerrak would then be based purely on the EU's decision on the TAC level fixed for the WBSS stock, and on the split between the management areas. During meetings in the CS, the fishery stakeholders from the MS settled on a 50:50 split of the TAC between Division IIIa and Sub-Divisions 22–24, mainly in order to achieve much desired stability in the advice given and management of the herring stock(s). The stakeholders underlined that it was of utmost importance to maintain relative stability and that the share for Norway should be fair in relation to what is allocated to the MS. Many were in favour of the 50:50 split, some because this was favourable for their own fishing activities and others because having such a share gave a predictable and semi-transparent management decision. This split is then combined with an optional transfer of the C-fleet TAC from Division IIIa to the North Sea, with a maximum 50% transfer. The science stakeholders agreed to use this approach in the short-term prediction methodology of the assessment, on the basis that the fishing industry will provide an estimate of the transfer at least two weeks before the assessment group meets and by using a three-year average for medium-term predictions. The final management rule was implemented during the Assessment Working Group for herring (HAWG) in ICES in March 2013 (ICES 2013).

The CS, by means of meetings and correspondences, has informed the participating representatives from EU involved in the more closed negotiation process on how to deal with the management challenge of sharing TACs for Division IIIa. This can be considered an extra element for a MAMP for the herring fishery and is historically where negotiations between the EU and Norway have been strenuous. The formulation of a model for TAC splitting, which was adopted by the 'Bergen technical WG' (a joint-working group created by the EU and Norway) and processed further, has now reached a point where the EU and Norway have sent a joint request to ICES for evaluation. The successful joint approach by the BSAC and PELAC, which represent several fleets in the EU utilising different parts of the WBSS, reached a mutually agreed solution for TAC sharing, something which has been valuable input to the participating representatives from EU involved in the more closed negotiation process.

After deciding upon the contents of the MAMP for the WBSS, the more technical details were agreed upon in the CS (by web-based meetings or by email). The general setting for the Management Strategy Evaluations (MSE) was approved by all and the CS ended up with a very detailed scheme for running several simulations within the MSEs based on identified scenarios. The officially suggested approach to predicting stock mix of TACs in Division IIIa in the EU-Norway request to ICES was according to the CS analysis the most optimal way to do so given the current assessment modelling set-up and its limitations. To come up with a full multi-stock assessment model, including a prediction model, and thus enabling MSEs where the full dynamics of NSAS and WBSS are included is now thought to be beyond what the CS can achieve. The focus of the CS will be on the more technical parts of running MSEs such as predicting stock shares with the use of single stock modelling. These have yet to be finalised, so the results cannot be described here. But it is worth noting that the CS created an arena where details, including the uptake of the bycatch ceiling TAC, stability clauses and possible TAC transfers between the areas, could be discussed and agreed upon.

6.6 Discussion

The CS set out to explore how to simplify and rationalise the highly complex management regime of the WBSS herring stock. The biology, the fisheries, the management and the politics this CS had to deal with were very complex. Communicating these complex issues, therefore, was the key to the success of this CS. Although science-related ecological issues of the fishery had been addressed by scientists, the results and TAC settings for these analyses have been discussed repeatedly by all CS members and modified with a common understanding of objectives, evaluation criteria and final outcomes. Initial disagreements between stakeholders on management objectives and criteria for a MAMP were amended after the first couple of meetings, as it became obvious that the disagreements were not that big when put into context and debated on common ground. The fact that the third party (Norway) was not present presumably enabled the EU stakeholders to find common ground for a MAMP through collective efforts at informing policymakers what their stand was regarding facilitating broad representation at the negotiations. Coming to an agreement/understanding regarding EU interests on the splitting of fishing opportunities between the Baltic Sea and IIIa was easier in Norway's absence. Differences in perceptions between scientists and stakeholders of the WBSS stock's complexities and dynamics were also reduced through dialogue and mutual understanding of the limitations and sharing of knowledge on this stock.

The CS has managed to strengthen the bridge between scientists and other stakeholders from around the EU, finding a common approach and position for the BSAC and PELAC based on trust and communication. This strengthened relationship was captured well by a participant from BSAC at the penultimate meeting of the CS: 'You (the group) have done very well; a trust has been built up amongst us and we are working towards a common goal now'. The MAMP has been tested in MSEs and can now be presented to managers in a refined format. It is expected that this will happen in the very last months of GAP in a workshop organised by ICES. However, the outstanding question is whether this MAMP is simple and yet comprehensive enough to match the needs of both the EU and Norway in relation to an agreed management approach for WBSS. Answering this question was beyond the scope and capability of the CS and will happen through the more closed processes of policy making between EU and Norway.

Getting the PELAC and BSAC to arrive at a common ground for the MAMP was not as difficult as anticipated. The stakeholders knew each other very well and were on good terms, which was an important reason for successful collaboration in the CS. The gap between science and stakeholders was bridged by a mutual exchange of knowledge. The stakeholders (finally) gained insight into the somewhat non-transparent advisory process and their views and expertise were taken up, discussed and aligned with science. Science was informed by and matched with what was observed in the fishery; the population diversity of the WBSS and sampling was optimised after taking aboard advice of the stakeholders. Finally, knowledge gained from stakeholders, who have years of experience in the fisheries management policy process, proved very valuable for the CS. An AC secretary put it like this: 'I think it is important to notice that knowledge has been shared. After all, scientists have information that fishermen don't and vice versa and by combining this knowledge people in general got a better understanding of the stock and the fisheries'. The CS managed to strengthen the bridge between scientists and stakeholders from around the EU, finding a common approach and position for the BSAC and PELAC. It successfully created an environment of trust. Co-management has moved from being an extractive exercise (getting data from industry and giving it to scientists) to a consultative one that involves co-management.

Although some involved have not always distinguished clearly between a research project and real negotiations, the gap among science, stakeholders and policy was reduced, because the CS informed policy members of the nuanced nature of problems, enabling a more integrated approach. Stakeholders and scientists both feel they are better represented in more closed processes through a CS like this. This has in turn increased the sense of inclusion in decisions made outside the remits of the CS. The CS has enabled stakeholders and scientists to inform the formal negotiation process between EU and non-EU countries, despite this process being rather closed and only certain policy makers being involved. A representative from the commission expressed it like this: ‘I would say that the GAP study has first and foremost been educational. It has brought stakeholders to a common understanding of the problems and enabled possible solutions to be discussed constructively’.

Collaborative approaches of formulating the settings of a MAMP increase the chance of the resulting regulations being accepted by everyone involved. One stakeholder expressed it like this: ‘Being at the table when the inherent trade-offs in management of a resource are discussed and decided is the difference between being ‘told to’ and having an informed ‘ownership’ of the results (the MAMP)’. Figure 6.6 illustrates how collaboration and dialogue in the CS have had an effect.

In Fig. 6.6, the ‘TAC wheel’ illustrates the annual process where scientists collate data in a stock assessment, then provide advice to policy makers which is used to decide upon the TAC for the fishery, leading to an effect on the fish stocks. This cycle does not necessarily imply any interaction among the three human parts of the

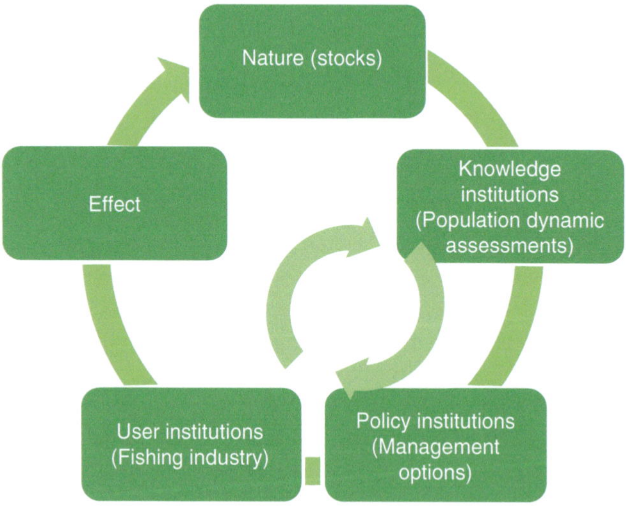

Fig. 6.6 A schematic overview of where the ‘GAP method’ has been making a difference in the CS: the smaller circle within the larger annual ‘TAC wheel’

cycle. Having a recursive cycle inserted, such as the CS, in the annual cycle of managing the resources in the sea allows for direct communication and immediate mitigation of problematic issues among all those involved.

Even though the CS faced problems in getting all parties involved, as already described, policy makers were able to take into account the practical experience of stakeholders. Stakeholders and scientists were also informed by policy makers, thereby getting an understanding of the interaction between trade-offs. So, although the overall inclusiveness of bringing everyone together in the CS was not achieved, mutual learning and sharing of problems related to management of herring in the study area has taken place among science, stakeholders and policy makers, albeit only on the EU side. The distance between policy makers and stakeholders within the EU was reduced.

The gap between management and the remaining participants was a bit more challenging, given the political sensitivity of the CS. This gap was labelled 'The Management Wall', since it appeared impossible to reach all potential stakeholders and managers of the CS. The Norwegian part of the CS was never represented and this caused some frustration. However, this 'Management Wall' was over time somewhat overcome. The 'breakthrough' was when two experienced managers and policy makers outlined the field in which the CS could operate and how this could then inform the more 'closed' process layers. This made the 'Management Wall' appear less daunting. The bridging among stakeholders, science and policy can be seen as moving from a situation where no dialogue was involved, to a situation where stakeholders and science sat around the table, explored options and discussed trade-offs inherent in management.

ICES has played a valuable role in providing independent scientific advice on fisheries for over a century. However, the scientific process behind the formation of the fisheries advice and limited opportunity to review other management aspects may have made the process quite closed both to independent observers and to stakeholders affected by the assessments. Scientific working groups have advised managers after a (scientific) peer-review process. It has been a rather non-transparent process with respect to the terms and scientific background of the advice provided and the nature of management. The CS has strengthened trust between scientists and stakeholders and shown that dialogue and information sharing is the key to getting operational and comprehensive advice.

A MAMP would have to respect relative stability both in the North Sea and the Skagerrak. If it is not based on solid undisputable science, then the political reasoning needs to meet Norway's demand to be able to manoeuvre in the negotiations of the TAC in Skagerrak and ensure that the MS maintains relative stability. If the collaboration is not transparent and open to all potential participants, then the risk of it being just as closed as the negotiation process between EU and Norway is high. As long as it is clear how results are reached and that they are open for review, then a collaborative approach to fisheries research can be credible.

Involving stakeholders in an "extended peer review" process may act as a positive driving force to change the whole approach to fisheries management, from a top-down, short-term approach to a bottom-up, long-term one. It is necessary to justify and explain the reasoning behind scientific models, outcomes of which will directly impact the livelihoods of stakeholders involved, as this will lead to an auto-evaluation of the quality and soundness of the scientific knowledge. This in turn will focus attention on the most uncertain, but at the same time most important factors. It drives a natural and shared understanding that these factors should be accounted for in the models, but with large confidence intervals around parameter values and related natural processes. This mutual understanding should extend throughout the governance process, so that the policy decisions should account for the potential risks linked to the inherent uncertainty. If scientific uncertainty cannot be resolved, then management must adapt to such a scenario and be precautionary. This is best communicated to, and acknowledged by, the stakeholders when they are part of the process. This can solicit input from a wide diversity of stakeholders; facilitate the creation of a shared vision of complex problems among scientific experts, policy-makers and stakeholders; and help maintain substantial, structured dialogue between these groups.

6.7 Conclusion

At the heart of fisheries management is expert advice. Scientific advice is required on the state of fish stocks and the impact of fishing on fish stocks. A lack of transparency with respect to uncertainties in fisheries science can damage legitimacy and credibility. In recognition of that, management can be made more credible if science is discussed on a common platform and then based on shared understanding used for management decisions.

It is clear that there is a need for more dialogue among scientists, politicians and stakeholders with regard to the precautionary approach and how it is applied. However, transparency is not only needed with respect to fisheries science for assessments and advice. How TACs are set and management decisions arrived at must also be understandable for the stakeholders involved. Vice versa it is imperative that the motives for sharing knowledge by stakeholders are transparent. This CS has clarified how the management process works, starting from the biological sampling over assessment and advice to the final political process. It has led to a mutual understanding of the limitations in the entire process. The CS has led to the establishment of a platform of communication that was needed to get the MAMP off the ground and underway. The ongoing development of the benchmarking system in the ICES community where scoping exercises with scientists, stakeholders and managers are part of the benchmark would facilitate such platforms of communication across the advisory process.

References

Baron N (2010) Escape from the ivory tower. Island Press, Washington, DC

Duke RD, Geurts J (2004) Policy games for strategic management. Dutch University Press, Amsterdam

Gröhsler T, Oeberst R, Schaber M, Larson N, Kornilovs G (2013) Discrimination of western Baltic spring-spawning and Central Baltic herring (*Clupea harengus* L.) based on growth vs. natural tag information. ICES J Mar Sci 6(1):1108–1117. https://doi.org/10.1093/icesjms/fst064

ICES (2010) Report of the Workshop on procedures to establish the appropriate level of the mixed herring TAC (Spring Western Baltic (WBSS) and Autumn Spawning North Sea (NSAS) stocks) in Skagerrak and Kattegat (Division IIIa). ICES CM 2010/ACOM:64. Retrieved from www.ICES.dk

ICES (2013) Report of the herring assessment working group for the area South of 62N. ICES CM 2013/ACOM:06. Available at www.ices.dk

ICES (2015) Report on the Workshop to evaluate the TAC calculation for herring in IIIa and management plan for herring in the North Sea. ICES CM 2015/ACOM. Retrieved from www.ICES.dk

Poulsen EM (1975) Fiskene i de frie vandmasser, Silden. In: Nørrevang A, Meyer TJ (eds) Danmarks Natur, bd3: Havet. Politikens forlag, Copenhagen, pp 372–378

Chapter 7
Aiming for By-Catch: Collaborative Monitoring of Rare and Migratory Species in the Wadden Sea

Kai Wätjen and Paulina Ramírez-Monsalve

Abstract The underlying motivation for this case study was to find ways to support the existing scientific monitoring system of the Wadden Sea by drawing upon fishermen's experiences, expertise and the information they collect. It was intended that this knowledge would contribute to a species inventory, and the designation of rare/migratory species of fish in particular, thus making a valuable contribution to monitoring the environmental status of the Wadden Sea. Collaborative working between fishermen and scientists was effective at closing important knowledge gaps about the prevalence of rare species and ways to reduce by-catch. At the same time, the project contributed towards raising fishermen's awareness on different nature conservation aspects of the Wadden Sea. The preliminary results of this study demonstrate a potential approach that could be used to establish a more comprehensive monitoring program. Other positive results include the design of a "different type of science", as well as gaining of fishermen's trust, which has been important in sustaining continued close collaboration in new projects such as "Sustain Seafood".

Keywords Monitoring by-catch · Brown shrimp · Ecolabel · Marine protected area · Cooperation

K. Wätjen
Alfred Wegener Institute Helmholtz Centre for Polar and Marine Research, Bremerhaven, Germany
e-mail: kai.waetjen@bremen.de

P. Ramírez-Monsalve (✉)
Innovative Fisheries Management (IFM), Department of Planning, Aalborg University, Aalborg, Denmark
e-mail: paulina@ifm.aau.dk

P. Holm et al. (eds.), *Collaborative Research in Fisheries*, MARE Publication Series 22, https://doi.org/10.1007/978-3-030-26784-1_7

7.1 Introduction

In June 2011, the main section of the brown shrimp (**Crangon crangon**) fishing fleet in the Netherlands and Germany went on a strike, which lasted for 5 weeks. The fishermen demanded fair prices from the wholesalers group controlling 80% of the market. The problem was that frozen storage was already filled up with shrimps from the previous year due to high landings in autumn and winter 2010, and as a result, the prices for shrimps in the new season dropped to a level where fishing was no longer economically viable for the shrimpers.

Although the crisis brought some fishermen close to economic ruin, it also served as a key to open new doors and to think about alternatives for the sector. For example, during the strike, fishermen recognized that alternative marketing structures had, and could be, reorganized to provide consumers with the opportunity of buying fresh shrimp rather than preserved shrimp products, which were peeled and processed in low-cost countries. Additionally, this facilitated fishermen's acceptance of researchers to help in working towards achieving a sustainable fishing business.

The catalyst for fishermen to work together with scientists started in 2008 with the GAP1 project, during which fishermen aimed to obtain the label for an ecologically sustainable fishery. In GAP2, fishermen and scientists agreed that a cooperative research project would help collect fishermen's information, experience and expertise about the recent fish fauna of the Wadden Sea, to complement the existing scientific monitoring system of the area. The need for more monitoring of rare and migratory species in the Wadden Sea has been called for by Jager (2015). Fishermen not only possess information that can add to the existing knowledge, but their continuous activity/fishing effort can also facilitate detection of rare and migratory species. Through this collaboration, fishermen's awareness of different nature conservation aspects of the Wadden Sea could be expanded.

The project also aimed to contribute to understand better the health status aspects of the shrimp population that affect fisheries sustainability. In particular, it included studies on micro-plastics in shrimp and monitoring of the black spot disease, although these were conducted in a more traditional, nonparticipatory approach where the fishermen provided samples to scientists who then did their laboratory work. These are not discussed further in this chapter.

The following sections give further insights into this case study. The first section below introduces the reader to some of the contextual aspects associated with the brown shrimp fishery in the North Sea, and particularly in Germany. In the second section, we report on aspects associated to the fishermen-scientist collaboration of this case study and more specifically, the steps which motivated shrimp fishermen to work with scientists and the expected benefits. Having established the contextual setting for collaboration, the third section focuses on the concrete scientific aims, followed by the process followed aboard the vessels for monitoring the species. The final two sections are reflections on those aspects, which are considered to be positive results from the project and on points, which should be taken into account for future development of similar projects.

7.2 The Fishery

The Brown shrimp fishery takes place along the entire coastline of the southern German bight, including the Wadden Sea. The Wadden Sea, a unique landscape with a distribution of more than 400 km between Den Helder (Netherlands) in the south, and Esbjerg (Denmark) in the north, is the largest unbroken system of intertidal sand and mud flats in the world (UNESCO n.d.), a biosphere reserve, a national park, and a world heritage site. It provides shelter to a number of juvenile fish species, is a perfect nursery and resting area for different kinds of bird species, and is important as a gateway (transitory water) for migratory fish species in the southern German bight (Fig. 7.1).

Fig. 7.1 Location of fishing ports Neuharlingersiel (Lower Saxony) and Hallig Hooge (North Frisia). (Adapted from Wadden Sea Secretariat)

The Wadden Sea is listed as European Marine Protected Area, a provision that falls under the Birds and Habitat Directives of the European Union. Under this classification, anthropogenic activity taking place in this area is authorized only on the condition that sufficient evidence can be delivered demonstrating that activities are not detrimental to the marine ecosystem (Council Directive 1992; MSC 2014a).

In addition to its rich biodiversity, the Wadden Sea also has a rich commercial interest. Along with tourism, shipping, renewable energy production and extraction of oil, gas and sediment, the brown shrimp fishery plays an important role. Landings are ranked fourth in value of all the North Sea Fisheries (STECF 2014), and the highest revenue for the German fleet is generated through brown shrimp (STECF 2018) (see Box 7.1). Commercial shrimp fishing is permitted in both seal and bird protection areas, as well as in the protection zone 1. Within the 12 nautical mile zone, shrimp fishery may only be taken with vehicles up to 300 hp.

Fishing for brown shrimp is regarded to be a traditional fishery. Although the German brown shrimp industry is considered to be a small-scale fishery, the sector plays an important role in employment and social development of the region, as well as for the leisure and tourism sector in disadvantaged coastal areas of northern Germany (see Box 7.2). The picturesque shrimp boats are a characteristic picture of the Wadden Sea coast and a symbol of the Wadden Sea tourism.

The beam trawl of the brown shrimp fishery is a light-gear, equipped with a bobbin trawl or bottom roller instead of heavy tickler chains like the beam trawls used in the flat fish fishery. Sieve nets are also suggested to be used instead of the traditional shrimp nets, mainly as one of the requirements for achieving a Marine Stewardship Council (MSC) accreditation. Although efforts have been made to reduce by-catch, the scope for further improvement is limited.

Box 7.1: European Brown Shrimp Fishery Overview (After Aviat et al. 2011)

- European brown shrimp production has been more or less continuously rising since 1990 and reached 33,000 t in 2009.
- The Netherlands, Germany and Denmark represent 95% of the total European production.
- The Netherlands is the largest producer due to having the most powerful fishing vessels that operate in various weather conditions, reaching thus highest landings per vessel.
- Denmark runs a very efficient fleet that is modern and manages catches through Individual Transferable Quotas.
- The German fleet counts a high share of smaller and older vessels with a strong seasonal fishing pattern and very little winter fishery, which results in lowest mean landings per vessel compared to the other fleets.

Box 7.2: German Brown Shrimp Fishery Overview (After Aviat et al. 2011)

- Germany accounts for more than one-third of the total landings, accounting for approximately 12,000 t annually. The turnover of the German shrimp industry is in the order of EUR 100 million, including the processing and distribution facilities with approximately 250 employees (Aviat et al. 2011).
- The German brown shrimp fleet consists of 212 vessels (2013).
- The crew consists of on skipper, operated by the owner of the vessel, and one or two deckhands. Most vessels of the fleet are between 16 and 20 m in length, with a draft between 1.5 and 2.5 m.
- The engine power is limited to 221 KW (300 hp) per vessel.
- Germany vessels generally do day trips within the vicinity of home ports. The more powerful vessels fish along part of the Dutch coast and all along the German and Danish coasts (in Danish waters outside the 6 nm zone).
- The fishing gear is a beam trawl with bottom rollers (7–9-m wide) mounted on both sides of the outrigger.
- The fishery effort is around 5,500,000 hp. days (2013).
- LPUE (landings per unit effort) [kg/horse-power days at sea] are approximately 2.4 Kg.
- The total annual catch value (all nations) varies with annual prices around 100 million € per year. Germany accounts for approximately 40%.

Once caught, the shrimps are typically cooked and cooled down on board of the vessels, then landed to shore and transported to the premises of two Dutch owned processors that specialize and dominate this trade. In addition, small quantities of shrimp from these fisheries are also handled locally at a small and artisanal scale, producing traditional products for local sale (MSC 2014b).

Generally, the fishery starts in early spring when the shrimps return from the deeper parts of the North Sea and lasts until December when the species migrate back into deeper and warmer water layers to stay there for the winter time.

The brown shrimp industry faces a number of challenges, including declining markets and constantly changing rules and regulations. The most important challenge, however, relates to the negative public image associated, mainly to the habitat destruction, and less so to the by-catch levels. NGOs suspect that fishing for shrimps had far-reaching ecological consequences decades ago and link the substantial disappearance of different species such as sea moss, sand coral (Sabellaria reefs) and other sessile living benthic animals as a result of the widespread beam trawling (Fischer 2009). However, the absence of other fish species – particularly the migratory ones – could also be explained by other factors as the decrease of suitable spawning habitats in the rivers.

Regarding by-catch levels, the brown shrimp métier is considered to have a high amount of by-catch, mainly due to the fishing method (beam trawling). The catch of unwanted species depends on the season, and a high number of by-catch consist of juvenile fishes like plaice and undersized shrimps. Ulleweit et al. (2010) found up to 10% of small fishes in the discard of the German fleet. Including juveniles, the brown shrimp fraction contributes between 50% and 80% to the total catch with strong monthly variations, but only 30% of these shrimps are marketable. The discarded undersized shrimps (40–50%) tend to have a high survival rate and could be easily discarded with high survival rates of approximately 80% (Aviat et al. 2011).

Since the 1980s, the brown shrimp market is controlled by two companies from the Netherlands. Most of the fishermen have contracts with these whole sellers and are obliged to sell them the whole catch. The price, which is determined by fishery auction on the free market, differ from week to week, depending on the demand on the market. This led to a monopolization of the business and a substantial dependency of the fishery sector concerning the whole sellers. In the shrimp fishery, landings and prices are loosely correlated, whereas poor catches might not necessarily be compensated by higher prices and vice versa (Aviat et al. 2011). However, since 2012 the market position of the fishermen has been strengthened by the establishment of the "Erzeugergemeinschaft Deutscher Krabbenfischer (EzDK)", a Producer-Cooperative of German shrimp fishermen, which brought together around 100 small-scale shrimp fishermen. Thanks to the establishment of the EzDK, fishermen organize themselves the classification of their goods and the prices are negotiated centrally through the EzDK.

Within the EU Common Fisheries Policy, the North Sea brown shrimp fishery is not considered a pressure stock fishery, and so it is not subject to management by Total Allowable Catch (TAC), that is, there is no legal catch limit for the shrimp fishery. However, the fishery is subject to ongoing scrutiny for stock assessment and management purposes, and such work is coordinated through a fishery-specific ICES working group (MSC 2014b). Difficulties have been encountered in relation to the stock assessment as it is quite difficult to assess the stock developments of a highly productive, fast growing and short-living resource like the brown shrimp. In addition to fishing mortality, abiotic factors, such as mild winter, which may contribute to a higher predator density, could also have an impact on the stock development. Although the fishery has no TAC, scientists agree that the present (pre-2011) stock sizes are very high and show no signs of overfishing (Aviat et al. 2011). However, ICES warns that the increasing fishing effort has led to growth overfishing. According to ICES (2014) a management system is required to prevent both: growth and recruitment overfishing (ICES 2014).

Having understood some of contextual aspects associated to the brown shrimp fishery in the North Sea, with particular focus on Germany, the chapter turns now into introducing the reader to the first steps that motivated shrimp fishermen to work together with scientists.

7.3 Fishermen-Scientists Collaboration

7.3.1 First Steps and Expected Benefits

Motivation for fishermen to work with scientists started with the fishermen's desire for obtaining the MSC accreditation for their fishery. During 2008–2009, a number of discussions took place between scientists and fishermen to discuss the pros and cons of accreditation. Looking at alternative fishing gears was one of solutions towards achieving accreditation. Contacts with NGOs were made, and informational events were jointly attended such as a trip to Oostende, Belgium (ILVO), in order to get knowledge about the pulse trawl development, one of the alternative gears that could be used.

In 2010, while exploring alternatives on what should the focus during GAP2, it seemed that the fishery would be certified by late 2011 – the fishery was finally certified in December 2017 (MSC 2017); therefore, the MSC issue was not considered a potential topic to explore further. Thus, evaluating the use of specific trawls to reduce by-catch was considered to be a good joint project. The plan was to play an active part in the development and implementation of the HOVERCRAN, a pulse beam trawl, which was tested in Belgium (ILVO). However, only one license was available for a German vessel and this was given to scientist from the Thünen Institute in Hamburg – not part of the GAP2 project.

Although at this point in time the GAP2 participants did not see their initial goals fulfilled (MSC certification and testing of the trawl to reduce by-catch), the experience showed that it was possible to work in collaborative projects such as GAP. After some deliberation within fishermen and scientists, it was decided that GAP2 project could be used to monitor rare and migratory species in the Wadden Sea and look at the incidence of black spot disease and micro-plastics in shrimp.

Scientists and fishermen involved in this case study have been working together for 7 years throughout GAP projects: 1 year in GAP1 (2008–2009) and 4 years in GAP2 (2011–2015). In between the GAP projects, the scientist was in regular contact with the fishermen. Collaboration has not only been about the development of the GAP projects but also about joint participation in diverse Wadden Sea-related fairs and events in topics such as sustainable fisheries, energy-reducing measurements for vessels and alternative fishing gear. Thus, although only two fishermen and one scientist have participated in this study, the level of participation has been deep and long-lasting.

Although not initially planned as one of the outcomes of this GAP2 case study, a concrete product of this collaboration has been the foundation of "Sustain Seafood" in 2013 by fishermen, marketing and food experts and a GAP2 scientist. The organization aims to address sustainability and traceability issues of the brown shrimp fished in the North Sea, and in an effort to cut costs and bring transparency back to consumers, the organization is encouraging consumers, for example, to peel the shrimp themselves. The Sustain Seafood organization is considered to be well established in the fishing community and backed by the management authorities as well as by the shrimper association.

Apart from knowledge on brown shrimp and rare/migratory species, fishermen have been providing scientists with additional knowledge, for example, about the occurrence and distribution of mussel beds, about the movement of tidal rivers and about the first sightings/increasing of grey seals. Furthermore, as the scientist says: "they gave me largely insights into the business and market rules" (Hadjimichael et al. 2015). Fishermen saw participation in this project as having a potential economic benefit in the long term since better public perception would translate into monetary benefits.

Having established the contextual setting of the fishermen-scientist participation of this case study, attention is now given to the concrete scientific aims, followed by a description of the process that took place on board of the ships.

7.3.2 The Focus: Monitoring Rare and Migratory Species

Not a lot is known about the occurrence of rare and migratory species in the Wadden Sea (Jager 2015). Due to the warming of the waters driven by climate change, southern species increasingly expand northward (Perry et al. 2005; Dulvy et al. 2008). For example, red/stripped mullet (Beare et al. 2004), although not common in the Wadden Sea, have been reported as by-catch (scientist and fisherman, personal observation). Lobsters actually need hard substrate; however, in recent years, fishermen regularly found a few lobsters each year in their catches (pers. comment Dirk dell Missier). The reason for that is until now not evaluated, so drawing upon fishermen's knowledge and expertise to complement existing scientific monitoring system under the Marine Strategy Framework Directive (MSFD) was considered to be one way to help fill this deficit. Until now, scientific surveys have been carried out twice a year, for around 2 or 3 days at the time. Fishermen on the other hand have the opportunity to record data on a day-to-day basis since they are fishing approximately 9 months a year, mostly in the same area.

The records of the incidence of rare and endangered species recorded in the project have been used so far by the fleet to demonstrate awareness regarding specific issues required in the MSC certification process. The records will also contribute to the inventory of rare and migrant species for the area, and to provide an open-access database about rare/migratory species for local fishermen. Of particular importance is the contribution it has made regarding the incidence of sturgeon in the Wadden Sea, which is listed as endangered on the IUCN Red List and is currently the target of concerted efforts by conservationists to reintroduce this species into Germany's waters. Responsible for the implementation of the aims is the Gesellschaft zur Rettung des Störs e.V. (society for the rescue of the sturgeon e.V). Just over one hundred years ago, sturgeons belonged to the typical fish fauna in Germany. But due to human activities, most sturgeon species are considered extinct or at least are at risk. Main threats for these and other migratory fish species are pollution of water and habitat destruction. Within the GAP project, fishermen are committed to inspecting their by-catch for Sturgeon specimens to help assess the species' population size as part of the Sturgeon Society's release and recapture programme. Sturgeons are strictly protected, and it is illegal for fishermen to be in possession of even dead specimens; thus, they must be discarded at sea. Sturgeons have a high survival rate after they have been accidentally caught in the nets. As one of the GAP fishermen said: "We are already looking for rare species as part of our GAP work, so why not support the Sturgeon project? It's interesting to find out about the unusual fish living in the sea we work on every day" (pers. com. Uwe Abken and Daniel Ahrends from Polaris vessel).

7.3.3 Process on-Board the Vessel

Three guilds of species were defined to be part of the study: (i) migratory species; (ii) southerly (exotic) species; and (iii) rare/decreasing species. This definition was the result of a collaborative process between fishermen and scientists who discussed mainly about which species could be referred in each one of the guilds, taking into account which species were currently being found as part of their by-catch. The monitoring of species was then deliberately restricted to (a) rare species, which are protected under the Habitat Directive, (b) uncommon and southerly distributed species, (c) IUCN listed fishes and (d) generally decreased species like cartilaginous fishes.

Although the term rare sometimes is primarily used to describe endangered species, in this case study, it refers not only to endangered species (sturgeon, sharks, rays, IUCN listed) but also to neoza (new) species, which are only rare in the Wadden Sea, although common in other parts of the Atlantic. Migratory species are also part of the study since they are considered also endangered mainly due to pollution and river constrictions as channelization and dams.

The fishermen were asked about their knowledge concerning the occurrence of the different identified species. It proved to be the case that most of the species were familiar to the fishermen from earlier times, but that the species' presence has decreased in the more recent times or even disappeared from the catches. The fishermen were informed about the special importance of the species, about their ecological role, life cycle, distribution and abundance.

Regarding the collection process on board of the vessels, it was agreed that the fishermen would collect only species from predefined areas, with the exception of species that were very rare and where the identification was very easily without too much effort for the crew on board (e.g. lobster, sturgeon, sharks and rays). Once caught, and during the sorting and cooking process of the shrimps, it was possible for the deckhands and skipper to search for certain species. Most of the species were either big enough or differed greatly from the rest of the catch (shrimps) and could be thus easily picked out from the funnel. If species were very abundant such as juvenile sea bass or twaite shad in certain month, it was agreed that the numbers are roughly estimated and not counted directly (<10/10–25/>25).

In order to ensure an accurate classification, identification cards (see Fig. 7.3) with the specific features were created and handed over to the fishermen. Potential misinterpretations with similar or related species were addressed particularly by means of using literature images from relevant key publications. In the cases where the crew was not able to recognize the species before entering the sieve drum, especially small pelagics and juvenile river lampreys, which can be difficult to spot, the individuals were collected, after the first sieving process in a basket, to be inspected later on. Very rare species or those which were difficult to identify were photographed by the fishermen and frozen for further laboratory analysis.

Furthermore, it was agreed that special findings were to be additionally noted into the protocol and stored on board until the next meeting between fisherman and scientist, for example, the mass occurrence of different species, algae blooms or cases where the shrimp catch had been very low or very high. Confidential information such as trawl distance, duration and fishing practice were not made public.

In summary, each fisherman was equipped with the following materials to document the information about the species, which were accidentally caught in the nets: identification cards, camera, labels, bags, protocol (see an example in Fig. 7.2) and a CTD mounted at the beam to measure temperature, depth and salinity. After recording the findings in the protocol, the species were discarded, or, if they required further investigation in the laboratory to avoid misinterpretations, the species were photographed, labelled and frozen. The CTD unfortunately was not accurate enough to calculate the different abiotic parameters, so no conclusions could be obtained from this process.

The method was designed to allow the fishermen to easily recognize certain fish species and quantify them, and was tailored to suit the work rhythm of the fishermen, aiming at not disturbing the fishing routine on board of the vessels. However, after an initial testing period, it was found that the sampling procedure was very time consuming and the fishermen were only able to count the species accurately if the total shrimp catch was not too big. The process was then revised to simplify the reporting duties. With the new procedure, the method changed from registering total amount of caught species, to register the data according to presence/absence. The new methods also aimed to ensure that the sampling would be easier for more fishermen in the future (Fig. 7.3).

To ensure the progression of the project, monthly visits were held on board of the vessels, either during fishing activities or at the harbour after the catch had been landed. The idea to meet on the vessels originated from both fishermen and scientist. From a scientific view, it was very important to observe and become familiar with the workflow on deck to consider the special requirements for a successful implementation of the documenting protocol. Furthermore, the shrimpers felt comfortable to meet on their boats. Furthermore, it was manifested that during these visits, there was time enough to discuss environmental and fishery policy and commercial issues of the brown shrimp industry, aspects that the scientists found very enlightened about.

Regular feedback, by phone or during the cruises, was also given to the fishermen on the findings from the data. According to the scientist, fishermen were sometime surprised about the life history (distribution/physiology/ecology) of these species.

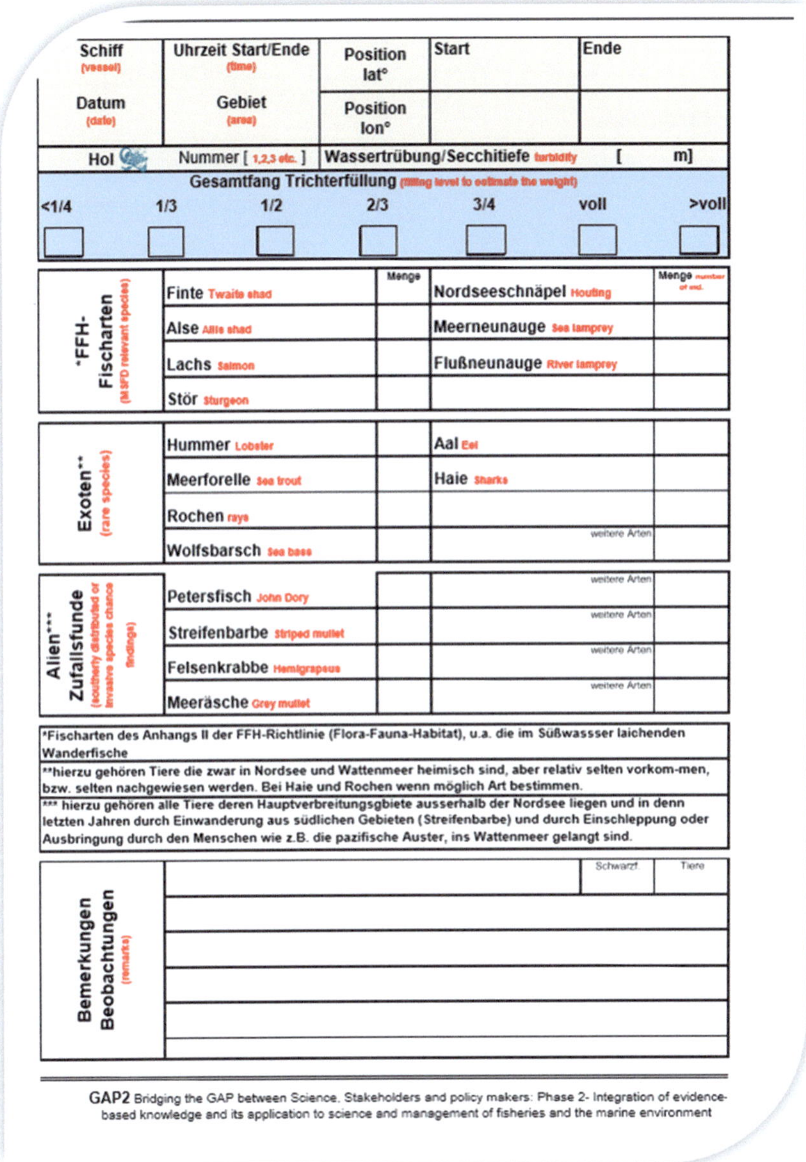

Schiff (vessel)	Uhrzeit Start/Ende (time)	Position lat°	Start	Ende
Datum (date)	Gebiet (area)	Position lon°		
Hol	Nummer [1,2,3 etc.]	Wassertrübung/Secchitiefe turbidity	[	m]

Gesamtfang Trichterfüllung (filling level to estimate the weight)

<1/4	1/3	1/2	2/3	3/4	voll	>voll
☐	☐	☐	☐	☐	☐	☐

		Menge		Menge number of ind.
*FFH-Fischarten (MSFD relevant species)	Finte Twaite shad		Nordseeschnäpel Houting	
	Alse Allis shad		Meerneunauge Sea lamprey	
	Lachs Salmon		Flußneunauge River lamprey	
	Stör sturgeon			
Exoten** (rare species)	Hummer Lobster		Aal Eel	
	Meerforelle Sea trout		Haie Sharks	
	Rochen rays			
	Wolfsbarsch Sea bass		weitere Arten	
Alien*** Zufallsfunde (southerly distributed or invasive species chance findings)	Petersfisch John Dory		weitere Arten	
	Streifenbarbe Striped mullet		weitere Arten	
	Felsenkrabbe Hemigrapsus		weitere Arten	
	Meeräsche Grey mullet		weitere Arten	

*Fischarten des Anhangs II der FFH-Richtlinie (Flora-Fauna-Habitat), u.a. die im Süßwassser laichenden Wanderfische

**hierzu gehören Tiere die zwar in Nordsee und Wattenmeer heimisch sind, aber relativ selten vorkom-men, bzw. selten nachgewiesen werden. Bei Haie und Rochen wenn möglich Art bestimmen.

*** hierzu gehören alle Tiere deren Hauptverbreitungsgbiete ausserhalb der Nordsee liegen und in denn letzten Jahren durch Einwanderung aus südlichen Gebieten (Streifenbarbe) und durch Einschleppung oder Ausbringung durch den Menschen wie z.B. die pazifische Auster, ins Wattenmeer gelangt sind.

Bemerkungen Beobachtungen (remarks)		Schwarzt.	Tiere

GAP2 Bridging the GAP between Science, Stakeholders and policy makers: Phase 2- Integration of evidence-based knowledge and its application to science and management of fisheries and the marine environment

Fig. 7.2 Example of protocol used to document the species. ∗FFH Fishcharten: Species of Annex II of the Habitat Directive (Flora-Fauna-Habitat), amongst others migratory species spawning in fresh water. ∗∗Exoten: these include animals that are native to the North Sea and the Wadden Sea, but are relatively rare, or rarely detected. If possible, determine sharks and rays to species level. ∗∗∗Aliens: this includes all animals whose main distribution areas are outside the North Sea and in the previous years by immigration from southern areas (striped mullet) and by introduction or application by humans, e.g. the Pacific oysters have reached the Wadden Sea

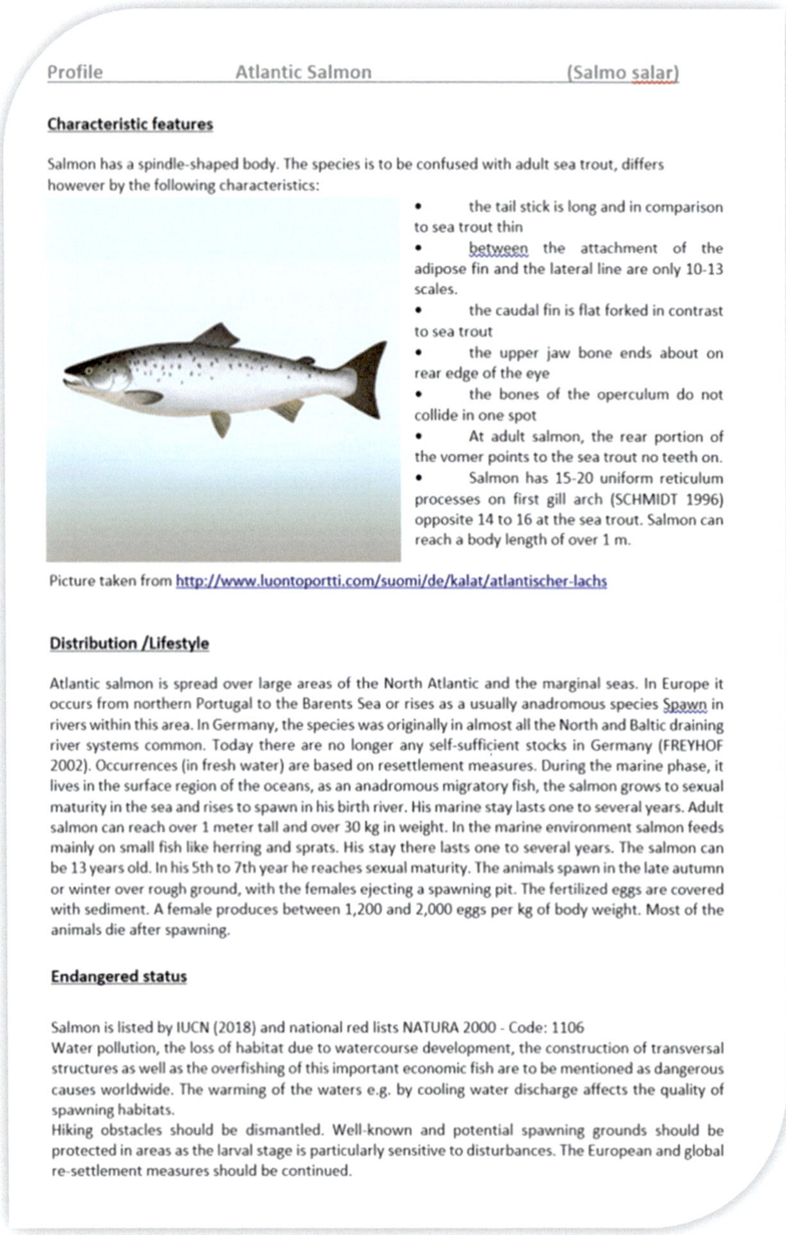

Profile Atlantic Salmon (Salmo salar)

Characteristic features

Salmon has a spindle-shaped body. The species is to be confused with adult sea trout, differs however by the following characteristics:

- the tail stick is long and in comparison to sea trout thin
- between the attachment of the adipose fin and the lateral line are only 10-13 scales.
- the caudal fin is flat forked in contrast to sea trout
- the upper jaw bone ends about on rear edge of the eye
- the bones of the operculum do not collide in one spot
- At adult salmon, the rear portion of the vomer points to the sea trout no teeth on.
- Salmon has 15-20 uniform reticulum processes on first gill arch (SCHMIDT 1996) opposite 14 to 16 at the sea trout. Salmon can reach a body length of over 1 m.

Picture taken from http://www.luontoportti.com/suomi/de/kalat/atlantischer-lachs

Distribution /Lifestyle

Atlantic salmon is spread over large areas of the North Atlantic and the marginal seas. In Europe it occurs from northern Portugal to the Barents Sea or rises as a usually anadromous species Spawn in rivers within this area. In Germany, the species was originally in almost all the North and Baltic draining river systems common. Today there are no longer any self-sufficient stocks in Germany (FREYHOF 2002). Occurrences (in fresh water) are based on resettlement measures. During the marine phase, it lives in the surface region of the oceans, as an anadromous migratory fish, the salmon grows to sexual maturity in the sea and rises to spawn in his birth river. His marine stay lasts one to several years. Adult salmon can reach over 1 meter tall and over 30 kg in weight. In the marine environment salmon feeds mainly on small fish like herring and sprats. His stay there lasts one to several years. The salmon can be 13 years old. In his 5th to 7th year he reaches sexual maturity. The animals spawn in the late autumn or winter over rough ground, with the females ejecting a spawning pit. The fertilized eggs are covered with sediment. A female produces between 1,200 and 2,000 eggs per kg of body weight. Most of the animals die after spawning.

Endangered status

Salmon is listed by IUCN (2018) and national red lists NATURA 2000 - Code: 1106
Water pollution, the loss of habitat due to watercourse development, the construction of transversal structures as well as the overfishing of this important economic fish are to be mentioned as dangerous causes worldwide. The warming of the waters e.g. by cooling water discharge affects the quality of spawning habitats.
Hiking obstacles should be dismantled. Well-known and potential spawning grounds should be protected in areas as the larval stage is particularly sensitive to disturbances. The European and global re-settlement measures should be continued.

Fig. 7.3 Example of identification card

7.4 Developing a Different Type of Science and Gaining Fishermen's Trust

Two aspects are presented here as being particularly positive results from the project, one of them being the different type of science that was created, the other gaining the trust of the fishermen and likewise, fishermen's trust of the scientist.

The method developed by scientists and fishermen during the GAP2 project allowed recording of information about rare and migratory species at a much faster rate and volume than what could have been achieved through conventional annual scientific surveys. A great deal of fishing effort is needed to spot rare and migratory species, and this effort is not feasible through scientific surveys alone. Currently, local authorities monitor the rare/migratory species within a relatively small time-frame window (twice a year between two or three or one to two weeks depending on the monitoring programme). Furthermore, fishermen fish at places where scientists do not usually fish, thus extending the sampling area and possibility of detection; as the scientist say "If you have some long-term monitoring stations and you're a scientist then you go every year to this point, but the fishermen go to other points, or other stations or other areas" (Hadjimichael et al. 2015). The few so far existing monitoring programs concerning the fish fauna are carried out exclusively in the Dutch and German parts of the Wadden Sea. For the Danish part, no data is available, and since 1960, no regular monitoring has taken place (Jager et al. 2009). A much better monitoring can be achieved with the help of fishermen who are at sea 9 months of the year, and this experience could be used to improve existing monitoring programmes.

The second positive point is related to gaining mutual trust. When talking about scientific data and cooperation – particularly in terms of implementation of protected areas or nature conservation – a slight negative note and fear of consequences (e.g. closure of fishing grounds) has been perceived as emerging from the fishermen (pers. comment Dirk dell Missier). This project provided a demonstration that fishermen do not have to be afraid to cooperate, and that there are positive advantages resulting from this cooperation. For example, increase in trust is identified by the fact that fishermen contacted the GAP2 team; in this platform, the fishermen can see that the participating scientists have a "shrimper" background and that scientist can do something for the fishermen. According to the scientist: "We reached with this mix of research and marketing a lot of more fishermen than we expected and we have reached more with our information campaign" (Hadjimichael et al. 2015). This could be seen as the start for future teamwork; connections have been created and there is potential for future research and collaboration, for example, two new projects have recently (2017) started: one bringing the human element behind shrimpers and their story of the last haul (Wattenmeerkrabbe); the second about creel fishing for brown crab (**Cancer pagurus**) in the offshore wind park area, as a strategy to diversify the shrimp fishery and to take fishing pressure from the Wadden Sea.

One specific point which was helpful in gaining the trust of the fishermen was related to the meetings on board of the fishing vessels. As the scientist commented:

"this is a good location where to talk about the project since the fishermen are feeling in their own area…and can make fun of the scientist when we get sea sick" (Hadjimichael et al. 2015). In addition, it was better to carry out meetings in conjunction with fieldwork, thus discussing the cooperative work on board during data collection or immediately after.

7.5 Reflecting Back

It has been wondered whether having other types of participants would have led to different results, for example, the involvement of NGOs or of fishery managers.

In the beginning of the project, contacts were made with the WWF. According to the scientist, this is an NGO that fishermen can cooperate with (Hadjimichael et al. 2015). Also, according to the scientist, an interesting conversation took place between one of the fishermen stakeholders and the WWF Wadden Sea when they, together with the scientists, were sitting at the table: "you can't discuss with every fishermen and with every NGO campaigner […] you need people who have a more open mind about the future" (Hadjimichael et al. 2015). Although there are good cooperative practices in the UK and Spain between WWF and the fishermen, this cooperation has, until now, been difficult in Germany. The cooperation with the NGO did not go further since there was the question of how environmental agencies or NGOs would deal with the sensitive data concerning catches of rare/migratory species and possible closures of certain areas. A point that the fishermen made clear to the scientist from the beginning of their participation was that the findings of the study should not to lead to closing of fishing grounds or further fishing restrictions. They were willing to help provide evidence on the incidence of rare and endangered species, but no to be punished for it.

In a similar way, the project had little involvement or cooperation with policymakers or managers. Their involvement was, afterwards, considered important as a means of perhaps having helped improved the participation of more fishermen into the project, because it was later noted that fishermen did not "see" the managers' support for the project. Thus, if there were to be another project, it would be very important to cooperate or at least have a better contact with the authorities.

Another point of reflection is that about fishermen's perception regarding the benefits gained from this GAP2 project. Post-project reflections include the fact that perhaps fishermen did not have a clear understanding of the potential benefits from this project. It is thus possible that perhaps the additional work and expected benefits were not made clear when the project was presented to the participants, and that this influenced the amount of fishermen who showed interest in participating in the project (Hadjimichael et al. 2015). While the benefits from an MSC certification and the implementation of the HOVERCRAN (although both are not supported by the whole fleet) were readily understood, it was more difficult to explain the benefits originating from increasing the scientific understanding of species composition in the Wadden Sea.

Acknowledgments This collaborative project took place, thanks to the engagement of several persons. Special thanks to Uwe Abken, Daniel Ahrends, Dirk dell Missier, Philipp Oberdörffer and Dirk Sander. Uwe Abken is the owner of "POLARIS", located in Lower Saxony. Daniel Ahrends, his decks hand who unfortunately died in a car accident in April 2014, was proud to be part of GAP2 and keen to collect the species and tried to find every individual. Dirk dell Missier is the owner of "∗De Liekedeelers", which means the crew of the pirate Klaus Störtebecker, "take it from the rich and give it to the poor". Notorious pirates – driven by the vision of equal shares for all – roamed the Baltic Sea in their sailing ships. They called themselves "Liekedeeler" or "equal sharers" and shared their booty equally with one another. Philipp Oberdörffer and Dirk Sander are managing directors of the EzDK (Erzeugergemeinschaft der Deutschen Krabbenfischer).

References

Aviat D, Diamantis C, Neudecker T, Berkenhagen J, Müller M (2011) The North Sea Brown Shrimp IP/B/PECH/IC/2010_102. Retrieved March 14, 2015, from http://www.europarl.europa.eu/RegData/etudes/etudes/join/2011/460041/IPOL-PECH_ET%282011%29460041_DE.pdf

Beare DJ, Burns F, Greig A, Jones EG, Peach K, Kienzle M, McKenzie E, Reid DG (2004) Long-term increases in prevalence of North Sea fishes having southern biogeographic affinities. Mar Ecol Prog Ser 284:269–278

Council Directive (1992) 92/43/EEC of 21 May 1992 on the conservation of natural habitats and of wild fauna and flora (OJ L 206, 22.7.1992, pp. 7–50)

Dulvy NK, Rogers SI, Jennings S, Stelzenmüller V, Dye SR, Skjoldal HR (2008) Climate change and deepening of the North Sea fish assemblage: a biotic indicator of warming seas. J Appl Ecol 45(4):1029

Fischer S (2009) Nicht nur Krabben im Netz. Der Beifang in der Baumkurrenfischerei auf die Nordseegarnele (Crangon crangon). Study on behalf of WWF Germany, 31 pp.

Hadjimichael M, Holm P, Ramírez-Monsalve P, Tveiterås K Framing participatory research. (2015) Analysis of the credibility, legitimacy and salience/description and evaluation of impact. http://gap2.eu/gap2wordpress/wp-content/uploads/2015/05/D4.3.1-Analysis-of-the-Credibility-Legitimacy-and-Salience-of-Cooperative-Research.pdf. Accessed July 2015

International Council for the Exploration of the Sea (ICES) (2014) Request from Germany and the Netherlands on the potential need for a management of brown shrimp (Crangon crangon) in the North Sea. Available at http://www.ices.dk/sites/pub/Publication%20Reports/Advice/2014/Special%20Requests/Germany_NL_Crangon_advice.pdf. Accessed July 2015

Jager Z, Bolle L, Dänhardt A, Diederichs B, Neudecker T, Scholle J, Vorberg R (2009) Fish. Thematic Report No. 14. In: Marencic H, de Vlas J (eds) Quality status report 2009. Wadden Sea Ecosystem No. 25. Common Wadden Sea Secretariat. Trilateral Monitoring and Assessment Group, Wilhelmshaven. Retrieved March 14, 2015, from http://www.waddensea-secretariat.org/sites/default/files/downloads/14-fish-09-12-22.pdf

Jager Z (2015) Conservation management of fish populations in the Wadden Sea – workshop report. Hamburg, January 2015. Retrieved December 5, 2015, from http://www.waddensea-secretariat.org/sites/default/files/downloads/news/wadden_sea_fish_ws_jan_2015_-_final_workshop_report.pdf

Marine Stewardship Council (MSC) (2014a). Marine Stewardship Council, from Request for variation to the MSC Fisheries Certification Requirement v2.0 – Implementation timeframes. Retrieved 19 December, 2014, from http://www.msc.org/track-a-fishery/fisheries-in-the-program/in-assessment/north-east-atlantic/germany-north-sea-brown-shrimp/assessment-downloads-1/20141219_VAR_RESP_SHR167.pdf

MSC (2014b) Marine Stewardship Council, from Germany North Sea brown shrimp. Retrieved 23 December, 2014, from http://www.msc.org/track-a-fishery/fisheries-

in-the-program/in-assessment/north-east-atlantic/germany-north-sea-brown-shrimp/germany-north-sea-brown-shrimp
MSC (2017) Assessment of North Sea Brown Shrimp. Retrieved 5 December, 2017, from https://fisheries.msc.org/en/fisheries/north-sea-brown-shrimp/@@assessments
Perry AL, Low PJ, Ellis JR, Reynolds JD (2005) Climate change and distribution shifts in marine fishes. Science 308:1912–1915
Scientific, Technical and Economic Committee for Fisheries (STECF) (2014) The 2014 annual economic report on the EU fishing fleet (STECF-14-16). Publications Office of the European Union, Luxembourg, p 363. https://ec.europa.eu/jrc/en/publication/eur-scientific-and-technical-research-reports/2014-annual-economic-report-eu-fishing-fleet-stecf-14-16-scientific-technical-and-economic, EUR 26901 EN, JRC 92507
Scientific, Technical and Economic Committee for Fisheries (STECF) (2018) The 2018 Annual economic report on the EU fishing fleet (STECF-18-07). Publications Office of the European Union, Luxembourg. https://doi.org/10.2760/56158. https://ec.europa.eu/jrc/en/publication/eur-scientific-and-technical-research-reports/2018-annual-economic-report-eu-fishing-fleet-stecf-18-07, JRC112940, ISBN 978-92-79-79390-5
UNESCO (n.d.) Wadden Sea. Retrieved 20 November, 2017, from http://whc.unesco.org/en/list/1314
Wadden Sea Secretariat Map Wadden Sea world heritage. Retrieved 14 November, from http://www.waddensea-worldheritage.org/sites/default/files/downloads/waddensea_worldheritage_en.jpg

Chapter 8
The Italian Job: Navigating the (Im)Perfect Storm of Participatory Fisheries Research in the Northern Adriatic Sea

Saša Raicevich, Mark Dubois, Marianna Bullo, Gianluca Franceschini, Monica Mion, Marco Nalon, Camilla Piras, Laura Sabatini, Tomaso Fortibuoni, Igor Celić, Adriano Mariani, Simone Serra, Andrea Fusari, Giovanni Bulian, and Otello Giovanardi

Abstract In fisheries management there is now a broad consensus on the importance of including and integrating fishermen and their knowledge. This shift reflected by the latest reform of the EU Common Fisheries Policy indicates, at least in principle, a move away from the traditional centralised 'top-down' model to a more decentralised 'networked' system of governance. What happens though to this ideal of participation when there is limited agreement about what the problems are and how to handle them? The GAP project case study in Chioggia, located in the Northern Adriatic, provides an opportunity to illustrate this question through exploring the differing perceptions and competing narratives surrounding 'la crisi' (the crisis) in the fisheries in question. Simply put, 'la crisi' represents a crisis in the sector, with stock collapse and the ruination of local fisheries-based livelihoods a

S. Raicevich (✉) · L. Sabatini · O. Giovanardi
Italian National Institute for Environmental Protection and Research (ISPRA), Chioggia, Italy

Institute for Biological Resources and Marine Biotechnologies (IRBIM), Ancona, Italy
e-mail: sasa.raicevich@isprambiente.it

M. Dubois
Innovative Fisheries Management, Aalborg University Research Centre, Aalborg, Denmark

Department of Fisheries, World Fish Myanmar, Yangon, Myanmar

M. Bullo · G. Franceschini · C. Piras
Italian National Institute for Environmental Protection and Research (ISPRA), Chioggia, Italy

M. Mion
Italian National Institute for Environmental Protection and Research (ISPRA), Chioggia, Italy

Institute of Marine Research, Swedish University of Agricultural Sciences (SLU), Lysekil, Sweden

M. Nalon
Italian National Institute for Environmental Protection and Research (ISPRA), Chioggia, Italy

National Institute of Aquatic Resources, Section for Ecosystem based Marine Management, Technical University of Denmark, Hirtshals, Denmark

P. Holm et al. (eds.), *Collaborative Research in Fisheries*, MARE Publication Series 22, https://doi.org/10.1007/978-3-030-26784-1_8

likely outcome should the status quo be maintained. This perspective is held by many fishermen and a number of scientists working with them and yet is opposed by a counter narrative of 'stability and ecosystem health', promoted by regional fisheries managers and affiliated scientists. The Chioggia case study, a participatory research initiative between scientists and local fishermen, seeks to bridge this gap in knowledge by mapping the spatial and temporal distribution of resources and fishing effort in the Northern Adriatic Sea. The paper emphasises the challenges and opportunities that this research collaboration entails and assesses its capacity to catalyse or inhibit the conditions necessary for mobilising collective action in fisheries management.

Keywords Participatory research · Knowledge systems · Fisheries management · Policy processes · Decentralisation

8.1 Introduction: A Sailor's Tale of Stormy Seas and Different Perspectives on the Weather

A shift in emphasis has occurred over the last two decades with top-down fisheries management approaches giving way to more direct and broader-based engagements (Symes 1997; Jentoft et al. 1998; Gutiérrez et al. 2011). One such approach to decentralised fisheries management is the EU GAP project, which takes a very practical approach to enhancing participation and broadening the policy arena and knowledge base used in fisheries management. It is this approach (termed the 'GAP method') that we describe here with a focus on how successful it has been to bridge the gaps between different groups of stakeholders, and particularly between scientists, managers and fishermen, and their knowledge systems.

We draw on the GAP case study in the Northern Adriatic focussing on a mixed fishery which operates in a complex governance context with multiple social, economic and environmental conflicts and concerns.

T. Fortibuoni
Italian National Institute for Environmental Protection and Research, Ozzano dell'Emilia, BO, Italy

Oceanography Division (OCE), National Institute of Oceanography and Experimental Geophysics (OGS), Trieste, Italy

I. Celić
Italian National Institute for Environmental Protection and Research (ISPRA), Chioggia, Italy

Oceanography Division (OCE), National Institute of Oceanography and Experimental Geophysics (OGS), Trieste, Italy

A. Mariani · S. Serra · A. Fusari
Consorzio UNIMAR Società Cooperativa, Roma, Italy

G. Bulian
Department of Asian and North African Studies, Ca' Foscari University of Venice, Venezia, Italy

In particular, the fishermen who took part in the project were part of the demersal trawler métier, i.e. otter-board and 'rapido' trawlers based in Chioggia, the most important fishing port in Italy. While this métier currently includes about 80 fishing vessels, the port of Chioggia also includes the following: mid-water pelagic trawlers (approximately 10 fishing vessels), hydraulic dredgers (approximately 70 vessels) and artisanal fishermen (10 fishing vessels) (Source: Coast Guard data). There is little confidence or trust within and between stakeholder groups and a widespread lack of agreement on even the most basic descriptors of the fisheries and the resources upon which they are based. This is a typical "wicked problem" in conservation, where different groups cannot agree on the problem or its solution. This of course makes it challenging for all those working on their management and those affected by management decisions.

A key issue for local fishermen, scientists and managers is whether the fisheries of the Northern Adriatic are sustainable. Some questions of particular importance include: what are the status of the stocks and are they over-fished or not? Are catches declining or stable? How does stock status affect fisheries and fishermen? In what follows, the views of various stakeholders are provided.

> My main worry is that I made a big mistake having two sons. Their life is lost because instead of suggesting they do something else, I convinced them to join the boat to work with me, but now there are no fish remaining (Chioggia fisherman, interview).

> I started working in this field in 1976–77. At that time, the fisheries were still expanding and good catches were available. So you know, the catch per unit effort has decreased a lot and even the total landings (Fisheries scientist, interview).

> Since the 90s, there has been a continuous decrease not only in revenues but also the condition of exploited resources... some species declined earlier and some species later whereas some disappeared or collapsed entirely. In the last (few) years, this crisis has become more evident (Fisheries scientist, interview).

> The problem is that stock assessments have rarely been applied in the Mediterranean, […]. Such assessments might have shown earlier that, […] stocks were over exploited (Fisheries scientist, interview).

These views, commonly expressed by local fishermen and scientists working with them, describe the fisheries of the Northern Adriatic and associated fishing communities, as a 'system in crisis'. This perspective often includes the view that fisheries have failed or that stocks are collapsing, overfishing is rampant and fishing livelihoods are under extreme threat.

By contrast, the managers of the Veneto Region (those responsible for local management of the region's fisheries) and the scientists that work with them have a different view, particularly with respect to the status of the stocks:

> In general, there is no significant problem in terms of stocks of the Northern Adriatic. Our data shows stability and in fact an increase in total production across species over the last several years (Fisheries economist & statistician for the Veneto Region, interview).

This is a view of stability and viability, at least concerning the fisheries resources themselves, and portrays an optimistic picture where the outlook is positive and business as usual can prevail. What this clearly demonstrates is a lack of agreement on the basic facts and indeed more broadly on the political descriptors used to justify specific courses of action or inaction, such as those related to 'la crisi' (the crisis). This is particularly true in areas where the knowledge base is limited or where the credibility and legitimacy of existing knowledge is contested.

> I know only this - in Italy reality is the way you look at things [...] To believe only in official statistics is difficult because there are a lot of things that are true but are not included in the statistics (Italian fisheries expert, interview).

However, it is worth mentioning that the Veneto Region has accepted certain problems, such as overcapacity, high fuel costs, marketing problems etc., for which the region has some capacity and interest to address. This is demonstrated by the set-up of a 'fishery crisis unit.' But tellingly, at present, the issue of fish stocks remains un-addressed, as management of fish stocks is far more difficult and Veneto Region does not have adequate capacity to address the issue.

Irrespective of whether the crisis is an 'actual or convenient truth', both framings are gross oversimplifications of extremely complex systems, and clearly demonstrate the gap that exists in thinking between proponents of the two framings (Friend 2009). Moreover, these competing positions and their story lines legitimise certain forms of knowledge and action and exclude others and are the means by which actors and institutions take political positions with regard to ownership over resources (Friend 2009).

> [the impression is that]... there is no political pressure to change, so a big crisis could be useful'... [it seems that]...' most are against change so it is good if there is a big crisis, a big collapse, rather than a gradual decline (Fisheries scientist, interview).

Finally, in addition to calling into question the credibility of the knowledge base, the legitimacy of policies and the capacity of policy-makers are also called into question:

> The rules from Brussels are made by incompetent people who do not know our real situation; they have wrong data, we should provide them the real data (Chioggia Fisherman, interview).

> Our fish is different from that of the Mediterranean Sea. We've been to Rome to protest and when we go back we will talk with the minister to find a solution, hoping that at least he understands the difference between a scallop and a tuna [...] the problem [will not be solved by] restrictions on fishing. The problem is that the institutions that make decisions do not understand the material on which to legislate. The Adriatic Sea is very different from the Mediterranean Sea (Chioggia fisherman, interview).

It is in the context of low levels of trust between stakeholders, the lack of credibility and legitimacy in the knowledge base for management, as well the different conceptualisations and narrative framings of fisheries management problems, that the tiny GAP 'life raft' plots its course across the great stormy seas of the Northern Adriatic.

This case study aims to determine if, in this complex scenario, the GAP method can be successful in bridging these differences by working collaboratively to produce knowledge according to agreed methods and shared platforms. The aim was to find common ground on basic management issues (such as the status of the stocks) and strategic actions necessary to rebuild them by forming alliances around mutual interests to engage with and shape the management discourse.

8.2 Planning the Voyage: Scale, Concepts and Methods

Our case study operates at lower spatial (and administrative) level from the Mediterranean Advisory Council, which was seen by the Common Fisheries Policy regionalisation process as the level to interact with fishermen representatives (EU Reg. no 1380/2013). Instead the case study operated at a regional administrative and ecological scale (Reid et al. 2006), i.e. the Veneto Region and the 'fishery system' includes fishermen and their community, fishermen representatives, managers and policy-makers. Focussing at this scale was considered to have two primary advantages:

- At this level, fishermen are able to represent themselves and their experience-based knowledge (EBK) directly rather than by proxy through their associations.
- This is the scale where administrations have some real power in terms of fisheries management.

Conceptually the chapter draws from Johnson and van Densen's (2007) framework for cooperative research processes and directly applies Raakjær's (2009) work on policy processes in fisheries governance. It does so to support an examination of how, in the knowledge production process, groups of actors, form alliances to pursue certain interests shaped by ideas and agendas, and the language of supporting narratives (Raakjær 2009). This approach is not applied to determine whether there is a crisis or not, should that even be possible in a system as varied and complex as the mixed fisheries of the Northern Adriatic, but rather to challenge these simple narratives and explore if and how they relate to policy process formulation. A key focus throughout is to analyse how scientists and fishermen cooperate in the co-production of knowledge and how this may shape conditions for collective action.

Our conceptual frame is 'operationalised' through incorporating a number of mixed social and natural science methods including field and participant observations, oral histories, semi-structured interviews and surveys conducted with 94 local fishermen. The research was carried out between 2008 and 2014.

Structurally, the chapter begins with a description of the origins and inception of the collaboration carried out under the GAP1 project, followed by details of the process of problem identification and the formulation of research objectives. We continue by describing the scale and approach, conceptual frame and methods used, including what we call 'critical signals' which emerged throughout the collaboration within GAP2. Finally, we outline and discuss the key results and outcomes of the participatory research process.

8.3 Setting Sail: Early Voyages

8.3.1 *The Ship, the Crew, and the Compass, Building Alliances in Participatory Research: GAP1*

Prior to the GAP project, which was conceived and initiated in 2008 by ISPRA (Italian National Institute for Environmental Protection and Research) researchers in the context of Chioggia, collaborative research with fishermen meant fisheries scientists were allowed on board fishing vessels during commercial trips and scientists able to hire vessels to conduct sampling exercises. Moreover, besides a degree of information sharing and occasional joint projects, no stable relationship existed between ISPRA and local policy-makers such as the Veneto Region fishery office and larger fishing associations.

The GAP1 project was an attempt to move away from such a conventional approach and establish a collaborative process between fishermen and fisheries scientists that was equitable in nature. Equitable referred to recognising fishermen's needs and capacities and then supporting these with scientific resources in a participatory way.

It could be said that GAP1 activities focussed mainly on building alliances with fishermen and their representatives (National Fishermen Organizations) in line with Johnson and van Densen's first two stages of a cooperative research process, i.e. problem identification and mapping out the research approach and design specification (Johnson and van Densen 2007). The follow-up GAP2 project (which commenced in 2011) focussed on implementing the last three stages of Johnson and van Densen's approach, specifically data collection, processing and analysis and the communication of results. We focus on both GAP1 and GAP2 so as to adequately analyse the initiation, implementation and evaluation of the GAP method as applied in the Northern Adriatic.

One of the first strategic alliances formed early on in the GAP process was with the UNIMAR consortium. UNIMAR represents a consortium of fisheries research cooperatives from the main national fishermen's organisations. Such a consortium was seen as key to horizontal collaboration between different entities, including in respect to their long-lasting experience in collaborating with fishermen. UNIMAR's relationship with both fishermen and the Ministry of fisheries was also seen as significant.

Several internal meetings were held in Chioggia by ISPRA researchers (also involving the stakeholder partner UNIMAR), before starting project activities, so as to define the possible route for GAP case study development. Personal and research aims were clarified as well as details regarding strategy and tactics necessary to ensure the success of the project. It is noteworthy that these initial consultations were held behind closed doors and did not include fishermen or their representatives.

A number of techniques were used during this phase including brainstorming and future scenario planning. In particular though, we would like to highlight the use of the "social-tachymeter", a method that allows participants to reflect on the range of stakeholders that could 'support or fight against' – to different degrees – a proposal. The method also allows for defining the tactics and strategies required to change the attitude of stakeholders, starting with the more neutral ones (Jelfs 1982).

It was decided to hold a start-up meeting with local representatives of National Fishermen Organizations (Federpesca, Federcoopesca, Legapesca and AGCI), delegates of local fishermen cooperatives and fishermen who were interested and influential in order to kick-start effective collaboration between stakeholders and fishermen in Chioggia. The latter also included fishermen who had already collaborated with ISPRA in the past.

The GAP1 kick-off meeting was held on the 6th December 2008 and was attended by 11 fishermen – most of them acting also as fishermen representatives – five ISPRA scientists and one UNIMAR representative. The meeting focussed on highlighting positive experiences of fishermen-scientists' collaboration and outlining GAP aims. ISPRA's role was recognised as important by participants because of its long history in Chioggia while UNIMAR's presence was considered strategic in terms of allowing for better communication with National Fishermen Organizations. All participants were in favour of an open and collaborative process.

8.3.2 *Charting the Waves: Creating a Common Ground*

Having identified the main players, six meetings followed that put forward the most important concerns of fishermen in relation to their activities and fisheries management (Table 8.1). Participatory research activities that could provide (scientific) evidence to foster solutions were also discussed. The emphasis from the outset was to produce knowledge collaboratively that could be considered scientifically credible and secondly to ensure legitimacy as a consequence of producing knowledge collaboratively with fishermen. The issues of primary importance to the process are presented below.

Table 8.1 Main fishery-related issues as identified by fishermen during GAP1

1. The minimum landing size regulations for some species were considered by fishermen to be too big
2. The regulation on the mesh size (EC Reg. no. 1967/2006; with new rules to be adopted on the 1 June 2010) and the need to test the selectivity of fishing gears
3. The progressive reduction of trawling areas due to other economic uses of the sea such as mariculture, sand extraction, off-shore LPG terminals, etc.
4. The effects of pollution and other anthropogenic pressures on fisheries resources
5. The duration of summer trawling closure (for some fishermen a management measure that was not effective as it was too short while for others it came at the wrong time)
6. The need to reduce fishing effort (e.g. hours of fishing per week)
7. Problems related to the facilities and functioning of the Chioggia wholesale fish market (e.g. trade and value of fish)

Discussions with local fishermen showed that whilst fishermen were willing to collaborate with scientists, there was a considerable lack of trust between the two groups. This was apparent in terms of the desire for privacy of information from fishermen, something exacerbated as a result of earlier collaborations where information was not shared back with participating fishermen.

> I offer my-self to collect and provide data provided that such data from my fishing vessel is not shared….fishermen working in restricted inshore areas would not be willing to share such information (Chioggia fisherman, interview).

> I would be happy to collect data again, but doing so manually is time consuming, it would be better to use electronic devices. But I think a degree of privacy should be ensured and only aggregate data used. Moreover, I never saw the results of the data analysis (Chioggia fisherman, interview).

Other fishermen expressed similar concerns:

> We are not sure whether our data will be used against us or not (Chioggia fisherman, interview).

There was at least one fisherman whose main concern was the necessity to act quickly:

> It is necessary to take immediate action. I use legal fishing gear and mesh size, but I accidentally catch undersized species..., I could be fined despite respecting the law. I would be happy to host scientists who could show data on such issues (Chioggia fisherman, interview).

Discussions and rich dialogue continued, and an attempt was made by scientists to raise awareness with regard to how decisions are made in Brussels. The debate was also enriched by scientists who gave several presentations on the main issues of interest raised by fishermen, specifically on minimum landing sizes, the spatial use of the Adriatic Sea, fishing gear selectivity and discards. Over time, and in response to these discussions, two meta-issues were prioritised:

- Selectivity of fishing gear in relation to minimum landing size and mesh size obligations.
- The spatial use of the sea and management proposals to reduce conflicts and promote sustainable fisheries (e.g. revision of summer trawling closure; reduction in fishing effort).

As these issues were quite broad, discussions continued with ISPRA scientists who suggested a sampling approach that could provide the information required to describe the status of fisheries in relation to the above concerns and influence current management. Estimates on possible costs, protocols and timeframes were also provided. Accordingly, a draft scheme was put forward that included:

- sea trial experiments on fishing gear selectivity;
- log-book installation for self-sampling by fishermen;
- on-board observations by scientists pertaining to biological data on major target species in the Northern Adriatic Sea.

This scheme was fully debated with regard to costs, sharing of money between research institutes and fishermen, expected results and time frame. It is worth noting that fishermen discussed even the general methodological approach and the need for data collection from fishing vessels of different sizes and using different fishing gears. This showed the fishermen empirical knowledge included complex sampling concepts such as what scientists call "random stratified sampling".

The fishermen involved stressed the need to establish such research activities immediately. Unfortunately, this proved difficult as GAP1 was largely a preparatory initiative aimed at defining and outlining the project scope and direction for GAP2. In addition to this, there was a gap between project phases and no guarantee over future funding. Therefore, the collaborators agreed on submitting a proposal to the Veneto Region fishery office to support selectivity trials and reducing undersized fish catch. The remaining issues, principally that of the spatial use of the sea, self-sampling and the fishing ban, were incorporated in the GAP2 proposal.

Fishermen decided that "it is necessary [that] the proposal includes all fishermen organizations" (Chioggia fisherman), and attempted to convince fisheries officers of the importance of this. Notwithstanding the effort put forward by scientists and fishermen, the proposal to the Veneto Region was rejected. The official position taken was that there was no financial instrument available to support it.

Despite the frustration of a negative outcome and a lag in activities, contacts with fishermen were maintained even after the end of GAP1. When the GAP2 proposal was accepted, activities started again with the full participation of all involved.

8.4 Sailing Together: Putting Collaborative Research into Practice, the GAP2 Experience

GAP2 focussed only partly on the objectives identified in GAP1 since the most urgent issues at that time (e.g. selectivity trials) had become outdated. The Chioggia case study under GAP ended up taking an approach which focussed on building a common knowledge base to describe the dynamics of fisheries resources and fishing effort with a view to providing management recommendations. The main goals of the case study related to the formalisation of a proposal for rethinking the duration and enforcement of the summer fishing ban and a proposal for regulating fishing effort in terms of hours at sea at the end of summer fishing closure. Field activities were designed to suit this purpose and coordinated, discussed, and framed in a series of meetings that were aimed at verifying the progress of the project and sharing information and knowledge in an open and friendly environment. Field activities included the collection of catch data by scientific observers during commercial fishing, the use of electronic logbooks to allow fishermen to self-sample spatially explicit data on commercial catches, and the surveying of trawling in the Veneto Region's administrative waters. The various activities were integrated together in the course of several open meetings including mutual learning events and exchange activities (e.g. an exchange to visit and learn from fisheries operations in the Netherlands). In addition, a detailed survey was carried out with a hundred fishermen and skippers which focussed on collecting opinions about current fisheries management practices (and in particular the summer fishing ban). The overall approach adopted within GAP was a participatory one with specific attention paid to getting fishermen to contribute their experience-based ecological knowledge (i.e. in trawl-survey design and data interpretation and in corroborating scientific results [see Sect. 8.4.4]) as well as through discussions on possible management alternatives aimed at addressing biological and economic needs.

Given that this chapter focuses on the process of collaboration, how it was instigated, how it operates and what we can learn from the process, we do not describe the scientific rationale/background and methodological approach adopted for sampling. Instead we focus on how the participatory approach was developed and implemented in each collaborative activity.

8.4.1 A Time for Dialogue: The Beginnings of a Shared Narrative

Critical to the participatory process was ensuring regular open meetings between collaborators. Informal meetings were held on a monthly basis and were open to all fishermen who wished to participate, although over time mainly GAP fishermen participated. They were notified of meetings mainly by phone and through the GAP case study Facebook page.

Timings of the meetings were important to ensure a good turnout. Meetings were mainly held on non-fishing days and at the fish market hall, a place considered more neutral and easy to reach by fishermen. Occasionally, when the fish market hall was unavailable, meetings took place at the ISPRA Chioggia branch. Usually scientists prepared a draft agenda for the meetings which also outlined issues which required a decision by the group. Short presentations were made detailing on-going activities, results and any other relevant issues for the project, including controversial matters and problems that had to be solved.

Meetings had an open structure, starting off with a session that allowed fishermen to discuss issues arising from their fishing activities. New agenda items were also raised when needed for discussion.

Participation initially was poor due to the mistrust of fishermen in the real capability of the project to change the 'business as usual' attitude of fisheries management authorities in the area. Fishermen also showed a lack of confidence in researchers and their activities. It was not uncommon to hear them say: "you [researchers] always say the same things' or 'you talk too much, but do too little" (Chioggia fisherman).

Gradually, by working with researchers, fishermen realised that collaborating researchers were sincere and were working to improve the sustainability of the fisheries and in this regard prioritised the interests of fishermen. Through this collaboration, researchers also learned about the problems of fishermen and better understood their mindset. Over time a "stable" group of fishermen – from 5 to 10 – participated regularly in the meetings. Although they still complained at times about the lack of tangible results, they nonetheless actively engaged in GAP activities and over time had more trust in the project.

> By participating we hope to change things', 'We invest time in meetings because we believe in this working group', 'It will take time to change things for the better' (Chioggia fisherman, interview).

There were still fishermen who remained skeptical. One fisherman said: "It has been years that I have been working with scientists but I do not see any results" (Chioggia fisherman). This fisherman stopped participating for several months but in the end other GAP fishermen convinced him to keep working with the team and show camaraderie and commitment to the overall cause.

We feel that the approach taken, that is, the on-going series of meetings, debates and planning discussions, maintained momentum, prevented or quickly addressed problems from emerging in the group and contributed to building a cohesive unit and developing mutual trust and credibility between stakeholders and their knowledge systems.

8.4.2 *Fishing for Trust: Scientists and Fishermen Working Together to Collect Catch Data*

The collection of fishery-dependent data on-board of commercial fishing vessels by ISPRA scientists was the first field activity undertaken by the group. Eighty-nine fishing trips with on-board researchers were carried out during 2012–2014 to collect biological data for eight of the most important commercial species. These target species were selected together by fishermen and researchers on the basis of their economic value and ecological importance. Discards were also analysed and served as a proxy for benthic assemblages (Piras et al. 2016). Field work was carried out across seven fishing vessels (five otter-trawls; two rapido trawls), and involved interactions with members of fishing vessels belonging to fishermen participating in the GAP project. Information on fisheries resources and their biological cycles and status were also collected. As a result of the data collection process, better two-way communication and trust were established between researchers and fishermen. Fishermen welcomed researchers to their fishing grounds and showed them their fishing gear and equipment (usually a well-kept secrets). They also let them know whether they comply with regulations or not and shared their own traditional knowledge and experience with researchers.

The reason for the latter was, as one fishermen explained: "you [researchers] have so much to learn from us fishermen, who go fishing every day" (Chioggia fisherman).

While fishermen welcomed researchers, they also questioned their understanding of the sea. Being at sea together reduced the distance between researchers and fishermen and, facilitated discussions about the fisheries 'crisis' and management problems/solutions. Fishermen increasingly valued the work of scientists and adopted their methodological approach, noting that scientists were also hard workers: "you are not only sitting comfortably in front of your computer" (Chioggia fisherman).

In other words, fishermen no longer viewed researchers as people who collected data for their own research purposes only (c.f. Jacobsen et al., 2012), but rather as people who collected biological data together with fishermen in an effort to address fishermen's needs in a bottom-up manner. Collaboration on board was crucial to stimulate exchange of knowledge and experiences on both sides.

8.4.3 Creating a Common Platform for Collecting and Storing Fishing Data: Electronic Logbooks and Self-Sampling

Based on earlier experiences developed in ISPRA, GAP2 scientists engaged fishermen so as to adopt electronic logbooks to record high frequency (i.e. haul by haul) geo-referenced data on catches. These devices were installed on March 2012 on five otter-trawls and two rapido trawls so that fishermen could collect data for the same eight target species selected for the on-board activities. This approach is, essentially, a self-sampling activity (Kraan et al. 2013), where fishermen record data themselves during commercial fishing. It results in a sharp increase in the number of records made available to scientists as compared to those that would be collected by on-board observers, in this case self-sampled data from 4800 hauls. A statistical analysis of the preliminary data showed that fishermen's self-sampled data were almost perfectly aligned with data independently collected by scientific observers, thus confirming their accuracy (i.e. fishermen are as accurate as scientists in their data collection or put another way, make similar mistakes) (Mion et al. 2015). This is a significant step towards a shared perspective on the credibility of self-sampled data.

That said, one of the main concerns of fishermen regarding their participation in this activity was related to privacy of data and lack of trust, namely the worry that data generated could somehow be used "against them". An example of this would be using the data to restrict fishing areas. After a year and a half of collaboration, one fisherman stopped collecting logbook data when he was fined because he sold some gastropods whose collection was forbidden in some areas. The fisherman accused GAP scientists of giving the GPS data that showed he was fishing in a restricted area to the control agency. Although the scientists said that they had not passed on such data, the fisherman stopped using the logbook and also decided against hosting observers on board. However, he continued to participate in meetings (being in fact one of the most frequent attendees).

This particular incident led the GAP group to agree that use of data by scientists would be agreed upon by the whole group (scientists and fishermen) and that data could be published only when aggregated and according to specific agreements.

Whilst the original misunderstanding was unfortunate, it ended up being a catalyst for ethical use of data, i.e. how and under what conditions data would be used, by whom and in what format.

It is also noteworthy that the data collection software is being further developed thanks to ongoing requests and suggestions from fishermen to improve usability. Additional functions have been added to allow fishermen easier access to their data and its interpretation (i.e. GIS map interpolations of haul coordinates, histograms displaying timeline catch per species etc.). In addition, and importantly, self-sampling activities have had the additional merit of demonstrating to fishermen that scientists trust them to collect crucial data and for fishermen that the data being used for assessment purposes is 'locally owned' and credible.

8.4.4 Trawling for Knowledge: The Summer Fishing Ban

Fishermen have repeatedly expressed concerns about the appropriateness of the summer fishing ban as a management measure. Common concerns relate to the timing and duration of the ban. The temporary trawl fishing closure (initially enforced in Italy in the late 80s and usually lasting about 30–40 days from late July to early September) was adopted as a management practice to reduce fishing mortality of juveniles of (some) target species. In order to test the effectiveness of such a measure and identify the spatial distribution of demersal resources, in particular of juveniles, at the end of the fishing ban, a fishery independent survey (i.e. otter-trawl survey) was carried out in 2012, 2013, 2014 and 2015. In the beginning, some fishermen had difficulty understanding the usefulness of the survey and the methodological difference between fishery-dependent and fishery-independent sampling. One fisherman wondered: "why are we going to fish where we know that the catch will be scarce?" (Chioggia fisherman).

As an offshoot of these discussions, a big debate ensued regarding involvement of only one fishing vessel to guarantee standardisation of sampling activities (a typical methodological problem that was put forward by scientists). At the end of a long discussion, a compromise was made by selecting two GAP fishing vessels with similar features (overall length, tonnage, engine power, fishing gear, etc.). The decision to involve only two vessels resulted in "jealousies" among fishermen who were either involved or excluded from the activity. Over time, however, fishermen accepted the decision without too much complaint.

The sampling scheme represented a compromise aimed at integrating scientific investigation with fishermen's knowledge and expectations and resulted in choosing 21 sampling stations in the Veneto Region administrative waters (from 4 to 18 NM from the seashore). Fishermen's contribution and experience was fundamental in defining sampling site allocation as well as timing and movements to optimise sampling activity. Fishermen also fully supported researchers during the sorting, sampling and measuring of catches so as to ensure that activities were carried out within the scheduled time and to an agreed standard.

Fishermen were paid (i.e. for renting their fishing vessels) for such activities, but the price they received was very low, covering just their expenses, as the main purpose was to support collaborative actions as opposed to simply buying fishermen's time. The resulting experiences at sea further promoted trust building and the exchange of opinions beyond those that focussed on professional fishing issues more commonly discussed.

At the end of each annual survey, results were presented in open meetings at the Chioggia town hall, which were also attended by (non-GAP2 participating) fishermen, local and regional administrators, policy-makers, coast-guard officers, scientists, journalists, etc. Some fishermen criticised the data during the presentations and were sceptical about its usefulness, claiming that in the few days between the survey and the beginning of fishing, the situation at sea might have changed sharply. On the other hand, a fisherman who participated in field activities praised the group's results saying:

> Congratulations to all for the work done in so short a time. I hope you were satisfied, even though I think my colleagues (beyond the GAP team) are not very satisfied. I think this may be because they do not fully understand what our purpose is in doing this (Chioggia fisherman).

Interestingly, the debate often heated up when the discussions moved away from perspectives on the credibility of the biological data, i.e. technical contestations over knowledge, to their management implications. This might suggest that rather than 'not fully understanding the purpose' of the collaborative research, fishermen were aware of the use of its results and their possible implications for access and use of the fisheries.

Over time, the authors observed a gradual decrease in the number of participants in these open meetings (from about 100 people in 2012 to 60 in 2015), yet they continued to attract the interest of all fishermen in Chioggia town. For instance, fishermen (not part of GAP) who participated in the meeting took pictures of species' distribution maps and shared them in real time to colleagues on social media. Presentations were also an occasion for the GAP group to socialise: after the presentation of the results, the fishermen-researchers continued to interact "among the stoves and pots," and each year a dinner was organised to cook and eat fish caught at sea. Moreover, every year, catches were shared with charities identified by fishermen and scientists together. The authors feel these are small but significant advances in building relationships, mutual understanding and trust between stakeholders.

8.4.5 Applying a Survey Approach to Broaden the Understanding of Fishermen Perceptions around the Fishing Ban

As already alluded to, the midsummer fishing ban - its duration, application, and regulations particularly those pertaining to effort - was the issue of highest priority in the GAP group. This theme was also somehow the least controversial, since there was a general agreement (both amongst fishermen and scientists) on the need of extending its duration and reducing fishing effort for the following fishing season. One of the reasons this was not as politically charged as it could have been has to do with the provision of subsidies to fishermen over the duration of the closed season. Yet, it should be noted that some fishermen stated that this would also be their opinion even without additional subsidies.

However, as the fishing ban would have far-reaching implications, it was considered necessary to open the discussion to those fishermen outside the GAP collaboration, particularly those discussions which would deal with potential conflicts as well as the general vision around the purpose and efficacy of the ban. Accordingly, taking advantage of a parallel project aimed at assessing the impact of European regulations in the Northern Adriatic Sea fisheries, an ad hoc survey was carried out. The survey involved nearly a hundred fishermen and skippers from the Veneto Region, who used different fishing gear, with fishermen having an average fishing experience of 28 years. Questions on the fishing ban were included in the survey.

A clear majority of those interviewed (70%) stated that they did not agree with the current duration of the summer fishing ban, most of them commenting on the need to increase its duration. In addition, 95% of those interviewed suggested an alternative period for the ban, although the periods suggested varied greatly amongst fishermen depending on the fishing gear used. This highlighted a strong and shared belief that the timing and duration of the ban was inappropriate. The consensus on the ban and potential fishing effort restrictions and the need to enforce them helped the GAP2 group to finalise their proposal to revise the fishing ban. This exemplifies the importance of developing a shared understanding and narrative to bridge the knowledge gap and create a common platform upon which to base management discussions and decisions. Furthermore, it also highlighted the need to build upon, where possible, the rather limited resources and reach of the GAP project, by engaging with similar or related initiatives.

8.4.6 *Navigating Stormy Weather*

Whilst encountering some rough waters, so far the description of the GAP process could suggest that the process was largely smooth sailing. Problems that have been described so far are primarily seen to have emerged from within project activities or as a result of outside events linked to fisheries dynamics and other management processes. This is possibly due to the need to represent in a coherent manner what was, at some points, a turbulent affair. In reality, whilst collective action did take place, there were significant periods of tension, as described below, which had to be resolved.

One on-going area of difficulty was in accessing funds from the Veneto Region for priority collaborative initiatives. For example, a proposal was drafted to monitor and raise awareness about heavy metal contamination in gastropods as it had considerable economic implications. Despite the project being 'accepted in principle', it was not funded, apparently due to an overall shortage in funding available for the region. Another project proposal related to experimenting with the "pulse trawl" in Chioggia (an idea put forward by GAP fishermen after a presentation made in Chioggia by the Dutch GAP fisherman, J. Bajii), had the same outcome. Informally, Veneto Region officers mentioned to us that there would have been a greater possibility of both projects being funded if the proposals themselves were supported by all fishermen cooperatives and organisations of the Veneto Region and not just a restricted number. In other words, they would have liked to see a much more involvement of fishermen cooperatives. Other proposals put forward by GAP affiliated fishermen and scientists, for example one to study the sand smelt fishery, were also rejected.

These failures of GAP scientists and fishermen working to establish a collaborative platform for monitoring aimed at broadening the knowledge base to solve specific resource problems. Yet they served to demonstrate a difference in political/research priorities between local scientists and those of the Veneto Region. This in itself highlighted the limitations of working through a small project like GAP with limited resources available to tackle large, complex problems.

Despite the lack of success accessing funds and the resultant frustration that emerged across stakeholders, such efforts were still very important for the following reasons:

- They allowed the group to act proactively to try and solve problems through maintaining the participatory nature of the GAP approach to solving fishermen's and fisheries-related problems.
- They allowed the group to increase collaboration with the Veneto Region administration which facilitated a better understanding of the latter's priorities and to participate in meetings from which the group was previously excluded.
- They illustrated the relevance of group activities and helped policy-makers better understand the nature of collaborative research.
- They showed that in order for GAP proposals to be successful, a higher degree of involvement of regional fishermen's organisation was required.

Moreover, such activities allowed us to maintain a certain momentum in a situation where tangible outputs from project activities were few (due to the need to collect data for at least 2 years to make it scientifically robust enough for managers and policy-makers). This was important given our experience (as scientists) at higher management levels (e.g. the EU Common Fishery Policy), which places a clear emphasis on direct evidence and credible scientific knowledge in decision-making. As such, we consider it important to see a similar approach taken in local management decisions in the Veneto Region.

8.5 Sighting 'Terra Firma': Mirage or Reality?

8.5.1 From GAP Life Raft to Flotilla

Empirically, the participatory research activities provided substantial evidence that it was necessary to rethink the summer fishing ban. In order to promote this change, it was necessary for the GAP group to become a bigger player in the management arena, not only at the Veneto Region administrative scale but also, and more prominently, at the Northern Adriatic Sea scale. To this end, field results (GAP logbook and survey data, as well as on board observations) were presented at a meeting of the FAO Adriamed project in late 2013. The presentation resulted in the establishment of a working group (WG) on the effectiveness of the summer fishing ban. In October 2014, the Adriamed WG met and several presentations were given, including one summary of GAP results and the outcomes of the survey that involved fishermen of the Veneto Region and focussed on the appropriateness of the summer fishing ban. In preparation for the meeting, and in order to set the agenda and the contents of the communication, the GAP team conducted a focus group on the summer fishing ban. Fishermen were asked to provide insights and ideas and make requests regarding changing the duration of the fishing ban, all keeping in mind the outcomes of the survey. The discussions were passionate and direct as can be seen below:

> Croatian fishermen should stop fishing as we do in Italy. Not only trawlers should be stopped, but also fishing with set nets and pots... During the ban Croatians enter the markets with their fish – this is not fair. It is not only Italians that should protect biological resources: the Adriatic is like a bathtub, all activities have an impact on resources (Chioggia fishermen).

Fishermen also told scientists: "Listen, you have to report exactly what we said, i.e. all aspects, even the most controversial ones because they are very relevant for us" (Chioggia Fisherman).

The detailed report of the focus group was well received at the FAO-Adriamed project WG meeting. The GAP participatory approach and the data collected were also commended. Adriamed agreed to put forward a detailed report highlighting the effectiveness of and possible revisions needed to the fishing ban that would be discussed in a forthcoming FAO GFCM plenary session. The GAP group also made further steps forward. In two different meetings, the proposal for the revisions to the summer fishing ban were drafted, proposing revisions to the regulations adopted in 2014. The new text proposed that the summer trawling ban should commence from the 15th of July to the 15th of September and it should be such for all trawling activities conducted within the Northern Adriatic Sea Fishery District. In addition, it was suggested that for the 8 weeks following the ban, fishing effort was to be restricted to a maximum of 60 h per week. Even after that eight-week period, fishing effort should not exceed 72 h a week. In order to generate support for the extended fishing ban, informal contact with different fisheries officers and politicians of the Northern Adriatic Sea fishery District were made. These processes compliment and add value to the GAP experiences, outcomes and visions. However, it should also be noted that the proposed revisions to the summer fishing ban rules was not accepted, as a number of issues need to be tackled first. These include: (i) the economic cost of the extended ban as fishermen receive compensation (subsidies) during the ban period, thus extending the ban will increase costs (despite some fishermen agreeing to a no cost extension; a key question is who will pay for this?); (ii) the need for a formal agreement among different regions (and thus fishermen) along the Northern Adriatic Coastline; and (iii) the adoption of a common strategy with other countries such as Slovenia and Croatia which is not straightforward as there is no binding obligation to use similar management approaches in the Adriatic Sea across the EU countries.

It is worth mentioning, in this context, that the GAP group also attended, in early March 2015, a meeting of a Pan-Adriatic project dealing with a common approach to fisheries management. The approach, outcomes and management proposal developed in GAP were welcomed by participants. This illustrated not only the value of the GAP case study generally but also, specifically, of building new knowledge and narratives under a participatory process involving fishermen and policy-makers. Given the positive feedback, the GAP group was invited to give a talk in another meeting that was to be coordinated by the fishery officer of the Emilia Romagna Region (a region neighbouring the Veneto Region) who was also the coordinator of the Northern Adriatic Sea Fishery District. This was another occasion to promote the GAP management proposal and approach (Raicevich et al. 2015).

An important point to make at this point is that though the GAP project had a limited reach, influence on decision-making was greatly increased through the use of the broader collaborations.

8.6 Making Port, Dropping the Anchor: Journeys End?

The Chioggia case study is a unique attempt in the Northern Adriatic Sea to operationalise the 'GAP method' in a complex and politically charged governance context. It has been successful in building a strong alliance of local fishermen and government scientists around a set of shared interests and values, and through the use of a common narrative (Raakjær 2009). This Chioggia case, which has focussed on sustainable fisheries management, has through the co-production of knowledge created to a certain extent a common platform from which both local fishermen and scientists have been able to reach an understanding over the current status of fisheries management and the need to extend the fishing ban in the Northern Adriatic. This has been achieved through an on-going participatory process of problem identification, research design and co-production of knowledge, in the process also bridging different knowledge systems (Reid et al. 2006). In doing so, it has demonstrated the potential of working together from the 'bottom up' in the pursuit of a broader more credible and legitimate knowledge base for fisheries management. Given the scale of the problem, however, and the fact that vested interests will try to ensure the maintenance of the status quo, in this case a centralised, top-down approach to management, GAP is a good beginning in ensuring broad and long lasting change. There is a need for others to build upon it as a viable approach of including fishermen and fisheries scientists in setting management objectives for addressing some of the problems faced by the fisheries sector. An example of where the approach gained some traction was the work with the Adriamed WG in relation to the summer fishing ban in the Northern Adriatic.

Whilst there were considerable successes, there remain serious institutional and practical challenges to fisheries management in the region (i.e. given the multiple administrative and ecological scales involved, sensu Reid et al. 2006), especially in relation to the limited resources available in GAP. One such challenge identified by the case study team is bringing fishing associations and Veneto Region managers together in support of more evidence-based decision-making. Our results demonstrate that influencing the policy agenda and ensuring appropriate policies is not simply a matter of providing knowledge up the management chain, but also instilling a participatory research (and the knowledge it produces) approach as a long term tool to include fishermen at the decision-making table. This then becomes a political issue, in the sense that knowledge becomes a political currency, and thus has implications in terms of enabling fishermen's' associations to represent fishermen and their interests. It also has implications in terms of the changing roles and responsibilities of scientists, fishermen and other stakeholders in the contemporary management of fisheries resources. Accordingly, collaborative research is political. Recognising that this is so affords a very real opportunity to work with fishermen, scientists and other stakeholders using collaborative research in a more politically nuanced and targeted way. If we seek to make explicit the interests and rationalities with which all stakeholders operate, how alliances are formed and discourses and policies shaped, we would expect to see greater traction for mobilising collective action. This is no easy task, nor short trip, but should this voyage continue, we hope and expect that the fisheries management challenges of the Northern Adriatic Sea will be better addressed.

References

European Commission (2007) Council regulation (EC) no 1967/2006 of 21 December 2006 concerning management measures for the sustainable exploitation of fishery resources in the Mediterranean Sea, amending regulation (EEC) no 2847/93 and repealing regulation (EC) no 1626/94. Off J Eur Union L 36:6–30

European Union (2013) Regulation (EU) no 1380/2013 of the European Parliament and of the council of 11 December 2013 on the common fisheries policy, amending council regulations (EC) no 1954/2003 and (EC) no 1224/2009 and repealing council regulations (EC) no 2371/2002 and (EC) no 639/2004 and council decision 2004/585/EC. Off J Eur Union L 354(22):22–61

Friend RF, Arthur RI, Keskinen M (2009) Songs of the doomed: the continuing neglect of capture fisheries in hydropower development in the Mekong. In: Molle F, Foran T, Käkönen M (eds) Contested waterscapes in the Mekong region: hydropower, livelihoods and governance. Earthscan, London, pp 307–331

Gutiérrez NL, Hilborn R, Defeo O (2011) Leadership, social capital and incentives promote successful fisheries. Nature 470(7334):386–389

Jacobsen RB, Wilson DCK, Ramirez-Monsalve P (2012) Empowerment and regulation – dilemmas in participatory fisheries science. Fish Fish 13(3):291–302

Jelfs M (1982) Manual for action. Action Resources Group, London

Jentoft S, McCay BJ, Wilson DC (1998) Social theory and fisheries co-management. Mar Policy 22(4):423–436

Johnson TR, van Densen WLT (2007) The benefits and organization of cooperative research for fisheries management. ICES J Mar Sci 64:834–840

Kraan M, Uhlmann S, Steenbergen J, Van Helmond ATM, Van Hoof L (2013) The optimal process of self-sampling in fisheries: lessons learned in the Netherlands. J Fish Res 83(4):963–973

Mion M, Piras C, Fortibuoni T, Celić I, Franceschini G, Giovanardi G, Belardinelli A, Martinelli M, Raicevich S (2015) Collection and validation of self-sampled e-logbook data in a Mediterranean demersal trawl fishery. Reg Stud Mar Sci 2:76–86
Piras C, Mion M, Fortibuoni T, Franceschini G, Punzo E, Strafella P, Despalatović M, Cvitković I, Raicevich S (2016) A photographic method to identify benthic assemblages based on demersal trawlers discard. Fish Res 178:142–151
Raakjær J (2009) A fisheries management system in crisis. The EU common fisheries policy. Dissertation, Aalborg University, Denmark
Raicevich S, Bullo M, Sabatini L, Giovanardi O (2015) Un futuro per la pesca in Alto Adriatico. Risultati e proposte del percorso partecipativo GAP2 tra ricercatori e pescatori di Chioggia. ISPRA, Quaderni – Ricerca Marina, Rome. n. 7/2015
Reid WV, Berkes F, Wilbanks TJ, Capistrano D (2006) Introduction. In: Reid WV, Berkes F, Wilbanks TJ, Capistrano D (eds) Bridging scales and knowledge systems – concepts and applications in ecosystem assessment. Island Press, Washington, DC, pp 1–20
Symes D (1997) Fisheries management: in search of good governance. Fish Res 32(2):107–114

Chapter 9
Trapped in the TAC Machine: Making a Fisheries-Based Indicator System for Coastal Cod in Steigen, Norway

Petter Holm, Asgeir Aglen, Maiken Bjørkan, and Jan I. Andersen

Abstract The GAP project is concerned with the lack of stakeholder participation in providing knowledge for management. This creates a legitimacy problem, as the fishermen do not get to share the knowledge basis for and justification of management decisions. In addition, fishermen's experience-based knowledge remains unused. GAP intends to bridge the legitimacy and knowledge gaps, finding practical ways for fishermen and scientists to co-create the knowledge basis for management. In this paper, we explore whether and how such co-creation can be achieved. We do that in the context of one of the GAP case studies, which set out to establish and test a fisheries-based indicator system for coastal cod in Steigen, Norway. The inspiration for the case study was the Norwegian Reference Fleet, in which commercial fishing vessels are used as platforms for data collection. In GAP, however, we did not simply aim for a replication of the Reference Fleet model. We were much more ambitious, dreaming of a project that would allow deeper engagement of fishermen. Our strategy was to break out of the strict framework of top-down management and create a cooperative platform that would also open up for fishermen's own insights and experiences. Inspired by collaborative research ideals, we sought equal partnership and true collaboration, where science and experience-based knowledge together would pave the way to a sustainable future. Based on 4 years of work, the conclusion is that we failed. In a technical sense, we did what we promised, setting up a fisheries-based indicator system for coastal cod. We ended up exactly with the type

P. Holm (✉)
Norwegian College of Fishery Science, UiT The Arctic University of Norway, Tromsø, Norway
e-mail: petter.holm@uit.no

A. Aglen
Institute of Marine Research, Bergen, Norway

M. Bjørkan
Norwegian College of Fishery Science, UiT The Arctic University of Norway, Tromsø, Norway

Nordland Research Institute, Bodø, Norway

J. I. Andersen
Norwegian Fishermen's Association, Steigen, Norway

P. Holm et al. (eds.), *Collaborative Research in Fisheries*, MARE Publication Series 22, https://doi.org/10.1007/978-3-030-26784-1_9

of project we tried to transcend, a mini version of the Reference Fleet. In this paper, we tell the story about how and why we got trapped in the TAC machine.

Keywords Participatory research · EBK · Fisheries management · TAC Machine · Coastal cod

9.1 Introduction

The GAP project is about the absence of cooperation among fishermen, scientists and managers and about how this – which is acknowledged as a major barrier to sustainable fisheries – can be remedied. The days when fisheries scientists would explicitly reject fishermen's knowledge as useless for management purposes are gone. Instead, we have seen a marked turn in fisheries governance, by which stakeholder participation in general and fishermen's knowledge in particular are accepted as important and valuable. At least, this is how it appears. Whereas this turn no doubt is an improvement in itself, questions remain about what it means and how far it goes. While bridges may have been built, we know less about the traffic they carry and the extent to which they have contributed to closing the underlying gaps.

One important expression for the participatory turn in fisheries governance is the emergence of Fisheries Dependent Information (FDI). FDI is a relatively recent approach for providing knowledge for fisheries management purposes (Graham et al. 2010; Hind 2014). As the label indicates, FDI stands in contrast to a previous approach, by which fisheries scientists would seek data involving fishermen as little as possible, for instance by relying mainly on scientific surveys for input in assessment models. FDI, then, involves research in which fishermen and commercial fishing vessels are involved in some way or another. The emergence of FDI raises a number of interesting questions concerning its origin, purposes and variations. There is little doubt, of course, that FDI does include fishermen in research for management purposes. But does it allow the establishment of a common knowledge basis? The reason for asking this question, which perhaps suggests some skepticism, is that the degree of fishermen involvement in FDI is typically on the low side, doing basic data collection and sampling, and providing their vessels as research platforms. What fishermen are actually allowed to do here is very much under the control of scientists. With Hoefnagel et al. (2006), we can say that FDI is based on a deference model, in which scientists make the rules of engagement and fishermen leave all important questions to be decided by scientists. As this indicates, FDI may work towards a common knowledge base, but primarily by allowing fishermen a peek into the world of science. This is not unimportant in itself. If fishermen learn to talk the talk of science, they may be able to express their insights and concerns in a way that could be more persuasive to scientists and managers (Hartley and Robertson 2009). This may be a step in the right direction, but remains quite far from the ideas of equal partnership usually promoted by the champions of collaborative research (NRC 2004; Haggan et al. 2007; Johnson and van Densen 2007; Mackinson et al. 2011).

These are the issues we want to explore in this paper. Would it be possible to go beyond the deference model, constructing a collaborative project that would allow deeper engagement of fishermen? Could we break out of the strict framework of top-down management and create a collaborative platform that would also open up for fishermen's own insights and experiences? Inspired by collaborative research ideals, we sought equal partnership and true collaboration, where science and experience-based knowledge together would pave the way to a sustainable future.

We did that in the context of one of the GAP case studies, aiming for the establishment of a fisheries-based indicator system for coastal cod in Steigen, Norway. The inspiration for the case study was a typical FDI project, namely the Norwegian Reference Fleet (NRF) (Bjørkan 2011). The NRF is a project run by the Institute of Marine Research (IMR), the Norwegian variety of the state-owned, marine research institutes in Europe, authorised to provide and certify knowledge for management purposes. In the NRF, commercial fishing vessels are used as platforms for data collection, and fishermen are engaged to collect data and samples under instruction from IMR.

As we designed the Steigen CS, the GAP team already knew the Reference Fleet quite well. IMR, the Reference Fleet owner, was included as a partner in the GAP project, represented by scientists who knew it intimately. In addition, one of the fishermen in the GAP project had previously owned and operated a vessel participating in the Reference Fleet. Finally, one of the researchers, working full time on the GAP project, had written her PhD thesis on the Reference Fleet (Bjørkan 2011). Nevertheless, in GAP we did not simply intend to do a Reference Fleet-type project all over again. On the contrary, we were much more ambitious, aiming for a project that would allow more and deeper engagement from the fishermen. Inspired by the GAP ethos, we imagined a fully developed collaborative project. In the Steigen project, then, the fishermen themselves would be able to take an active part, not leaving all important decisions in the hands of the science partner. We did not want a project where fishermen would collect data as prescribed by science, to feed into existing assessment models. Our strategy was to break out of the strict framework of top-down management – the TAC machine (Holm and Nielsen 2004; Schwach et al. 2007; see Chap. 2) – and create a cooperative platform that would also open up the fishermen's own insights and experiences. Inspired by the collaborative research ideals, we started out with the dream of equal partnership and true cooperation, where science and fishermen's experience-based knowledge together would pave the way to a sustainable future.

After 4 years of work, we conclude that we failed. It was not a spectacular failure. We have carried out what we promised, setting up a fisheries-based indicator system for coastal cod. During the project period, fishermen collected data and samples on a regular basis, providing information that is useful for management purposes. If the project had been allowed to live on after the end of GAP, it could have produced an indicator series useful for assessing the status of coastal cod. In some ways, the fishermen were allowed more responsibility than in a typical FDI project. For instance, the fishermen succeeded in getting extra funding for the project, allowing us to extend the sampling program. All in all, however, the ambition to have the fishermen take control, to provide better access for and usage of fishermen's experience-based knowledge and to break the privileged position of science did not succeed. On the contrary, we ended up exactly with the type of project we tried to transcend, a mini version of the Reference Fleet.

In this paper, we tell the story about how and why this happened. In the first section below, we describe the fisheries and management context for the project, focusing on the fascinating story of Norwegian coastal cod. In the following section, we describe the practical work of carrying out the project and how we ended up where we did not intend to go. This leads to a final discussion, where we summarise our experiences and try to account for the reasons why we remained trapped in the TAC machine.

9.2 Norwegian Coastal Cod

In order to achieve our goal – a project that would give fishermen a strong hand in knowledge provision as well as a platform for integrating science and experience-based knowledge – the choice of research setting was crucial. The basic framing for the GAP case studies was participatory research for management purposes. Ideally, we wanted our project to come up with knowledge products with a potential to affect management decisions. This meant that we would have to start with some management problem, for instance a fish stock for which management measures were in place, but perhaps not working as intended. Nevertheless, it did not seem like a good idea to focus on the main commercial stocks, already caught up in the routine stock assessment procedures. For such stocks there is already a well-developed knowledge infrastructure in place, and it would be little hope that our tiny project could contribute anything useful. On the other hand, it was not tempting to go to the other extreme, focusing on marginal species like mussels or ling. Then there would be no demand for the knowledge products, and the project would not be able to prove anything one way or the other. What we needed was something in between the centre of things and the margins.

It was not really difficult to find our target: Norwegian coastal cod (NCC). NCC is a sub-category of Atlantic cod, *Gadus morhua*. In North Norwegian waters (above 62°N), there are two main types of cod. In addition to the NCC, which mostly remain along the coast and in the fjords throughout the year, there is the North East Arctic cod (NEAC). The NEAC is a dominant natural resource in the Barents Sea. The stock is managed jointly by Norway and Russia through the Norwegian-Russian fisheries commission. In historical, economic and political terms, the NEAC is big and important, whereas the NCC is small and relatively insignificant. What is crucial, however, is that the NCC and the NEAC stocks are harvested in mixed fisheries. The two stocks have been (and still are) managed under a common quota regime for the whole area north of 62°N. For the NEAC, this works great. There is a well sophisticated assessment system, a comprehensive management plan, and the stock status is well within precautionary approach criteria (ICES 2008: 24–26). For the NCC, in contrast, the situation is different. While this stock is included in the management regime, the main focus remains with the NEAC. The minority status of the NCC means that it usually cannot sway the decisions in its favour. As a result, this stock has decreased considerably over the period 1997–2005 and has remained low since then. ICES has recommended zero catch for the years 2004–2011. These recommendations have not persuaded management authorities, mainly because it would be difficult to implement without shutting down important parts of the NEAC fisheries. Instead, a less ambitious rebuilding plan (ICES 2008: 38; ICES 2013) has been adopted. The management of NCC north of 62°N included several technical regulations, stricter inside fjords than outside. There were no regional differences in the regulations. Two important spawning areas, one close to Svolvær in Lofoten (68°N) and one close to Ålesund (62°30′N) are closed during the spawning season. While the rebuilding plan may have prevented a stock collapse, it was not immediately successful in rebuilding the NCC stock. The stock has been reasonably stable since 2005, but has remained at a very weak state and has not responded well to the management measures in place. In its 2014 advice, ICES stated: "The survey indicates that the SSB is close to its lowest level" (ICES 2014).

The NCC assessment is widely considered – also among scientists – to be inaccurate. This is partly linked to the question of stock structure and the difficulties in differentiating between NCC and NEAC. In addition, the methodology in the standard coastal cod survey is to a large extent dependent on trawling (Aglen et al. 2008). This means that the bottom topography represents a challenge, since trawling is not possible over large areas of the Norwegian coastal zone. The way stock assessment is carried out is also an issue for debate due to the limited number of sample events over the course of the year. Such knowledge gaps have been recognized both by the management authorities and the fishermen. As part of an effort to improve the management of coastal cod resource, the Norwegian Ministry of fisheries and coastal affairs in 2005 established a task force to evaluate the existing knowledge base for the management of NCC. One of the recommendations was, besides indicating a need for more research, that models for geographically differentiated management regimes should be considered (Anon 2005).

From the perspective of the GAP project, the NCC provided an ideal context for a case study, with the absence of a solid and shared knowledge base for management. Although we can take note of the divergence views on stock development between the fishermen and the scientists, there was no deep and systematic polarization. The absence of a solid knowledge base naturally established a situation where different stakeholder, according to the interests at stake, could emphasise different viewpoints. Nevertheless, there was actually an agreement among managers, scientists and fishermen that the knowledge base on NCC is weak and should be improved (Anon 2005). In this situation, a GAP case study on NCC could offer exactly what the situation demanded.

Furthermore, since the knowledge problem related to NCC was intimately tied to the unresolved question of stock structure, a relatively small and localised case study, as available funding suggested, would be appropriate. From an IMR perspective, such a case study was considered to be a good opportunity to learn what could be achieved by collecting more local information and local scientific data.

As the geographical location for our case study project, we quickly settled on Steigen, a fishing community and municipality in Nordland County. The reason for this had more to do with the early involvement of local fishermen from Steigen than with the strategic location of Steigen in relation to the research issue per se. Nevertheless, the rich natural environment together with a complex socio-economic setting makes Steigen a good location for the project. In Steigen, more than a thousand small islands are scattered in this large fjord area. Here we find spawning and feeding grounds for several important species of fish, of which cod is the most important. Steigen has known spawning grounds for NEAC as well as stationary and migratory stocks of NCC.

9.3 The Project in Practice

The aim of the project was to set up and run an information system sophisticated enough to produce data acceptable for management purposes, while robust enough to be operated as an integral part of on-going fishing operations. Since the project was intended to build a common knowledge base for fishermen and scientists, it was important to have active fishermen participation in all phases of the project. Keeping with the guidelines on participatory processes (Johnson and van Densen 2007; Mackinson et al. 2011), fishermen should participate in the planning and method development, do data collection and contribute with local knowledge. That was the plan. Here we report how it worked out in practice.

9.4 Finding Common Ground

In contrast to some of the other case studies (CS) within GAP, the Steigen CS was not anchored in collaborative networks established prior to GAP. In practical terms, the first time the would-be partners met in the context of GAP was in a meeting at the Fishermen's Association's offices in Trondheim in 2008. Initially, the UiT partner had started with an idea to work with the model developed within IMR's "Coastal Reference Fleet", in which selected vessels are engaged in systematic collection of data for assessment purposes (Bjørkan 2011). However, this was just one possibility and a starting point for the discussion in Trondheim. The meeting here was one in which all ideas were welcomed. We discussed central notions such as fishermen's Experience Based Knowledge (EBK) and how it differs – or not – from scientific knowledge. Another issue was how EBK could be used for management purposes. The fishermen representative, Jan Andersen, was very engaged and eager to test some of his experience-based knowledge. One of the issues he drew attention to was that of moon phases and if they affect the catch rates of fish. The UiT scientists, Petter Holm and Maiken Bjørkan, talked about their research on knowledge production for fisheries management. At this meeting, agreement was reached that the Reference Fleet model would give a too-restricted role for the fishermen. It was decided instead that we would develop a fisheries-based model targeting coastal cod. A key point was to make sure that the project would allow the fishermen to take an active part in the design and implementation of the project.

During the second project meeting in 2009, also in Trondheim, it was decided to focus on the Steigen area. The partners had agreed that a regionally based project concerning the use of the coastal zone would be desirable. The home area of Jan Andersen, Steigen, has important spawning grounds for several ground fish stocks and there is known spawning sites for the Norwegian Coastal Cod. At the same time, this is also a crucial area for the coastal fisheries, occupational as well as recreational.

The third meeting was held in Steigen in 2009, and engaged with the practical design and planning of the project. At this meeting, the IMR was drawn in for the first time and presented their viewpoints on how the GAP project could be undertaken. Knut Sunnanå, the IMR scientist, emphasised the limitation of the existing knowledge on NCC, since the current surveys only cover areas where trawling is possible. These areas, however, only constitute a minor part of the coastal zone. In addition, the catch statistics are not accurate enough to give a precise measure of the fishing pressure in the fjords. Together this implies that getting a full overview of the stock situation is difficult under the current survey regime. A better resolution could be achieved if local fishermen were engaged in data collection, as intended by the GAP CS. The IMR also pointed out that there were no stock abundance indexes for gear types other than trawl. Methodological development was therefore of great importance. There was a need to develop standardized gear for measurement purposes.

On the basis of the IMR perspectives on the possibilities, we went on to the key issues of developing a specific project idea for GAP. First, we identified possible issues and areas of concern. In Steigen there are several marine species that could become interesting subjects to further research, for instance redfish (*Sebastes marinus*), halibut (*Hippoglossus hippoglossus*) or common crab (*Cancer pagurus*). Still, it was agreed that the NCC would be of greater interest, particularly because of its relevance to ongoing matters in the Norwegian fisheries management sector.

Second, we discussed the relevant research question our project could engage with. This revolved around the many puzzles regarding the different types of cod in Steigen. What is the abundance and patterns of local cod resources in Steigen? Where do we find the stationary fjord cod, and where is the migrating coastal cod dominant? How does the NEAC mix with the NCC? How separate are the different stock components? What are the migration patterns for the different kinds of cod?

Third, we discussed the actual design of the project. It was agreed that we should select three areas in Steigen where the fishing activities are carried out in different ways. From each of these areas, at least two fishermen should participate in the research. The following methodological approaches were considered relevant for our study:

- Developing measures designed to improve the accuracy of catch statistics. Draw up a detailed code system for the different catch areas;
- Collecting samples. Otoliths. Genetic material. Measure length/weight, estimate degree of sexual maturity. In the winter season: State whether catch is believed to be NCC or NEAC;

- Using echo-sounder and Olex. These devices can be used to provide data to IMR concerning the abundance of cod on the spawning grounds;
- Tagging experiments to determine migration patterns;
- Methodological development: Calibrating the use of gear types like gillnets and/or fish pots as standard devices for assessing fish abundance.

At this meeting, it was agreed, both among scientists and stakeholders, that our planned study would be both relevant and interesting. We looked forward to its implementation and we agreed to work hard to get it funded! At the end of this process, the fishermen and scientists had finalised a collaborative project proposal entitled "Developing a fisheries-based resource monitoring system: The case of Norwegian coastal cod".

In the project description, it was made clear that the project would pursue two different objectives with different types of outcomes. On the one hand, the technical-scientific objective was to deliver, by way of a fisheries-based project, "reliable and timely information" on the status of the cod resources in Steigen. On the other hand, the project emphasised that the development and testing of the cooperative research design in itself constituted an important objective. To fulfil this objective, the project description promised to deliver a range of indicators on the project performance:

- A detailed description of the monitoring system, including main objectives and functions, equipment, personnel, training, organisation, information flow, cost, etc.
- An analysis of the performance of the monitoring system with regard to reliability, effectiveness, cost and legitimacy.
- An evaluation of the institutional and political challenges such as a monitoring system would have to meet if it were to function as part of the ordinary information infrastructure for marine resource management.
- A handbook of cooperative research with particular emphasis on how to facilitate effective fishermen participation in research.

In passing, we can note that the chapter you are now reading is what came out of these promises. Besides the tendency to be over-ambitious with regard to the level of detail in documenting the project, we can note that the optimism with regard to the project was strong at the time. Nevertheless, the high hopes at the start of creating a platform that would enable equal partnership between scientists and fishermen had already been significantly toned down at the end of the project development phase. In the project description there was still a strong commitment to include fishermen in active and responsible positions in the project. The monitoring system "will be built on cooperation between local fishermen and scientists". While the project was organised with a clear commitment to active fishermen participation, it is also obvious that the overall project design was attuned to scientific requirements. Under the section specifying what was in it for different stakeholders, the main benefit for the fishermen was to be educated in the ways of science:

- To participate actively in the development of information systems.
- To gain hands-on experience in the processes of information generation.
- To gain insights in the way science works.
- To become skilled in working with and communicating with scientists.

The list of benefits for science emphasises the access to new systems and sources of knowledge and new models to engage fishermen. This is all fine, of course. But we note the absence of fishermen's experience-based knowledge. While the project would allow fishermen to learn about science, there is no mention of the ways science may get to learn about and utilise what the fishermen know. In practical terms, the project had been redesigned, now aiming for a system collecting scientific information, operated by fishermen under scientific supervision, not for connecting to and collaborating with fishermen as knowledge agents.

9.5 Implementing the Project

The implementation phase of GAP started in April 2011. In order to kick off our case study and get all the partners together, we met in Steigen in May 2011. Hector Peña, the scientists picked by IMR for GAP, came up from Bergen. Maiken Bjørkan, who was engaged by UiT as a researcher working full time on GAP, came in from Tromsø. In Steigen, they met with Jan Andersen and Jon Erik Pettersen, who represented the local Fishermen's Association. Since it was important to generate an environment where everybody felt that they could share ideas, views and questions, we started off by an informal chat during breakfast at the local hotel, where the entire meeting took place.

Making sure that everybody was on the same page about what GAP in Steigen should be, we talked about our expectations. For IMR, it was important to develop a monitoring system adapted to the local resource system. Steigen was a good place to make a pilot project. The local fishermen were happy to hear this and underlined the project's importance of building up the fishermen's trust in the science.

The methodological issue was in focus at this meeting. Based on the previous discussions in GAP, the fishermen started out with a preference for a methodology on Catch Per Unit Effort (CPUE), using standardised gillnets in different areas. However, when Peña took out his laptop and presented several examples of how modern echo-sounders can be used to collect data, he immediately had their interest. With the particular equipment Penã demonstrated, it is possible to generate videos instead of still pictures, and Hector showed us a number of possibilities with the software. When he started the slideshow, the fishermen leaned in and understood immediately what information they contained (Fig. 9.1). Fishermen and IMR scientists obviously have some experiences and knowledge in common that could serve as a platform for collaboration.

After Penã's presentation, the fishermen were exited to follow this methodological idea, partly because it offered a way to collect necessary scientific information without compromising their fishing activities. One benefit would be that the echo-sounder could collect information of the same quality as that routinely used by IMR, and thus the data could be integrated with and complement the data series IMR is collecting themselves. Jan Andersen, himself a fisherman and the local coordinator of the GAP project, underlined several times how important it was that the data collected by the fishermen actually would be useful to the scientists.

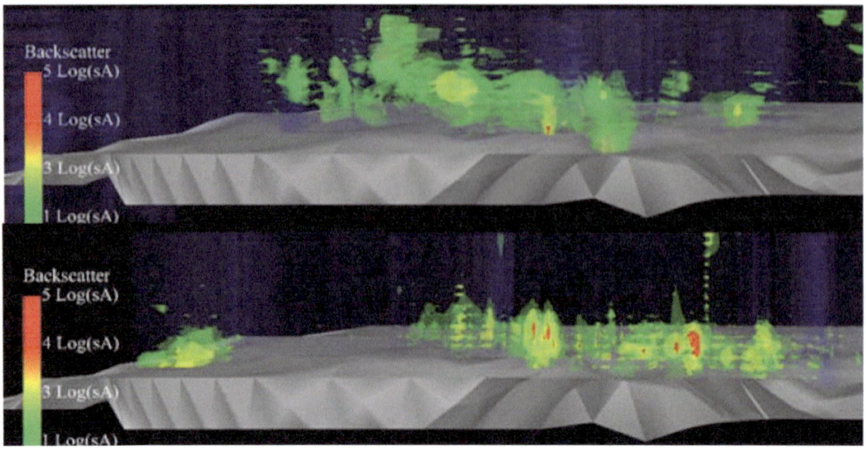

Fig. 9.1 Two of the echograms Penã showed the fishermen at the kick-off meeting

The fishermen and scientists then discussed what problems could be generated for the fishermen with a different type of echo-sounder. For instance, the acoustic gear suggested by Peña required some data skills and it was expensive. The IMR has already installed this gear in larger fishing vessels, but never in a small coastal vessel. Hence, we were entering unknown territory. A lot of the discussions focussed on what had been done earlier, what was realistic and what it would take to implement this methodology in the Steigen project.

In short, the meeting agreed that the objective was to assess the state of the local coastal cod resource in Steigen, and this would be done using acoustic technology installed on six selected vessels and also biological monitoring of the catch.

While fishermen and scientists were satisfied and confident with the new approach, several challenges and questions became evident soon after this meeting. First, the equipment was more expensive than we had planned for. With the available EU funding, the project could only afford to buy one echo-sounder and hence only engage one vessel in the project (MS Fix), instead of six as planned. Second was that Penã did not speak Norwegian, and so the fishermen doing the sampling had to be able to speak English.

9.5.1 The Echo-Sounder

By the end of 2011, we ordered the echo-sounder. The installation was a three- step process. One company (Simrad) delivered the equipment, the Simrad EK60 Scientific Echosounder (Fig. 9.2). The price tag was NOK 500,000.

Following installation and calibration, the IMR partners went out to test the equipment and train the fishermen in data collection and data storage. The following is an extract from an e-mail reporting how it went down. Figure 9.3 is a screen shot from the trial run mentioned in the e-mail.

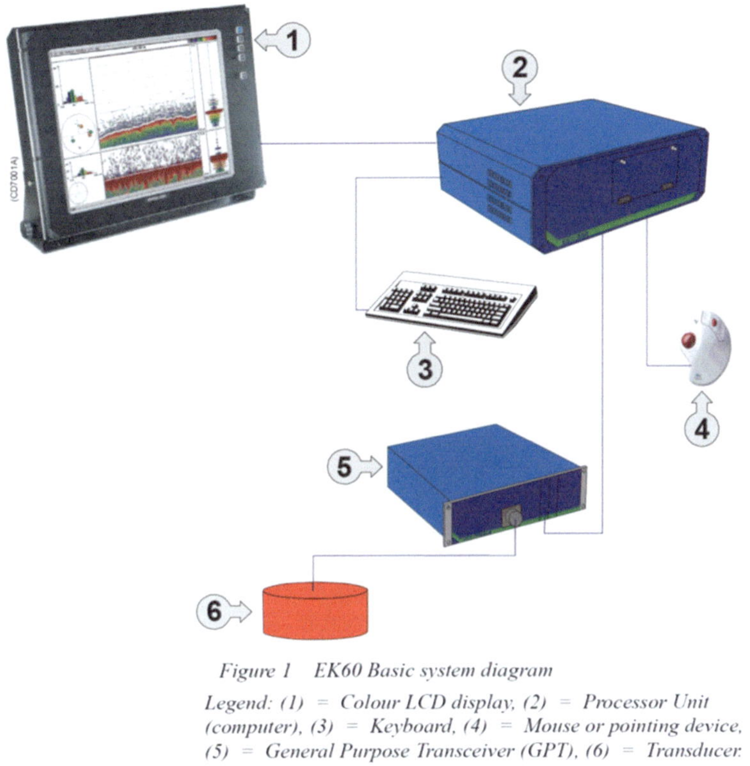

Fig. 9.2 Schematics for the echo-sounder Simrad EK60

On Thursday we went on a fishing trip to record data and observe the performance of the echo-sounder, using my notebook to avoid the problem previously mentioned. The fishing ground was ca. 1 h from the port and several other vessels were in the area fishing. I attach one echograms of the conditions in the fishing ground, where cod and sei were together feeding on a large herring school. The echogram is very good, with no interference, being able to identify the single targets of the larger fish in the surroundings of the large herring school in the center (email from Aglen 26.01.2012).

9.5.2 *Designing the Data Collection Program*

Having successfully installed and tested the echo-sounder, we needed to design a survey plan and a sampling program for the MS Fix. In order to do that, Asgeir Aglen from IMR traveled to Steigen. Here, he sat down with the fishermen, a map between them, to decide exactly the survey transects between the islands, making sure the important areas were covered. This is how Aglen described this step in the process:

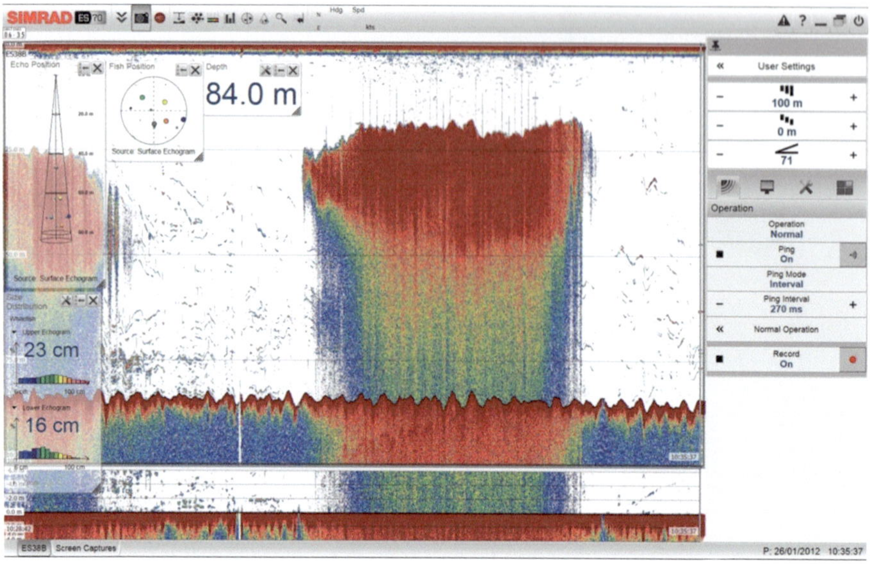

Fig. 9.3 Ecogram taken from the test run

> During the planning of the project the scientists got access to mapped data on fishing areas and spawning areas, developed by Steigen municipality based on information from fishermen. These maps were important as basis for the planning meetings with the fishermen and fishermen's organization (e-mail from Aglen, 26.01.2012).

Due to the relatively high investment in project equipment, it was decided that surveys and sampling would be undertaken the whole year, not only in the traditional cod season from January to April. This would allow us to collect data on other fish resources in Steigen besides the NCC and NEAC stocks. We also had to agree on what and how to pay the fishermen for doing the survey and sampling. The fishermen partners underlined that it is important that fishermen are loyal to the project. Committing to undertake the surveys and sampling represents a substantial cost for the fishermen involved. This does not only involve the time and effort involved in collecting samples. In addition, the pre-determined acoustic surveys ties the vessel up, taking time away from active fishing. In order for the fishermen to commit to such a project, there needs to be some financial compensation. Also, in the Norwegian context, the Reference Fleet has set a precedent, indicating the level of payment the fishermen can expect for this kind of work. For the Steigen project, the fishermen suggested a model by which the participating vessels would get a fixed sum for every month of data collection. In addition, each sample collected should be paid according to the rates used by the Reference Fleet. The other project partners agreed that this model made sense, and the fixed monthly payment was set to 17,000 NOK.

After several months of preparation, we were finally ready to begin the actual data collection. Trygve Skogheim and MK Fix started in March 2012, reporting the daily catches by species. Nevertheless, this was not based on an agreed sampling program. In March 2012, we received the following e-mail from Jan Andersen, insisting that such a program must be made:

> When it comes to getting the Steigen project started there is no lack of local quota, expertise or willingness. I have at an early stage, already in November 2011, proposed to make a written agreement with Trygve [the skipper on s Fix] about how data collection should be done: how long periods he should fish in Steigen and periods he would be able to fish elsewhere. This is so that he would be sure to be able to finish his quota. The remuneration he receives from the Steigen project [100.000 NOK + samples] will be little to live on. This got little response, and still Trygve does not know how sampling should be conducted. The IMR must as soon as possible to give us a plan of how the research will take place, so that Trygve can start fishing in Steigen with the quota he has left on the " Fix " (…) (email from Jan I. Andersen, 23.03.2012).

As a response, Asgeir Aglen (IMR) and Maiken Bjørkan (UiT) went to Steigen in April to meet the fishermen and agree on a sampling program.

9.5.3 Money, Money, Money

In trying to keep with the original plan, a key challenge throughout the project was to find extra funding. We looked into numerous possibilities to get extra funding during the first 2 years, which was time consuming and a source to frustration, especially since efforts were unfruitful.

The first attempt, made in the fall of 2012, was to get extra funding from the Norwegian Seafood Research Fund (FHF), the seafood industry's own instrument for industry-based research and development. Initially, we were quite confident that our project could get funding here, since the Steigen project, as we saw it, is innovative and extremely important to develop a new methodology for cooperative research (www.fhf.no), but turned out to be a huge disappointment. While the presentation from Jan Andersen went well, the feedback was entirely negative. As it turned out, the project we proposed did not fit into the FHF strategy, since they saw it as related to fisheries management, which FHF defined this as a government responsibility falling in the Ministry budget. Our argument of the importance of mobilising the fishermen in knowledge provision, the need for finer resolution, the importance of collaboration and of developing methods for participatory research did not find an interested audience in FHF.

The second attempt to find extra funding was initiated in the beginning of 2013, to Råfisklaget, one of the fish sales organizations. As such it is one of the truly important and powerful and in normal circumstances quite wealthy organisations in the industry. The team was invited to submit a concept note. Under the project title "Center for Experience Based Knowledge", we described a project that would collect and systematise the fishermen's own experiences and stories about the fisheries and the local resources (Andersen and Holm 2013). We were careful to mention the ongoing data collection in cooperation with IMR. The point of the project we now applied funding for, was to add an extra layer of information. Using fishermen to undertake systematic acoustic and biological sampling and develop indicators that could be accepted and used by the science, was a good starting point, we claimed. Despite only applying for a small amount of money, there seemed to be little interest in the type of concern for which the project was set up to address.

In the core institutions of Norwegian fisheries, we gradually began to realize, there was no active discourse regarding a knowledge gap between fishermen and scientists. While the fishermen representatives would acknowledge, in principle at least, the importance of fishermen's knowledge, they were not ready to allocate their own hard-earned funds to collect and use it.

The third attempt finally met with some success. It started already in January 2012, when Jan Andersen arranged a meeting in Steigen:

> I had a meeting today with the political leaders in the Steigen Municipality, the head of the development section and the Steigen Seafood company regarding the fisheries in Steigen. Here, I informed them about GAP and asked for funds. They found it very interesting and were impressed that the research is focused on Steigen. I think there is a possibility here (email from Jan I. Andersen, 27.01.2012)

As it turned out, Steigen municipality, which is very small and not at all wealthy, had little money available. However, as part of this process attention was directed towards the funding possibilities at the County level. Andersen and Bjørkan went to Bodø to present the project for the County administration for Nordland and were well received and guided through the formal application procedure. While the whole process took well over a year to complete, we did in the end receive 250,000 NOK. Since the technical development of acoustic equipment apparently had been rapid, the price of an echo-sounder of acceptable quality now was only 220,000 NOK, including mounting and calibration. The extra money from the County thus allowed the project to include one more vessel in the sampling program. The equipment was installed in the vessel Økssund – almost identical to Fix – owned and operated by Inge Wilhelmsen. With this vessel, data collection started from October 2013.

9.6 Results

As indicated in the project description formulated in the early phase of GAP, we aimed for two kinds of objectives. The first kind, which is the main focus of this chapter, is about the process of collaboration itself. The second type, to which we now briefly turn, concerns the actual data and knowledge generated by the project.

For both vessels, Fix and Økssund, the skippers collected, in addition to the echograms, weekly samples of cod and/or saithe, measuring length, weight and taking otoliths for age readings. They have also collected some genetic samples of cod. The otoliths are used to determine the age of the cod as well as the age at first spawning. In addition, they are used for separating coastal cod from NEA cod. The genetic samples are used for verifying the stock separation by otoliths and for investigating the relationship to coastal cod in other areas. Both vessels have been fishing in other areas during the peak spawning fishery for NEAC, which is mid February to end of March. The observations from cod fishery in Steigen during the spawning season in 2012, 2013 and 2014 have confirmed reasonable concentrations of spawning cod in those spawning areas highlighted in the planning. The acoustic observations also showed good recordings of cod in those areas where catches were obtained. The exception was at Englevær where high catch rates were obtained (up to 90 kg per net), but very little fish was observed on the echo-sounder. The fishermen claim that this is an area where the fish migrate through rather quickly (towards other spawning areas). Thus, the cod could already have moved away from the area when the nets were pulled in and the acoustic observations were made. When revisiting Englevær in the 2014 spawning season, some patchy distributions of cod were observed on the echo-sounder. Both in 2013 and 2014, the fish in this area had running gonads, indicating that some of the fish would have spawned there. Since the cod is spawning in batches at 1–2-day intervals, these observations could indicate that some fish might choose to spawn in several spawning areas during the spawning season (Fig. 9.4).

These observations represent glimpses of the collected data and only serve to illustrate the kind of insights the work in Steigen could produce. The way the Steigen CS project was set up, the project only had resources to collect the data; putting them to use depended entirely on the interest and resources of our science partner. The otoliths sampled, together with length and weight measurements, have been included in the IMR data collection program. By the end of the GAP project, no systematic attempt at including the echo-sounder data was apparent, however.

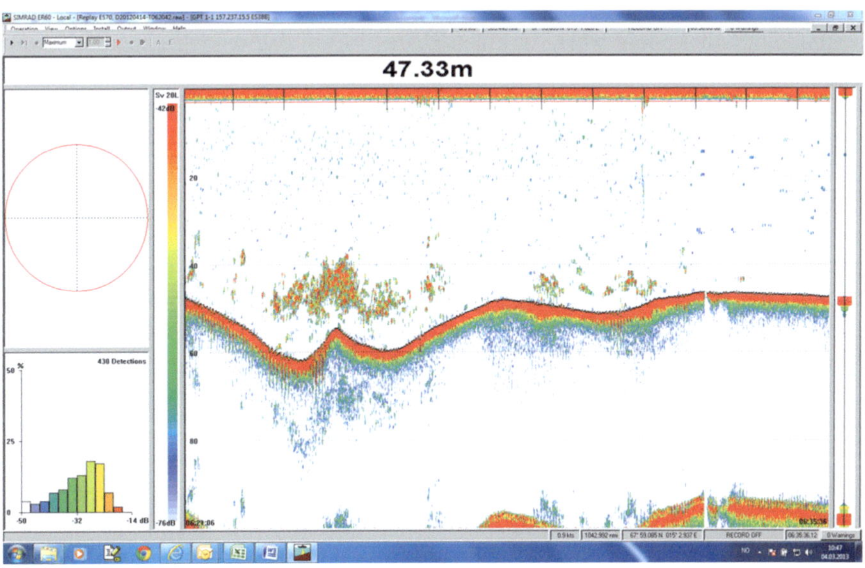

Fig. 9.4 Echo-sounder records of spawning concentrations of cod at Bøvika, Steigen, March 2013

9.7 Trapped in the TAC Machine

When we started GAP, the premise was a project conducted on the fishermen's terms, rooted in the fishermen's experience-based knowledge. The result would be a method to bridge the gap between scientists and fishermen in the estimation of fish stock size and development. This method was meant to be applicable anywhere. The project has not achieved this. Instead, the objective of the project changed along the way. Losing sight of the original ambition, we quickly moved on to become part of the established Norwegian marine research regime. Here, all important tasks are undertaken under instruction by scientists, and the fishermen use their experience-based knowledge to catch fish.

We are, at the end of the day, not too disappointed. We are confident that new knowledge has been generated and that the collaborative model we have developed will work. Nevertheless, the type of knowledge produced through the project and the methods we have applied remain scientific. As for the gap between scientists and fishermen, it is surely reduced by the project simply because of the confidence built up through the practical collaboration and joint goal of sustainability. What the project has not achieved, however, is to develop a model for utilising the fishermen's experience-based knowledge.

There are several reasons why we were not able to realise our original intentions. One important factor is related to the limited funds relative to the high cost of the equipment. This effectively turned the GAP in Steigen into a mini-project. We may have underestimated the practical challenges this represented in developing the project. A project like ours, with limited funding and without precise, formal objectives for guidance, easily becomes vulnerable to external forces. In this case, we became very much dependent on the scientists from the IMR to guide us towards a methodology that could work. In practice, this meant that we were trapped in the TAC machine.

We do not mean to blame our science partner. Quite on the contrary. Without the valuable inputs from IMR, we would have achieved nothing. From the start of the project, all partners were in full agreement that we wanted the project to make a difference. This meant that we had to deliver knowledge products that would count in the existing management system. While we wanted to go beyond the Reference Fleet model, involving fishermen in more responsible roles, we did not want to do that in a way that would reduce the usefulness of the knowledge products. When we invited the IMR to take part, this was exactly to make sure the knowledge output from the project would stand up to the relevant requirements. In hindsight, it seems clear that we underestimated the degree to which this would reframe the project. Accepting the IMR as the authority on what would count as credible and salient knowledge, the GAP project in Steigen turned into a mini-version of the Reference Fleet. While the project group of course was still in command, the expertise of the IMR scientists in practice came to define the terms of the project. Through the careful guidance from the IMR science, we realised that the options they proposed actually were in our best interest. Or rather, given the premise that the knowledge generated in the project was going to fit into and be used by the existing management system, there simply was no alternative but to go with the IMR suggestions.

At the outset, we had aimed for a project where we wanted, as much as possible, to use relatively cheap and simple equipment and procedure. This would not only allow us to include more fishermen in the project, but was also important to be able to integrate data collection and fishing operations. Or so we had thought. In practice, however, we had to invest in a science-grade echo-sounder in order for the data to be useful for IMR. While this reduced the number of participating vessels to one, it did solve the problem of combining data collection and fishing. With the chosen methodology, however, the task left to the fishermen were not particularly interesting. All the fishermen had to do was to turn on the machine when they left the harbor and, from time to time, follow pre-defined survey transects. At the time, however, we did not see any alternatives but to go with this plan. Collecting data acceptable for IMR was an absolute requirement. The fact that the proud notion of active fishermen participation had transformed through the implementation process and that our project now began to resemble the Reference Fleet model that we originally had intended to transcend somehow escaped our attention at the time.

References

Aglen A, Berg E, Mehl S, Sunnanå K (2008) Akustisk mengdemåling av sei, kysttorsk og hyse, Finnmark – Møre, hausten 2008. Toktrapport. Working paper 5/2008. Havforskningsinstituttet, Bergen

Andersen JI, Holm P (2013) Kartlegging og bruk av erfaringsbasert kunnskap. Prosjektsøknad til Norges Råfisklag. Prosjektdokument. NFH, Tromsø/Steigen

Anon (2005) Rapport fra Kysttorskgruppen. Fiskeridirektoratet, Bergen

Bjørkan M (2011) Fishing for Advice. PhD Dissertation. University of Tromso, Tromsø

Graham N, Grainger R, Karp WA, MacLennan DN, MacMullen P, Nedreaas K (2010) An introduction to the proceedings and a synthesis of the 2010 ICES symposium on fishery dependent information. ICES J Mar Sci 68:1593–1597

Haggan N, Neis B, Baird IG (2007) Fishers' knowledge in fisheries science and management. UNESCO Publishing, Paris

Hartley TW, Robertson RA (2009) Stakeholder collaboration in fisheries research: integrating knowledge among fishing leaders and science partners in Northern New England. Soc Nat Resour 22:42–55

Hind EJ (2014) A review of the past, the present, and the future of fishers' knowledge research: a challenge to established fisheries science. ICES J Mar Sci 72(2):341–358. https://doi.org/10.1093/icesjms/fsu169

Hoefnagel E, Burnett A, Wilson DC (2006) The knowledge base of co-management. In: Motos L, Wilson DC (eds) The knowledge base for fisheries management. Elsevier, Amsterdam, pp 85–108

Holm P, Nielsen KN (2004) The TAC machine. In: Report of the working group on fishery systems, WGFS annual report. ICES, Copenhagen, pp 40–51

ICES (2008) Advice 2008, book 3: 24–26. ICES, Copenhagen

ICES (2013) Advice 2013, book 3: 1–7. ICES, Copenhagen

ICES (2014) Advice 2014, book 3: 1–4. ICES, Copenhagen

Johnson TR, van Densen WLT (2007) Benefits and organization of cooperative research for fisheries management. ICES J Mar Sci 64(4):834–840

Mackinson S, Wilson DC, Galiay P, Deas B (2011) Engaging stakeholders in fisheries and marine research. Mar Policy 35:18–24

National Research Council (2004) Cooperative research in the national marine fisheries service. National Academy Press, Washington

Schwach V, Bailly D, Christensen AS, Delaney AE, Degnbol P, van Densen WL, Holm P, McLay HA, Nolde Nielsen K, Pastoors MA, Reeves SA, Wilson DC (2007) Policy and knowledge in fisheries management: a policy brief. ICES J Mar Sci 64(4):789–803

Chapter 10
When Fishemen Take Charge: The Development of a Management Plan for the Red Shrimp Fishery in Mediterranean Sea (NE Spain)

Maiken Bjørkan, Joan B. Company, Giulia Gorelli, Francesc Sardà, and Conrad Massaguer

Abstract This chapter describes how a small fishing community in Mediterranean Spain has taken charge over the management of the local red shrimp fishery. In Catalonia, the deep-sea red shrimp (*Aristeus antennatus*) is the most important commercial species, targeted by trawler fleets located in the numerous harbours along the Costa Brava. In recent years, it has become increasingly obvious that the heavy exploitation of the shrimp populations is no longer sustainable. As they faced decreasing catches and overcapacity, the fishermen in one of the fishing communities, Palamós, joined up with marine scientists from Barcelona to discuss the crisis. Together, they developed a management plan to secure a sustainable fishery. In this chapter, we describe the inspiring story of a bottom-up management initiative and discusses the mix of factors that made it into a success.

Keywords Tragedy of the Commons · Overfishing · Bottom-up management · Long Term Management Plan

---

M. Bjørkan (✉)
Norwegian College of Fishery Science, UiT The Arctic University of Norway, Tromsø, Norway

Nordland Research Institute, Bodø, Norway
e-mail: maiken.bjorkan@nforsk.no

J. B. Company · G. Gorelli · F. Sardà
Institut de Ciències del Mar (ICM-CSIC), Barcelona, Spain

C. Massaguer
Confraria de Pescadors de Palamós, Palamós, Catalonia, Spain

P. Holm et al. (eds.), *Collaborative Research in Fisheries*, MARE Publication Series 22, https://doi.org/10.1007/978-3-030-26784-1_10

10.1 Introduction

In his essay on "The tragedy of the commons", Hardin (1968) pointed out that some types of problems defy repair even when they are perfectly understood by those who are stuck with them. Open access resources, like marine fish stocks, are good examples. To the extent that there are no limitations on who has access and how much can be caught, there is a danger of overfishing. Even when fishermen realize this danger, they may neither be able to agree about the main problem nor to impose the regulations required to solve it (ibid).

Hardin has been criticized for his neglect for the commons as a social institution (McCay and Acheson 1987) and for his lack of appreciation of the commoners as community members and institutional entrepreneurs (Ostrom 1990). Both points are important to have in mind when we explore how the heroes of our story, the shrimpers of Palamós, went about to solve their overfishing problem. Even so, Hardin's basic insight rings through: The challenges involved in reaching agreement on what the (open access) problems are and how they can be fixed. When such initiatives succeed, as in the case of Palamós, it is well worth to pay attention to understand and learn about how challenges can be overcome.

Before we go into the details of the case, however, there is another issue we need to keep in mind. Since Hardin published his paper in the late 1960s, a new oceans regime has been adopted. A basic principle of this regime is that of coastal state responsibility for sustainable fisheries management within 200-mile Exclusive Economic Zones (EEZs). While most ocean resources used to be open access commons, this is no longer the case. Nevertheless, the implementation of the new regime varies. Even for countries with the highest ambition in fisheries management, only the most important commercial stocks receive full management attention, including scientific stock assessment, licensing schemes, quotas, etc. This does not mean that the situation remains unchanged for the less important fisheries, however. Fishermen who find that existing regulations are inadequate, like the shrimpers of Palamós, now have two challenges to face instead of just one. There is still the basic problem of agreeing on a common understanding of the overfishing problem and a plan to fix it. In addition, they now must find a way to make their plan fit into the national management system and have it sanctioned by the authorities. If this succeeds, of course, the implementation problem is drastically reduced, since, under the new regime, state resources are to be allocated for implementation purposes. For that to happen, however, government representatives must be convinced that this is in the common interest. Somehow, the local fisheries problem has to be transformed into an issue of national priority.

In this chapter we describe how the shrimpers of Palamós successfully solved these two interconnected problems. In short, they first agreed among themselves on something fishermen often dispute, namely that the resources on which they depend were overexploited and that regulations were in their best common interest. Then they managed to convince the Spanish government that their plan to solve this problem was sound and important enough to be made into official policy. The lowly shrimpers of a small Catalan village had accomplished a remarkable feat, turning the mighty Spanish government into their own willing instrument. Hence, in Palamós, the normal order of things was turned up-side down. In the following, we explore in detail how that was made possible.

10.2 The Red Shrimp Fishery

10.2.1 The Resource

The red shrimp (*Aristeus antennatus*), is a decapod crustacean with a high commercial value. It is intensely exploited along the continental slope of the western and central Mediterranean Sea. Here it is one of the most abundant species in the deep-water fishing grounds (Sardà et al. 2004; D'Onghia et al. 2005). The populations of the red shrimp show seasonal movements in and out submarine canyons (Sarda et al. 1994; Tudela et al. 2003). Sudden intra-annual fluctuations in the landings have been correlated with the downslope shelf water cascading, a climate-driven oceanographic process (Canals et al. 2006; Company et al. 2008). In fact, these cascading effects are believed to influence negatively the abundance of the red shrimp in the fishing grounds. In addition, the fishing conditions are difficult due to strong currents and the changing bottom conditions. In order to work under such conditions, the fleet is highly specialized. Only the skippers that have sufficient knowledge of the bottom conditions and the currents are able to exploit the red shrimp resource effectively.

In Northeast Spain, the Girona region is the most important area for the fishery of the red shrimp, with Palamós as the central fishing harbour (Fig. 10.1). During late spring and summer, the grounds mainly visited by fishermen are the open slope grounds north and south the submarine canyons (Fig. 10.2) and the catches are mainly comprised by large adult females (Sardà et al. 2003). During winter, the main fishing grounds are located at the canyon head and walls as illustrated in the map of Fig. 10.2. Here, aggregations of juveniles form in winter, when they are actively targeted by fishermen (Sardà et al. 1997).

There are seven main fishing grounds located along the submarine canyon of Palamós: Sant Sebastià, Rostoll, Els Clots, Abisínia, Gamba de Llevant, La Malica, and Candelero. Two fishing grounds are shared with the neighboring fleets of Blanes (La Malica, south of the submarine canyon) and Roses (Gamba de Llevant, north of the submarine canyon) (see Fig. 10.2).

10.2.2 Management Context and Regulations

The red shrimp fishery largely takes place within the 12 nm zone of Spain. This is of significance since the management jurisdiction resides with the Spanish government rather than with the EU, who has jurisdiction beyond 12 nm, or the regional

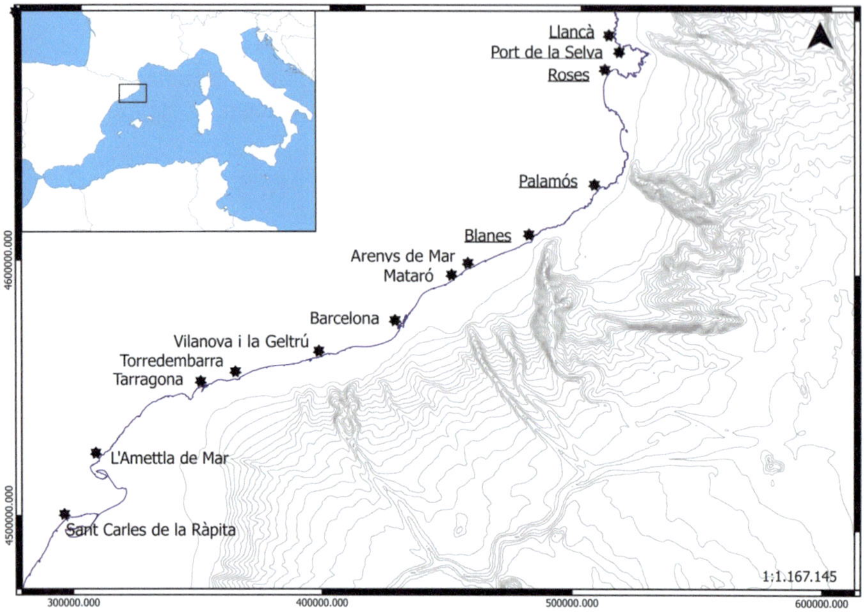

Fig. 10.1 Map of the Catalonia region, NE Spain (northwestern Mediterranean) indicating the main fishery harbours along the Catalan coast. The fishing harbours of the Girona region are underlined. Bathymetry lines every 100 m depth

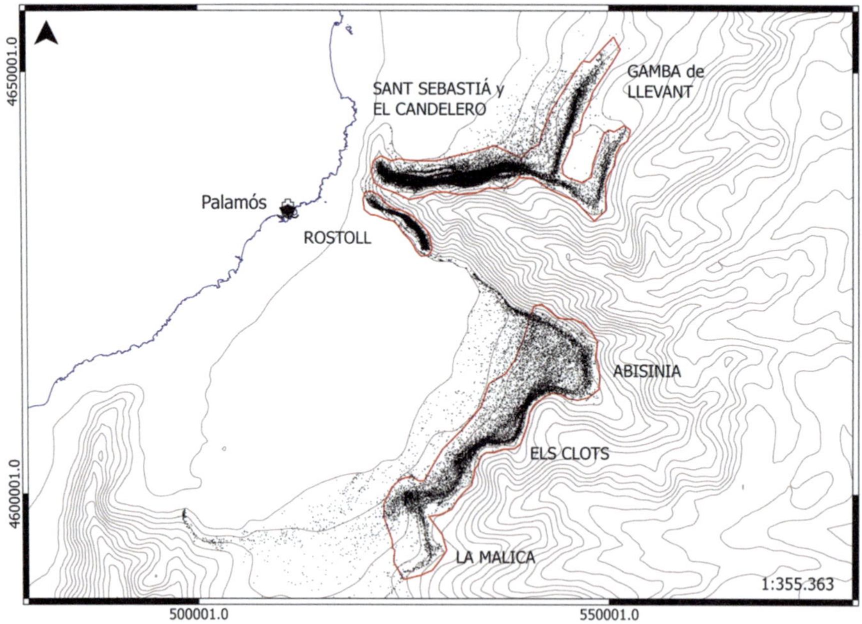

Fig. 10.2 Vessel Monitoring System (VMS) data points for the Palamós trawlers targeting the red shrimp, with indication of the main fishing grounds on the margins of the submarine canyon. Bathymetry lines every 100 m depth

government of Catalonia, who only has jurisdiction in the waters between land Capes. In Spain, it is the Secretary General for Fisheries of the Spanish Ministry of Agriculture, Rural and Environment who are in charge of fisheries management. The state delegates the execution of some regulations and inspection to the Regional Governments. In Catalonia it is the Department of Fishery and Maritime Affairs that is given this responsibility. In the red shrimp fisheries, as in all fisheries under national management authorities, access to fisheries is regulated through vessel licenses. Fishing effort is regulated through a maximum number of vessels that can have a license for different types of fisheries. In addition, there is a maximum limit of engine power per vessel (500 horsepower, with scarce compliance), as well as a limit on active fishing time (5 days and 12 h daily). There are also specific gear regulations in place.

10.2.3 Palamós

Palamós is a community with about 18,000 habitants (IDESCAT 2013) and strong tradition for fishing (Alegret and Garrido 2004). While the 24 red shrimp trawlers in Palamós are outnumbered by other vessels in the local fishing fleet, they are by far the largest vessels and land the most valuable catch. Over the years, the red shrimp fishery has become the "golden resource" and the economic cornerstone for this harbour. The red shrimp serves as a symbol for the community and is for instance the main character in their Christmas decoration and for other commercial products (picture 1). The red shrimp from Palamós is marketed under the quality label of "Denominación de Origen", and it is presented by fishermen, scientists and managers as a success story with quality at the center stage. The label "Shrimp from Palamós" is a brand to certify that this shrimp is caught by fishermen from Palamós and landed in the harbour of Palamós. It is a hallmark to identify and differentiate it from red shrimp caught and landed in other ports. Even though this is the same species that landed also in almost all Spanish Mediterranean harbours (Aristeus antennatus), it is highlighted by fish mongers and restaurants in Costa Brava that this is "the shrimp of Palamós".

The Palamós Cofradia comprises 73 vessels, targeting different fisheries (Bargalló 2010). The cofradias are organizational bodies for consultation and collaboration with the government on issues of interest to fishermen, including crew as well as ship owners. The cofradia in Palamós, which was founded in 1948 (Alegret and Garrido 2008), performs a number of services such as running the auction and economic accounting, providing ice, cleaning of equipment, and a clothing and equipment shop for fishing. In addition to being members of the cofradia, the ship owners also have a separate organization where they meet and discuss issues of common interest.

The red shrimp is not protected by a minimum landing size. Nevertheless, the size of the shrimp affects the market price. For a kilo of small shrimp (comprising almost only juveniles, below 26 mm CL) the fishermen receive a yearly average of about 10 euros. If they wait a year to catch them, these shrimps triple their weight and are classified as medium size shrimp (between 28 mm CL up to 35–38 mm CL) and costs 30 € per kilo. A shrimp classified as large costs on average 60 € per kilo (over 38–39 mm CL). The price of large shrimp can go up to 200 € per kilo euros in special dates such as summer or Christmas (numbers based on selling bills from the Fishery Association of Palamós). This means that it is far more profitable catch the larger shrimp than the smaller ones. As we shall see, this is a feature that plays a significant role in fishermen's interest in establishing a management plan.

While Palamós has a thriving tourist industry and through that is affected by urbanization and cultural change, agriculture and fishing still are important, retaining the characteristics of a close-knit traditional community. In most cases, the vessel owner and crew are close relatives, living in or close to the old village of Palamós. This does not necessarily mean that there are no conflicts of interests between vessel owners and crew. In Palamós there is a system called "reballa" (Bargalló 2010). The "reballa" originated as a way to distribute a part of the catch among the crew to ensure food for their families. In the case of the red shrimp, the crew were given the small-size shrimp (ibid). Over the years, the fishermen started to sell this catch on the black market, until the "reballa" was accepted as part of the formal wage system (ibid). Today, the small shrimp and the small individuals of other species of invertebrates and fish are sold through the Cofradía auction, with the main share of the income belonging to the crew rather than the vessel owner. While the economy of the red shrimp fishery as such will strongly improve by catching larger shrimp, this is not necessarily the case for the crew, whose income are partly tied to the catch of juveniles. In the struggle to agree on a management plan, this was one of the thorny issues that had to be settled.

10.3 The Crisis

In 2005, the red shrimp fishery along the Costa Brava was in trouble. As the skipper and president of the cofradia put it: "in 2005, there were no shrimp".

Natural fluctuations in the red shrimp abundance have always occurred, and as such did not come as a surprise (Company et al. 2008). Captures show a marked pattern of intra-annual variability with higher landings in spring-summer and lower landings in autumn, which has been attributed to reproductive displacements of the population. However, as Fig. 10.3 shows, the 2005 downturn was deeper than usual (ibid). At the same time, the disappearance of the shrimp hit harder than usual since it coincided with the removal of state subsidies on fuel (Bargalló 2010) (Fig. 10.4).

Faced with a severe crisis, the fishermen looked for an explanation. They explored the reasons for the disappearance amongst themselves: Why did the shrimp disappear and where did it go? As the President of the cofradia explained, "The old ones, old fishermen from Palamós with a lot of prestige, they did not know what was going on with the shrimp. There were a lot of theories" (interview 09.01.2014).

Fig. 10.3 Specialty of the Palamós Chocolate shops: chocolate in the form of shrimps. (Photo by Joan B. Company)

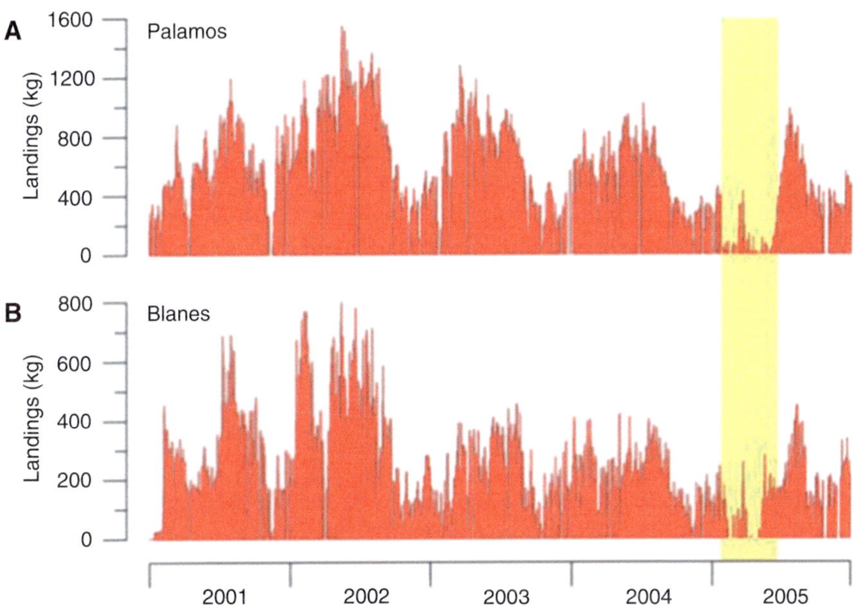

Fig. 10.4 Intra-annual variability of daily catches. (**a** and **b**) Daily landings of Aristeus antennatus since 2001 at Palamós (**a**) and Blanes (**b**), the most representative studied harbours. The pale-yellow band shows the period affected by the temporary fishery collapse caused by the winter 2005 dense shelf-water cascading event, disrupting the intra-annual variability of precedent years. (Courtesy of Company et al. 2008, used with permission)

The shrimp disappearance was a mystery for the scientists as well. The red shrimp had been studied by biologists from the 1960s, without solving the question about the periodic fluctuations of the stock (Bargalló 2010). While the regular disappearance of the shrimp itself was a well-known phenomenon, the reason for such fluctuations was not known. It was only on the basis of the event in 2005, which was a "very dry windy and cold winter (…) when cascading is exceptionally intense" (Company et al. 2008: 1) that scientists were able to find an explanation (Company et al. 2008; see also Bargalló 2010). A set of oceanographic instruments was deployed near the bottom. Current meters were placed 12 m above bottom and sediment traps 22 m above bottom. With these instruments, the team of scientists was able to record extreme downwards speed of the water masses. During cold and windy winters, like that of 2005, the cascading effect would be intense, carrying the small shrimp off from the fishing grounds down to the deep-water canyons. With the discovery of the cascading effect, the scientists had arrived at an explanation for the periodic disappearance of the shrimp (Company et al. 2008).

10.3.1 Spreading the News

The findings above were published in 2008 (Company et al. 2008). Before that, however, the scientists engaged in the task of informing the fishing communities dependent on red shrimp. This started when the scientists went to all the fishing harbours to explain their findings in meetings organized by themselves. This process took about a year. First, there were very few fishermen that came to listen, but at the end, there were more than 40 people attending. At these meetings, the scientists presented a hypothesis to the fishermen: Three years after the disappearance, there would be a "babyboom"; an unusual abundance of juvenile shrimp. The fishermen were skeptical. In 2008, however, the prediction came through. The scientists showed that they could provide knowledge about the shrimp that was valuable to fishermen, and this was a key element in building trust between fishermen and scientist.

According to both fishermen and scientists, finding the answer to the shrimp-riddle and predicting the babyboom changed the relations between fishermen and scientists. The President of the cofradia said:

> I met CSIC scientists. They gave me information that was beneficial for the fishing. We had to do something to safeguard. We had a general meeting with presentations. The young ones are the key, the elder ones are sceptical (President of the cofradia, Interview 09.01.2014).

According to one of the vessel owners, "The biologists came, and we learned from them and they from us. It was an interchange" (Interview 08.01.14). Another vessel owner said that "We do have more trust in biologist now" (Interview 08.01.2014).

While the fishermen's characterization on their new trust in science is perhaps somewhat guarded, it remains a fact that the establishment of such trust formed the basis of cooperation that had not been there before. After the crisis, in 2007, the fishermen in Palamós drafted a Management Plan for the Palamós red shrimp. The plan consisted of a number of different management options and was developed jointly by fishermen and ICM-CSIC scientists. While this draft was not adopted by the authorities, it formed the basis of the both the GAP case study project on Palamós red shrimp, which eventually would lead to the adoption of a management plan.

10.4 The Gap Process

Built on the strength of the relationship between fishermen and scientists forged by the shrimp crisis in 2005, Palamós red shrimp was developed as one of the case studies in the GAP project. The GAP1 project ("Common Grounds", 2008–2009), provided an opportunity for fishermen and scientists to continue to work together in order to improve the red shrimp fishery. This led to a second Long Term Management Plan (LTMP) draft, which was submitted in late 2009 but not adopted. The development of a third LTMP draft was undertaken with financial support from GAP2 in 2012–13. On this basis, a revised LTMP was officially recognized by the Spanish Government in mid-2013.

10.4.1 Step One: An Ambitious Plan Materializes

When the process of planning a LTMP for red shrimp during the first stage of GAP was initiated, it was a very ambitious plan. Originally, the intention was to include the whole region of Girona, which includes five fishing harbours (Port de la Selva, Llança, Roses, Palamós, and Blanes, see Fig. 10.1) and more than 150 vessels.

The idea behind this initiative was to devise stricter regulations than those currently implemented by the government. More specifically, it was the fishermen from the Cofradia in Palamós that came up with the idea and initiative of making a red shrimp management plan. Based on the fishermen's wish to establish an LTMP, a meeting was organized with all the different actors. Since the aim was to convince all five harbours in the area to follow the management ideas of Palamós, these were also included in the beginning of the process.

The fisheries sector, including the presidents of the three most important cofradias of the Girona province (cofradias of Blanes, Roses, and Palamós), a scientific team from ICM-CSIC and the Catalan Government worked together to generate a LTMP draft document. The discussions were centred on two key issues that were thought necessary to maintain a sustainable and viable plan for the red shrimp: to ensure a good protection of juvenile population and a reduction in fishing effort. In line with this, the GAP team designed a joint plan aiming to control fishing effort, reduce the catch of juveniles, preserve females and ensure a biological rest through closed seasons during main recruitment period. Also, the plan designed scientific monitoring and pilot studies of the red shrimp. This was a very ambitious and expensive plan. Estimated costs were more than eight million euros over 5 years. The LTMP was first submitted to the Spanish Ministry of Environment and Rural and Marine Affairs in September 2008 (see Box 10.1). Note that the red shrimp fishery has a number of features that make it suitable and urgent to establish a management plan: (1) Their economic value far exceeds 50% of all fishery products landed in ports of Girona. (2) In the recent years, this fishery maintains a modern and powerful fleet but could hardly survive without this resource. Because of this dependency, when the resource fluctuates downward, the fleet is in serious financial difficulties and exploits other fishing grounds in shallow water, generating widespread social conflict. (3) This fishery can be considered mono-specific due to the depth at which the target species, shrimp, is caught at. This optimizes a positive response to this appeal with a concrete plan of management. (4) The biology and ecology of this species are complex, but in Catalonia experts have been studying the species for years, which ensures a good and rigorous development of the management plan. (5) This fleet is particularly vulnerable to the current price of oil and rising foreseeable future. (6) The situation of joint arrangement between fishermen, managers scientists and the need for future regulatory and economic projection of this important resource are a challenge and an unprecedented opportunity in the Mediterranean area.

Summing up, step one in the process consisted of different suggestions for management measures taken by the fishermen themselves, advised by scientists at ICM-CSIC. After many meetings with Spanish and Catalan policy-makers, the final draft of a LTMP was almost ready late 2009. However, the severe economic crisis in Spain stopped all negotiations about its implementation. Given the economic downturn, the first plan was simply too ambitious both in terms of economic costs and the number of harbours included.

Box 10.1: Summary of the 2008 LTMP Proposal

(1) Evaluate the resource biomass and how it responds to a particular level of exploitation.
(2) Managing the resource: Evaluate those management measures that allow the maintenance of a sustainable fishery according to the sector.
(3) Preparation for the fluctuations: Given the impact the environment has on the shrimp population and hence on this sector's economy, it is essential to anticipate their movements.
(4) Pilot Actions: Given the urgency to take action for the sustainability of shrimp, the Pilot Actions are presented as those immediate actions to ensure minimum containment and protection of certain vulnerable phases of the shrimp populations.
(5) Budget LTMP for 5 years: 8,284,826 €.

10.4.2 Step Two: Scaling Down

With the start of the second stage of GAP in 2010, it was possible to continue the work done during GAP1. Given the experience from the first stage, it was important to reduce the complexity of the plan. In addition, the costs had to be reduced. In order to do that, the number of fishing communities was cut. Since Palamós had taken the lead in suggesting a LTMP, it was an obvious choice to include them in the process. Hence, the other communities were excluded in order to reduce the costs and social complexity. In addition, the ambitious ideas related to the biological and oceanographic data were reconsidered.

In order to ensure a bottom-up process and to get the project running, several meetings were held in 2011 and 2012, where those with a stake in the LTMP for red shrimp were present. This included the Palamós Fishery Association, Regional Federation of Fishermen Associations of Girona, Scientific Administration and the Autonomous Government of Catalonia (Generalitat de Catalunya). Although the Fishermen Associations of Girona such as Roses or Blanes were not a player on the new LTMP initiative, their regional federation (Regional Federation of Fishermen Associations of Girona) was also present during these meetings. Their support and acceptance on the management strategies planned by Palamós Fishermen were considered essential because of the sharing of some fishing grounds. First, scientists from CSIC and the Palamós red shrimp fishermen met to discuss the objectives for the GAP CS in June 2011. A workshop was arranged focusing on possible measurements for the LTMP. In October, the vessel owner of the vessel chosen to assist the sampling (of daily landings per size class, the size and abundance of the red shrimp to map and estimate the fishing effort, size-class of catches and the catch per unit of effort for this fishery) went together with the scientific team to Girona. Here, they attended a meeting with stakeholders and policy-makers at the Central Fishery Association in Girona for ongoing discussions on the LTMP. In addition to these formal meetings, the scientific team has been in contact with fishermen, vessel owners and the Administration on a regular basis with phone calls and e-mails. In summary, the meetings have been an arena to discuss and refresh the agreements already made for the first LTMP draft. Importantly, focus was on what options were available and the possibilities of establishing a LTMP for the red shrimp in Palamós. The following issues were given most attention: the need for effort reduction; the need for stricter gear regulations; and the need for closed seasons.

In the beginning of 2012, over 80 fishermen from 24 boats reached a unanimous agreement to close the deep-sea shrimp fisheries in Palamós. The decision was reached in an effort to protect juvenile shrimps which at the time made up 50% of the fishermen's catch and represented over 30% of their income. Hence, the fishermen made a significant economic sacrifice. The fishery closed for 3 months, from February to April 2012. At this period, it was expected that the juveniles migrate to deeper waters, out of the reach of the fishing boats (Sardà et al. 2004). Noticeably, this was before the official recognition of the plan. The hope of the fishing community was that the juvenile shrimps then would mature and boost the adult catch next year.

It was also decided that over the next 2 years, the CSIC scientists, would collect data to assess whether this strategy would benefit the fishery in the long-term. From 2012 to 2014, biological surveys (funded by GAP) was undertaken twice a month by a PhD student, funded by the Spanish Government (Gorelli et al. 2016).

10.4.3 Step Three: The LTMP as a Spanish Official Bulletin

The LTMP was officially published in the Spanish Official Bulletin in May 2013 (Spanish Government 2013). Hence, this is the third and final step in the process. In this final LTMP, focus is on protecting juveniles and reducing fishing effort through a number of measures listed below (see Box 10.2). The areas subject to regulation of this order are the seven grounds listed in the map. Below, we will first describe the process, before we turn to the plan itself.

Getting to the point where the LTMP was officially recognized required considerable effort. Even when building on the work carried out during GAP1, it had taken several years with meetings, negotiating drafts, details of the plan and its changes. The Catalan government's role has been to act as the main interlocutor between fishermen and Spanish Government. As the fishery takes place in open water beyond Capes, Spanish government is the officially responsible actor for the management measures and its implementation.

The plan establishes a number of technical measures, which have been developed and defined though participatory planning over years, building on the GAP process. The present LTMP only involves the waters belonging to the Palamós fishing grounds, i.e. the area reserved for fishermen from Palamós. In Catalonia, the fishermen usually conduct their fishing trips during a single working day. In Palamós, the vessels are allowed to leave the harbour at 6 am and must return by 6 pm. This means that the fishing grounds must be fairly close to the harbour. When a vessel leaves and when it comes back is easily monitored from the cofradia building. In addition, the vessels are tracked electronically by the "blue box" (the Vessels Monitoring System, VMS - also known as Blue box data). The total period of the closing season was finally established to 4 months (8 weeks), which in 2015 was whole January and 2 weeks of February. Additionally, 2 weeks are placed in the first half of March. These 2 weeks can vary from the beginning of March to end of March after the decision of a committee comprised by Spanish and Catalan Fishery Administration, scientist (ICM-CSIC), NGO (WWF-Mediterranean) and fishermen from Palamós. The decision is taken depending on the presence and abundance of juveniles of each year. The plan is designed to have some flexibility, to be decided by the committee, as for example to move two-weeks of the closing season from beginning of March to the end of March.

In terms of monitoring and enforcement in Spain, the authority is divided by several institutions. Besides the official monitoring by national and regional authorities, the president (and his committee) of each local fishery association has the right to implement economic fines or other kind of sanctions. For instance, he can ban for a period of time the fishing activity of a fishing boat that does not follow the agreed rules of the LTMP. The fishery administrations (Spanish and Catalan) also help and collaborate with the president of the fishery associations in sending the needed information for the implementation of the sanctions (controlled via VMS).

In summary, the LTMP measures include closed seasons, mesh size regulations (increased mesh size) and fishing effort (number of vessels) reduction in order to protect juveniles and reduce the overall impact of the fishery on the stock.

Box 10.2: Summary of the LTMP Accepted in 2013

(1) Vessels are only allowed to use their trawl gear armed using square mesh with a minimum opening of 40 mm, which may not exceed twine 3 mm in diameter.
(2) The maximum time spent may not be more than 11 h and 30 min, counting from the output to the input port when crossing the mouth of the same, and weekly activity period may not exceed 5 days.
(3) When vessels conduct fishing activities in Areas A and B, the maximum number of haul sets shall be limited to three on each day. In Area C, one cannot overcome two hauls per vessel per day.
(4) Closed period: A temporary closure of fishing is set during 60 days per year (see the extended version for a schedule).
(5) Fleet Capacity, structural program: The target is the reduction by 20% of the number of units.
(6) All vessels should have an engine with a maximum output of 500 HP.

10.5 When Fishermen Become Managers

In the GAP case study project in Palamós, the goal was to devise and implement a management plan for the local red shrimp fishery, aiming for sustainability both in biological and economic terms. With the official recognition of the LTMP in 2013, this goal had been achieved. Moreover, the fishermen of Palamós had been actively involved in making the plan and getting it accepted by the authorities. This is remarkable. Under a traditional top-down management approach it would NOT come as a surprise that such a management plan had been devised without much local involvement and implemented despite active resistance from fishermen. That the fishermen themselves have been the driving force behind the plan is unexpected and deserves an explanation.

In Palamós, the fishermen came to agree that stricter regulations would be in their interest and that they needed a management plan. In this sense, the 2005 crisis can be understood a starting point or window of opportunity for a new regime (see for instance Armitage et al. 2011 and Bjørkan 2009). An important basis for that is the price structure in the red shrimp market, where larger shrimp receives a much higher price than smaller ones. If a skipper can skew the composition of the catch towards the larger sizes, it will easily pay off. The practical mechanism for achieving that is a key part of the management plan, namely the 60-day closure during winter.

After the fact, when such a measure has been implemented and the effects documented, it is not very difficult to find out whether it is serving its purpose. For the fishermen to device such a measure, and even implement a closed season on a voluntary basis, is a different matter altogether. In Palamós, two important factors help explain why this was possible, in addition to the 2005 crisis. The first is related to their relationship with and trust in the ICM-CSIC scientists, which allowed them to establish a common understanding of the management problem and what was needed in order to achieve biological and economic sustainability. The second is related to the social and organizational structure of the Palamós fishing community, which established a sound basis for collective action.

10.5.1 The Issue of Trust: Making Music Together?

> If you only have the notes – science – and nobody to play - fishermen, there will be no music. The experience of the fishermen and the scientist is both important. (skipper of Palamós, interview 07.01.2014)

In addition to the crisis, there was already a basis for science/fishermen relations in Palamós. According to the fishermen in Palamós, their relationship to science and scientists was also affected by strong relationship with scientists from the University of Girona. Among other things, a fishing museum had been built and was used regularly to organize meetings between fishermen and scientists. According to the President of the cofradia, who is also a skipper, the relationship between fishermen and scientists before they started to interact with the Museum as a key arena, was a negative one: “We saw the scientific world as very distant. The only thing they [scientists] wanted, was to remove the fishermen” (interview 09.01.14). These relationships seem to have been important in order to draw on both scientific and experience-based knowledge for new knowledge. As one of the skippers stated:

> It is not by chance that it has developed this way. The establishment of the fishing museum by the social scientist Alegret [Dr. Juan-Luis Alegret, University of Girona], as well as the studies done by the biologists Sarda and Company [ICM] over many years, has been very important (Skipper of Palamós, interview 07.01.2014)

In this interview, the fishermen underlined the importance of the personal relations, rather than the institutions as such:

> One scientist has been working with me for 12 years. Another scientist has been here for many years with a shrimp project. The museum has facilitated the contact and collaboration. (…) it was a little bit of one and a little bit of the other one and a little bit of a third one (fisherman of Palamós, interview 07.01.2014).

As these developments reveal, a change was starting to take place in the relationship between fishermen and scientists. Before the discovery of the cascade effects, the relations between fishermen and scientists were more strained, but after the discovery fishermen started to trust biologists and their scientific findings.

In 2005, there was a combination of events that had an impact on the actions that were taken by the Palamós red shrimp fishermen. While the disappearance of the shrimp due to the effects of particular meteorological conditions was key, (Canals et al. 2006; Company et al. 2008), the sum of economic and biological factors forced the trawling sector to act. The main initiatives were based on the severity of the situation where the catch was going down fast (see Fig. 10.3 above); the petrol prize was high, and in particular scientists were able to answer the riddle of the disappearing shrimp that was fundamental to the fishing communities in the area, established a new trust in scientists.

Recall that the GAP project is in particular concerned with gaps in communication and understanding between fishermen and scientists. But in the case of Palamós, we did not find this to be the case. Rather, the fishermen and scientists seem to agree that they have different, complementary roles on the management stage: they seem to be more occupied with making music than that they are playing different instruments.

For instance, when we asked the fisherman about his view on the role of science and scientists, he simply answered: "They explain the reason why" (Fisherman 08.01.2014). Another fisherman elaborated more: "…they [scientists] are the ones that know. (…) We learnt a lot from biologists that we did not know. But they don't know how to fish" (Fisherman 10.01.2014). Yet another fisherman stated that: "We knew about the where and the size and so on, but we did not know the why" (Fisherman 07.01.2014).

In Palamós, fishermen did not hint in any way at the need for including fishermen's knowledge or the exclusion of fishermen in the management process. One of the interesting findings one can draw from this is that fishermen asked seem to find it unproblematic to simply divide fishermen and scientists' knowledge into two categories: fishermen know how to fish, e.g. where and when to find fish and issues related to bottom structure and gear, and scientists can explain the reason behind different phenomena, e.g. molting, changes in stock composition and of special importance here, why the red shrimp occasionally disappears. Importantly, one should note their knowledge overlaps when it comes to the red shrimp – they agree about the key issues such as the state of the shrimp (it is overfished) and the necessary measures (to reduce effort and implement size restrictions). This was also confirmed and fortified as the first results of data analysis were presented during a meeting between scientists, fishermen and regional administration in spring 2014. The data showed how the stock abundance and CPUE have declined over the years, stressing the need for a reduction in fishing effort. A scientist commented that "This result confirmed the fishermen's perception of the stock status, matching perfectly their personal experience as well as the stories told them by their parents" (e-mail 19.02.2015).

In this case study, the division of roles has been clearly defined with regard to who does what in the LTMP process. The fishermen came up with the idea and the initiative to establish a LTMP as they saw that the official management of the red shrimp was not enough to ensure a sustainable fishery in the future. Even if the low prize of the small shrimp in theory should motivate the fishermen to target the larger shrimp, this is not what is reflected in their practices. As mentioned above, the "reballa" is a possible reason for this. So it was the Palamós cofradia that contacted CSIC and asked for their help in order to design a LTMP. The function of the scientists has mainly been to document whether the LTMP achieves its goal or not. Their division of roles is well reflected by how the data collection is organized. The commercial vessel "la gaviota" has served as a data collection platform, and their tasks onboard are strictly defined. The fishermen take the scientists out to the predetermined fishing grounds which they have decided on together, and here they are strictly fishermen taking care of their catch, while the scientists are strictly measuring a part of the catch. The scientists' role is to collect data that can evaluate how the LTMP is working, and if the management measures are effective. The fishermen activities are related to the gear and the catch, while the scientists have a little station in the corner of the vessel, out of their way.

The point here was not to collaborate on research and data collection – knowledge production – as such. This is rather surprising since the rejection of fishermen's knowledge in management and the need to include other forms of knowledge than science has stirred heated debates (See for instance Chap. 9; Bjørkan 2011; Davis and Ruddle 2010; Holm 2003). While the details of what knowledge production involves in detail is not the issue at hand as such here, it is necessary to address some key aspects of participation. There are different ways of ensuring participation in management processes and several examples of how fishermen can be working together with scientists at different stages and with different levels of responsibility (See Bjørkan 2011 for a detailed description). Literature describe cases where fishermen are working together with scientists in research activities that can range from "chartering commercial vessels for research to the full integration of fishermen in all stages of research" (Johnson and van Densen 2007: 834). Importantly, not all of them involves input that have an impact on the knowledge production process as such (Bjørkan 2011). In the case of the red shrimp, the issue of including fishermen's knowledge seemed to be causing little, if any, conflict. Rather, the traditional roles where science are supposed to produce knowledge and fishermen fish (Holm 2003), seems to be embraced. This is contrary to the management slogan that gained force in the 1990; "ignore fishermen's knowledge and miss the boat" (Johannes et al. 2000).

Obviously, planning the design and organizing a vessel like described above is based on cooperation, and as the scientists underlined: "During the survey fishermen showed a deep respect for scientific research, actively collaborating with scientists' onboard" (e-mail 19.02.2015). The two groups also meet and communicate regularly through phone calls and e-mails. Importantly, the scientists and fishermen involved in the process have been collaborating closely on designing every step of the management strategy.

10.5.2 In the Common Interest: Organizational Aspects

Arguably, the capacity to take action for a sustainable future is related to their sense of belonging. According to the fishermen and vessel owners in Palamós, the community is able to make a local strategy because they have the same motivation for fishing – they share the same attitudes. For instance, a Palamós fisherman explained that:

> We think about the future, about tomorrow. I have a son and if he would like to go fishing, I would like him to. (Fisherman, interview, 07.01.2014)

When asked about the reason why he thought the Palamós fishermen were able to organize themselves and find common ground, a skipper said that

> In other harbours the owners are not skippers. They are in the fishing to get as much money as possible. They exploit the resource, and for them it is not important if it is sustainable. I don't care if my son wants to be a fisherman or not, but I want him to be able to choose. (Vessel owner and skipper, interview, 10.01.2014)

In Palamós, fishermen seem to feel included in the process of establishing the LTMP. This brings us to an important organizational feature of Spanish fisheries, the cofradia. The red shrimp fishermen of Palamós are members of the cofradia. All of the red shrimp fishermen interviewed were positive towards the cofradia and its role in ensuring participation of fishermen in the LTMP process. One skipper and vessel owner explained that:

> The cofradia is a democratic organization, with elected representatives. The President is on top, then two vice presidents where one represents the fishermen the other the vessel owners, then 15 representatives for the fishermen and owners. It has been a public process. If someone feels excluded it is because they wanted to be excluded. The cofradias doors are always open. It has been a long process. (Vessel owner and skipper, interview 08.01.2014).

And continues, "There are people that are not fan of the plan, maybe 10%. The majority is pro". In addition to the cofradia organization, there is another important organization for the vessel owners, the "asociacion de armadores". Hence, the vessel owners – who in Palamós also are skippers – have their own organization. Several of the vessel owners pointed to this as a key arena for the development of the plan. A former President of the cofradia, explained that:

> In addition to the cofradia, there is a parallel organization for the owners (Asociacion de Armadores) which is not really related to the cofradia as such. It was used as an arena to discuss the issues and give them information with time to process it on beforehand. (Former President of Cofradia, interview, 08.01.2014)

Some of the fishermen also related the success of the plan to the "owners' organization". According to one fisherman:

> The owners decided the LTMP on beforehand. Not everybody was for the plan. Of the owners, 80 % were for, they have a different mentality. Some of the fishermen are pro and some con. But they have seen that it can be good. It has been working for a couple of years now. (Fisherman, interview 08.01.2014).

Moreover, the Palamós fishermen argue that the owner structure is important to how the resource is exploited. If the vessel owners do not live in the community and do not fish themselves, this affects their strategies. For them, cash flow is more important than the state of the resource:

> Almost all the vessel owners are also the captains. In other harbours, there are few owners that go to sea. This is our home; we have been over 100 years in Palamós. An owner on land wants to take out cash. That is not the same vocation (Fisherman, interview, 08.01.2014).

The strong notion of Palamós as a community and the shared dependency on fisheries is central for agreeing on a common understanding for how to regulate the fishery. They have a common interest in avoiding the overfishing problem since the sense of belonging is strong; this is also reflected in the way that the "reballa" system has developed and how the family ties remain important. Of course, the organizational structures like the cofradia and the owners' organization are key to nit the often related fishermen together even further. The cofradia as an organization has also been an important tool to strategically organize their strategy of implementing stricter regulations and to face the challenge of fitting their LTMP into the national management system as well as convincing government representatives that this is in the common interest. As the above statements shows, their relations to scientists have a personal character and their trust is tied to some persons rather than institutions.

Importantly, the fishermen in Palamós are confident that the benefits from the management measures will fall back to the community. This is related to several issues, such as the organization of the fishery, the fishing methods, the biology of the shrimp and local geography: First, to the institutionalized system where certain cofradias have exclusive right to their fishing areas – in the case of the red shrimp, this involves seven fishing grounds, where two are shared with two other cofradias (see Fig. 10.2 above). In the LTMP, this aspect is strengthened and formalized, making sure that the gain is not appropriated to free riders – people that do not belong in Palamós. Second, the red shrimp fishery has some attributes that are important for where the gain will fall: this is a very specialized fishery where it is easy to control key aspects. The monitoring is easy compared to other fisheries: i.e. who is trawling in the seven "caladeros" is easy to monitor, the hours spent fishing is monitored through the blue box but also visually, and there is no by-catch hence it does not interfere with the interests of fishermen targeting other species.

10.6 Final Remarks

As described above, this case study is set apart from the other GAP2 cases, as its main focus is on management measures, while the knowledge question is secondary. While other cases, like the Norwegian Coastal Cod, is concerned with participation in research and knowledge production, the Spanish red shrimp study's main concern is on participation in planning. The scientists are doing the research and this takes place on board fisherman's vessels. But the research itself is done without any involvement of fishermen. The task of sampling the shrimp for instance, is done exclusively by scientists. Importantly though, when it comes to the development of the LTMP, fishermen's involvement has been broad in every phase and stage.

A number of factors facilitate a process in which the local ship owners together with scientists have been able to develop a management plan and get it authorized by Spanish authorities, with the collaboration of the regional government of Catalonia. Some of these are obvious reasons for succeeding, while others are more complex and generated in a mixed web of organizational frames and ad hoc approaches. First, the red shrimp is a clearly defined and selective fishery. Together with the economic value of the catch and the current situation and the positive attitude of the fishermen themselves, the red shrimp fishery is ideal for using special temporal regulatory measures that tend to ensure their conservation and sustainable development in the future. The fishermen were in a situation where they had to make changes, or they would disappear. Second, the Palamós vessels that target the red shrimp share several important features. They all fish in a limited geographical area (submarine canyons and nearby grounds), and they use the same gear (bottom otter trawls) and sell their catch at the Palamós Cofradia. Third, by means of the LTMP, they can ensure the benefits of their sacrifices for themselves, due to the above described characteristics. Turning to the organizational aspects, there are a limited number of interests – still, it is important to underline that the process has not been friction free. While the cofradia is an important institution that enables direct participation and a process that is conceived as legitimate and democratic, it is plausible that the additional organization for vessel owners has been key to the LTMP.

Finally, an important question is if it would be possible to copy the process that has taken place in Palamós. As a matter of fact, this process is already in motion. As described in the above section on "the GAP process", the LTMP originally aimed to include the whole region of Girona, with 5 local fishery harbours and over 150 vessels. When the first LTMP was turned down as it was too ambitious both in terms of costs and complexity, a new strategy was developed that only involved Palamós. However, in the long term, the idea was to include the other communities. According to the Palamós scientist:

> [W]e all together (including the fishermen from the other harbours) decided to first implement the idea in Palamós with the commitment that if the LTMP in Palamós succeed, the other harbours will also implement some measures in their fishing ground. In fact, we are now starting very promising contacts with fishermen from Roses (another important fishery harbour of northern Catalonia). We will start some management measures at the beginning of 2015 with the fishery of the Hake (Merluccius merluccius) one of the most important fishery in our waters together with the deep-sea red shrimp Aristeus antennatus. (Scientist, e-mail 11.12.2014)

This indicates that the neighbouring harbours are learning from the Palamós red shrimp fishermen and seem to be adapting a similar strategy where they are taking charge of their red shrimp fishery. It is important to underline though that none of the other communities have actually implemented a LTMP at the time of writing.

Whether the LTMP of the deep-sea red shrimp in Palamós will actually achieve its goals of protecting the juveniles as well as reducing the fishing effort or not is a question that needs time in order to be answered.

References

Alegret JL, Garrido A (2008) History of the Palamós Harbour: development, associationism and migration. AREAS Rev Int Cienc Soc 27:27–40

Alegret JL, Garrido A (2004) Història de la Confraria de Pescadors de Palamós. Confraria de Pescadors de Palamós (ed)

Armitage D, Berkes F, Dale A, Kocho-Schellenberg E, Patton E (2011) Co-management and the co-production of knowledge: learning to adapt in Canada's Arctic. Glob Environ Chang 21:995–1004

Bargalló M (2010) La Pesqueria de Gamba Rosada en Palamos: Nueva Propuesta de Plan de Gestion. MSc Thesis, Càtedra d"Estudis Marítims de Palamós de la Universitat de Girona

Bjørkan M (2011) Fishing for advice: the case of the Norwegian Reference Fleet. Norwegian College of Fisheries Science. PhD Thesis. https://munin.uit.no/handle/10037/3770?show=full

Bjørkan M (2009) Putting MPAs to work: a Mexican case study on community empowerment. MAST 8(1):11–31

Canals M, Puig P, Durrieu de Madron X, Heussner S, Palanques A, Fabres J (2006) Flushing submarine canyons. Nature 444:354–357

Company JB, Puig P, Sardà F, Palanques A, Latasa M, Scharek R (2008) Climate influence on deep sea populations. PLoS One 3(1):1431

Davis A, Ruddle K (2010) Constructing confidence: rational skepticism and systematic enquiry in local ecological knowledge research. Ecol Appl 20(3):880–894

D'Onghia G, Capezzuto F, Mytilineou C, Maiorano P, Kapiris K, Carlucci R, Sion L, Tursi A (2005) Comparison of the population structure and dynamics of Aristeus antennatus (Risso, 1816) between exploited and unexploited areas in the Mediterranean Sea. Fish Res 76:22–38

Gorelli G, Sardà F, Company JB (2016) Fishing effort increase and resource status of the deep-sea red shrimp Aristeus antennatus (Risso, 1816) in the Northwest Mediterranean Sea since the 1950s. Rev Fish Sci Aquac 24(2):192–202

Hardin G (1968) The tragedy of the commons. Science 162:1243–1248

Holm P (2003) Crossing the border. Maritime Stud 2(1):5–33

IDESCAT (2013) Official statistics website of Catalonia, https://www.idescat.cat/

Johannes RE, Freeman MMR, Hamilton RJ (2000) Ignore fishers knowledge and miss the boat. Fish Fish 1(3):257–271

Johnson T, van Densen WLT (2007) Benefits and organization of cooperative research for fisheries management. ICES J Mar Sci 64(4):834–840

McCay BJ, Acheson JME (1987) The question of the commons: the culture and ecology of communal resources. University of Arizona Press, Tucson, Ariz

Ostrom E (1990) Governing the commons. Cambridge University Press, Cambridge

Sardà F, Cartes JE, Norbis W (1994) Spatio-temporal structure of the deep-water shrimp Aristeus antennatus (Decapoda:Aristeidae) population in the western Mediterranean. Fish Bull 92:599–607

Sardà F, D'Onghia G, Politou CY, Company JB, Maiorano P, Kapiris K (2004) Deep-sea distribution, biological and ecological aspects of Aristeus antennatus (Risso, 1816) in the western and Central Mediterranean Sea. Sci Mar 68. (Suppl. 3:117–127

Sardà F, Maynou F, Talló L (1997) Seasonal and spatial mobility patterns of rose shrimp Aristeus antennatus in the Western Mediterranean: results of a long-term study. Mar Ecol Prog Ser 159:133–141

Spanish Government, Ministerial Order number AAA/923/2013 (2013, May 16) For the regulation of the deep-sea red shrimp fishery in Palamós area (NE Spain). Retrieved from http://www.boe.es/boe/dias/2013/05/27/pdfs/BOE-A-2013-5555.pdf

Sardà F, Company JB, Castellón A (2003) Intraspecific aggregation structure of a shoal of a western Mediterranean (catalan coast) deep-sea shrimp, Aristeus antennatus (Risso, 1816), during the reproductive period. J Shellfish Res 22(2):569–579

Tudela S, Sardà F, Maynou F, Demestre M (2003) Influence of submarine canyons on the distribution of the deep-water shrimp, Aristeus antennatus (Risso, 1816) in the new Mediterranean. Crustaceana 76(2):217–225

Chapter 11
Does Slow-Burn Collaboration Deliver Results? Towards Collaborative Development of Multi-Annual Multi-Species Management Plans in North Sea Mixed Demersal Fisheries

Steven Mackinson, Michael Park, and Barrie Deas

Abstract Embedded in the ecosystem approach, a transition is taking place in fisheries governance. This transition is underpinned by the need to work collaboratively in generating, mobilising and applying knowledge and data to solve problems of shared interest. This paper gives an example of this 'participatory research' process and is intended as a practical and critical reflection of learning-by-doing. The example is a science-stakeholder partnership on developing the knowledge and tools needed to support development of a multi-annual management plan for North Sea mixed demersal fisheries. We chart the process of defining the issues, establishing ways of working and developing a common language, and then assess how these were affected by the shifting political discourse on fisheries governance. The process revealed institutional barriers that risked efforts to achieve knowledge co-creation and much needed continuity. We describe approaches that were taken to keep moving constructively so that none of the barriers derailed the collaboration. Even though the work was a 'slow burner', which is only now on the point of delivering applicable outcomes, the collaborative process made a valuable contribution to developing the trust and relationships necessary for transition to more inclusive governance of fisheries. The value and utility of the collaboration are questioned by those directly involved and put in context of recent developments.

S. Mackinson (✉)
Centre for Environment, Fisheries and Aquaculture Science (Cefas), Lowestoft, Suffolk, UK

Scottish Pelagic Fishermen's Association, Fraserburgh, UK
e-mail: steve.mackinson@scottishpelagic.co.uk

M. Park
Scottish Whitefish Producers Association, Fraserburgh Business Centre, Fraserburgh, UK

B. Deas
National Federation of Fishermen's Organisations, York, UK

P. Holm et al. (eds.), *Collaborative Research in Fisheries*, MARE Publication Series 22, https://doi.org/10.1007/978-3-030-26784-1_11

Keywords Long-term management · Ecosystem Approach to Fisheries Management (EAFM) · Ecosystem model · Stakeholders · North Sea

11.1 Introduction

This article is part of a wider story about the transition taking place in fisheries governance towards an ecosystem approach to management (FAO 2011–2018). The story focuses on the emergence of partnership arrangements for tackling issues about the sustainability of fish stocks and the fisheries that depend upon them. At the centre of this story is the need for knowledge and data, and how a transition to partnership working is helping to generate, mobilise and apply it where it matters (see Chap. 1).

The transition presents challenges for management systems founded upon top-down and linear styles of management. Increasing demand for active participation of stakeholders (as a requirement of good governance) together with the increasing complexity of management goals – such as a move from single species to multi-species, mixed fisheries management plans – requires evolving research practices and advice frameworks that allow for the co-creation of common knowledge bases for management (Mackinson and Pastoors 2014).

Lack of trust and collaboration between the fishing industry, scientists and policy-makers/managers is a well-known barrier that has frequently undermined the effectiveness of management measures. Our case study sought to help alleviate this by strengthening collaboration through participation in relevant research actions. In particular, we worked together to evaluate how food-web interactions among species in the North Sea affect the performance of management strategies aimed at achieving Maximum Sustainable Yield (MSY) and eliminating discards. Specifically, we wanted to look at what alternative options under the reformed Common Fisheries Policy (CFP) may lead to better management of mixed demersal fisheries, and contribute this knowledge to the process of developing multiannual management plans consistent with the EU's CFP's commitment to an ecosystem approach (EU 2013).

This paper is a critical self-reflection of the process of research through a collaboration principally between scientists and stakeholders. As such, it is a practical examination of learning-by-doing. It does not get in to the technical details of modelling and management plans in mixed fisheries, which is the subject of separate scientific paper (Mackinson et al. 2018). With an idealistic commitment to collaboration, we strove to create an arena for knowledge co-creation, where the value of collaboration was obvious both in its utility for management and in making research better suited to stakeholder needs. The process served to test the desire and willingness to collaborate, and to illuminate barriers to developing common knowledge bases to support transition to more inclusive governance. By charting the process, we look at how and where we were successful and what it takes to improve.

11.2 Description of the Process

The following description draws loosely upon Johnson and van Densen's (2007) outline for organising cooperative research processes. Loosely because only some of what Johnson and van Densen describe is relevant to this case study and because much of what was learned emerged by adapting to case-specific needs rather than following a pre-scripted structure.

11.2.1 Genesis: Problems, Needs and Incentives

The main partnership involved scientists from the Centre for Environment, Fisheries and Aquaculture Science (CEFAS) in the UK, and the Demersal Working Group of the North Sea Advisory Council (NSAC)[1] whose members include a range of stakeholders from the fishing industry and non-governmental organisations. Policy-makers from the Department of Environment Food and Rural Affairs (Defra), UK, were closely involved only in the latter stages, but their support was present throughout in defining evidence needs for policy and providing financial support to match the EU funding of the work.

Our case study evolved from accumulated thinking about how single species management plans make a transition toward more holistic management plans aimed at achieving Maximum Sustainable Yield and taking account of the long-term sustainability of ecosystems and fishing businesses. Specifically, it was catalysed by previous work (Mackinson et al. 2009), funded by Defra in 2007 to support a NSAC initiative to establish dedicated scientific support 'to explore the potential short- and long-term advantages and disadvantages of different levels of fishing aimed at achieving MSY, the realism of the time-scale and finding out whether it can be achieved simultaneously for all species within a mixed demersal fishery'. This work was undertaken outside of the GAP1 project, but was the stimulus for the principal researcher (Mackinson) to approach the NSAC for continued support of a case study under GAP2.

The original premise of the work is now cemented in the new EU's CFP (EU 2013), which commits to taking multi-species and ecosystem impacts into account in managing fisheries (Articles 9–10 on Multiannual plans). This is of particular importance in mixed fisheries where the catch includes multiple species caught at the same time, a good example being the North Sea mixed demersal fisheries for cod, haddock, whiting and saithe. The historical 'problem' with these fisheries has been that for many years the total allowable catch (TAC) management system has failed to deliver intended reductions in fishing mortality, principally because setting TACs and quotas independently for each stock only limits the amount of the catch that may be legally landed. As long as the quota for at least one stock has not been exhausted, fishing continues and discarding of any catches in excess of quota of the other stocks occurs.

[1] Previously called the North Sea Regional Advisory Council (EU 2004 2004/585/EC)

Scientific research on mixed-fishery and multi-species effects in the North Sea have been undertaken for many years (see Pinnegar et al. 2008), channelled at the EU level through two International Council for Exploration of the Sea (ICES) working groups: ICES WGSAM and WGMIXFISH. The vanguard of research in this area and its application to management of mixed fisheries issues is discussed in a recent joint report (ICES 2014: WGSAM Annex 5). Of particular relevance here is the recognition that new demands for provision of integrated scientific advice on mixed fisheries and multispecies issues are intimately connected, leading to scientific challenges that equal the complexity of the management issues. For example, accounting for the different time scales that mixed fisheries and multi-species processes occur, defining optimum biomass and yields for all species when the biomass and yield of one species is affected by another, and defining in models an appropriate aggregation of fleets/gears so that the differential mortality effects are captured, while at the same time thinking about the level of aggregation relevant to manageable units.

With these kinds of scientific challenges in mind, the core objectives of the case study were to develop research outputs for application to scientific advice on multi-annual management plans and support the NSAC in preparing position and advice papers on issues pertinent to sustainability of mixed demersal fisheries. Particular needs were: (i) to consider options that comply with CFP regulations, both in relation to targets to achieve MSY and the obligation to land all (commercial) demersal catch (due for a phased implementation from 2016–2019 in the North Sea) and (ii) to evaluate the consequences of changes in the ecosystem arising from direct and indirect (food-web) fishing impacts on target and non-target stocks, as required by the Marine Strategy Framework Directive (MSFD).

Given such challenges, a key incentive for the scientists working on the case study was the opportunity to place scientific research on ecosystem modelling, which has hitherto been seen as a research novelty, firmly in the applied research domain. Other incentives were a passionate belief that collaboration in research leads to outcomes more relevant to society, as well as being enjoyable. For members of the NSAC, who have been advocating the strengthening and deepening of their collaboration with researchers (ACs 2012; ICES 2013), this opportunity resonated with their beliefs and specific interests to address a fishery issue which they felt typified many of the failings levelled at fisheries management institutions and the CFP itself. In particular, the need to do this work was a result of (i) dissatisfaction with the impacts on fishing opportunities for cod, haddock and whiting in the North Sea (NSAC 2012) and (ii) industry taking a long-term perspective that views fisheries management plans in the context of fishing business plans and thus includes consideration of socio-economic and environmental factors (Pope et al. 2006).

The work presented here was not undertaken in isolation. There were other complementary research activities on North Sea mixed fisheries issues feeding into the NSAC discussion. One example was the continued work on the evaluation of quota allocation options that lead to best use when fleets target multiple stocks, which forged the "f-cube model" (Ulrich et al. 2011). Another example was research using multispecies models to assess the stock biomass and yield trade-offs that arise when the predation of one species by another is considered together with the impacts of fishing (EU projects Myfish, Ecofishman, Mareframe). What is unique about the modelling work undertaken in the case study presented here is that it incorporates both multi-species and multi-fleet dimensions, enabling us to simultaneously evaluate ecological and fishery trade-offs.

11.2.2 Organisation: How we Organised Ourselves

The original proposal stated that the case study aim was 'to work with the NSAC in gaining the knowledge required to support the development long term management plans (= multiannual plan) based on an ecosystem approach'. It defined the focal areas of work as the development and application of ecosystem modelling tools and the distilling of information (and uncertainties) from models into readily communicable forms. It did not describe in detail how the work would be organised and who would be responsible for what. It was assumed that the work would build on previous arrangements where small informal meetings were undertaken.

The beginning of our technical work started in June 2011 with a presentation by Cefas scientists to the NSAC's Executive Committee. The purpose of this presentation was to revisit the ideas agreed during proposal preparation and discuss to what extent they still reflected the NSAC's need for scientific support and how to best match the skills at hand with their needs. It was also to illustrate how the case study fitted in with the overall aims of GAP2 and discuss how best to organise our working partnership. It took another 2 years to resolve the latter.

Aware that many NSAC members have a very busy travel schedule, the original approach to organising the working relationship was based on the need to minimise travel and maximise the chance of getting relevant people together at the same time. It made sense then that the scientists should travel to NSAC meetings, with formal slots on the agenda, and seek to extend meetings where needed (including paying necessary costs). What is more, the NSAC seemed to already have a near-perfect receptacle for discussions, namely, 'The whiting long-term management focus group'.

11.2.3 Roles: What Bits of Knowledge Did we Bring to the Table?

While our work, like that of all the GAP2 case studies, was predicated on the notion that developing a shared knowledge base will help solve problems and provide opportunities for innovation, we understood from the beginning that our roles would be different.

The anticipated roles of the scientists were: (i) data collectors and processors, (ii) developing a software tool that could show how interactions in a food-web might affect the outcomes of alternative management options for mixed fisheries, (iii) translating management options proposed by the NSAC in to specific model strategies that could be compared, (iv) learning how best to express and present results in a way that was both understandable and had utility for stakeholders and managers and (v) teaching and making transparent the model, its assumptions and limitations – thereby enabling the NSAC members to make an educated assessment of the utility of the work in relation to their needs.

Anticipated roles of the NSAC members were: (i) to identify their priority needs for scientific support of ecosystem considerations of long term management plans and define which fisheries in which region to address – thus guiding model development, (ii) to advise on specific management options that would be palatable to them – thus avoiding wasting time in evaluating unrealistic options, (iii) to guide us in jointly developing the outputs in a form that would help them to generate their advice, (iv) to be the vehicle to give science utility and impact by making it relevant to applied management needs and (v) to teach the scientists about the nuances of fisheries and what matters to stakeholders so that scientists could improve their comprehension of mixed fisheries issues and methods for helping to address them.

It was expected from the start that the co-creation of the results format and 'hands-on learning' would be essential for the success in developing a common understanding of the value of the work and in communicating it to others so that it had the desired impact.

11.2.4 Modelling Approach

After establishing the needs and finalising technical concepts, an in-depth approach to modelling work was adopted. This approach sought to evaluate realistic management options that would be identifiable and meaningful to NSAC members and policy-makers. An alternative would have been to take a gaming approach and explore hypothetical scenarios of the possible impacts of multispecies interactions on the performance of management options. Such approaches are useful teaching tools, but in this instance we felt that the level of knowledge of the NSAC members was high already because of (i) the high technical level of their knowledge with regard to stock assessment, harvest control rules and management measures and (ii) previous exposure to multi-species modelling work (e.g. Mackinson et al. 2009). For these reasons, we felt that a gaming approach would quickly be surpassed and thus not a wise choice in terms of use of resources.

11.3 Key Lessons on Making the Collaboration Work

Throughout the process, we found that a major barrier to overcome was the lack of time and attention that could be devoted to the case study, particularly but not only, by the NSAC members. It highlighted to us that the development of collaborative ways of working can be vulnerable because of the demands on time required to work in existing 'established' structures and modes. It was frustrating but also a critical learning outcome. It indicated how existing structures were not as well suited to establishing close collaborative interaction as we thought. It also created a dilemma, since trying to create an arena for more effective collaboration by creating additional meetings (e.g. as a new focus group) ran the risk of increasing demands on time. Recognising that there was no easy solution, we developed simpler strategies to maintain dialogue and evolve work so that it remained worthy of attention. These strategies included setting aside dedicated time in advance, being flexible and grabbing opportunities, connecting the value with relevant policy meetings, working one-to-one and adapting goals to opportunities. The following paragraphs illuminate these more clearly by describing in detail some important aspects of case study development.

11.3.1 Establishing Effective Ways of Working

Our plans to tap into existing NSAC WG structures to maximise opportunities for input and avoid duplication and fatigue never materialised as anticipated. While attendance of scientists at the NSAC's Demersal Working Group meetings were always welcomed, meetings never delivered the depth of engagement required to get into technical detail. A key reason for this was that the NSAC meeting agendas tended to prioritise issues based on their urgency, leaving little time to address important strategic interests. It is a familiar pattern for many people. In this case, meeting agendas beating to the drum of CFP reforms, annual TAC negotiations and pressing enforcement issues, all played a part in influencing the timing and order of priorities for the AC. On more than one occasion, the time originally allocated for discussion on the case study was cut short during the meeting. A good example of this was a meeting where an agenda dedicated to Long Term Management Plans (LTMP) was hijacked by presentations on the ICES stock advice, net measurements and landing obligations.

Reflecting on this, we discussed whether a dedicated focus group would be more appropriate, with GAP2 supporting some of the costs of these specific meetings. But that never occurred either. Furthermore, during the period of the project, neither of the existing relevant focus groups convened a meeting. Hence, neither the Demersal Working Group meetings nor its focus groups proved to be successful arenas for getting into the detailed work of the case study. The fastening pace of CFP reform during 2013, with a very public and political focus on banning discards, increased the demand on NSAC members in their day jobs and NSAC work. With urgent issues fully occupying their time, they were not able to think about long term management plans.

Unlike the NSAC members, the researchers involved in the project were not constantly in fire-fighting mode and therefore had time to work on strategically important issues. But the researchers' situation was not without problems either, and, perhaps unbeknown to the NSAC members, the dip in attention to LTMPs was perhaps fortuitous for the scientists. It gave them additional time to resolve complex technical challenges, which were compounded by loss of capacity and skills (three of the four scientists having left Cefas by the end of 2011), which threatened model development and also thwarted efforts to maintain the high level of interaction expected in 2011–2012. Although not without significant delay, subcontracting arrangements saved the day, enabling continued input from the model programmer and additional support from experts at the University of British Columbia, Canada. With additional expert input, technical issues were overcome. Foreseeing the priority need to provide scientific support on discard issues, new innovations to address this concern were also included. This additional work posed a significant risk in the development of the model because it meant breaking what existed to build new capability. The scientists were confident that this was a risk worth taking, but also aware that it would push back the timing of the work.

July 2012 marked a turning point in the organisation of interactions, with members of the Demersal Working Group nominating Michael Park as their representative to work closely with scientists on technical details. Yet, despite good intentions, the delays caused by the additional technical work on discard capability, as well as changes in the domestic circumstances of the subcontractor, resulted in 2013 being a year of low interaction. These setbacks tested the resolve of the collaboration, but we maintained a lifeline to the case study by taking opportunities for brief updates on the side lines of other meetings. Moreover, scientists provided written submissions for relevant NSAC meetings when unable to attend in person.

11.3.2 Common Language

Establishing a common language was a continuous learning process throughout the case study and essential to making results from a complex model accessible and useful (i.e. relevant and meaningful) to stakeholders and policy-makers. Given that NSAC members routinely use fisheries' scientific terminology, technical language barriers were assumed to be minor. The reality is that many of the scientific concepts have different meanings, or at least are spoken about in a different context among NSAC members and scientists, resulting in the knowledge co-creation process being complex.

Aware of this, scientists in their presentations paid close attention to ensuring that technical complexities of the model were understood, but not dumbed-down. Wherever possible, jargon was avoided and graphical explanations used to help establishing a common knowledge base. NSAC members were perhaps less aware of competing meanings to scientific language than their scientific colleagues, resulting in scientists getting lost in the NSAC members' use of fisheries language as it was often broad and deep. When it came to one-on-one technical meetings, things got a lot easier because we were able to take the time to ask questions and educate one another. The challenge with the language was thus wrapped up with the level of organisation. In small informal settings it was possible to take time and co-create language and meanings necessary for shared understanding. In larger formal groups, however, invited participants hitchhiked on the discourse, not catching all the intended meanings. This was perhaps one of the root causes of not consolidating an effective organisation early on; it is difficult to get formal endorsement from the NSAC if there is no shared understanding.

The need to get the language right was also an essential part of making the modelling tool fit-for-purpose. We strived to build shared understanding by avoiding scientific jargon in the model interface and its outputs. While much of the ecosystem modelling environment has a detailed 'scientific interface', here we opted for a minimalist interface with a step-wise logical sequence to guide users through. Focusing on the evaluation of management options, the interface gets straight to the necessary components that form the basis of discussion around management options. The interface design and format of the outputs were designed following discussion with NSAC members and policy-makers. While more work remains to be done to fulfil the potential of the analyses, the core outputs are pertinent to present needs of evaluating ecological and fishery trade-offs against management targets, including plots of stock biomass (relative to reference points), fishing mortality and fishery yields for specified fleets.

Specific changes were made to the model development to ensure that the outputs of modelling were readily identifiable to NSAC members based on their knowledge. Specifically, we converted stock densities (the modelling units) to total stock biomasses – thus allowing for quick 'sense checking' of outputs. We also incorporated existing management reference points so that the performance against known (familiar) targets was transparent. This made it possible to quickly spot weaknesses of the model. Moreover, exposure meant that modellers could not hide in the abstract, leading to a healthy conversation about confidence and utility of the modelling results.

11.3.3 Utility and Adapting to a Changeable Policy and Research Environment

Because this case study speaks to an issue that is tightly bound up with management needs, we thought that it would be easy to bring the work to the attention of policy-makers. Indeed, as a body having a direct-line for advice to the EU Commission (hereafter Commission), the scientists assumed that the NSAC itself would be a key vehicle for facilitating application and uptake of the results. In 2013, political uncertainty in the form of disagreement among the Council of Ministers, the EU Parliament and the Commission cast a shadow over progress by preventing clarity on which parties are responsible for initiating and developing multi-annual plans and the scope of their content. NSAC members were pushed to the side of this so-called 'trilogue' between the EU institutions, leading to further uncertainty about their remit in relation to engaging with scientific aspects of the development of multi-annual management plans. Combined with this, constitutional uncertainty of the NSACs put them in limbo during the middle to late stages of the CFP reform process – making it difficult for them to commit to the pursuance of particular work avenues of strategic importance.

Following implementation of the new CFP in 2014, dedicated one-to-one technical meetings took place between the author, Mike Park (NSAC) and Defra policy makers. These allowed for detailed technical discussions on methods and results, providing direct co-learning and knowledge co-creation opportunities. One of the important parts of the knowledge exchange that was influential in terms of the credibility of management options was the definition of plausible management strategies. The key word here is "plausible". In initial trials, scientists developed a raft of options based on published ranges of F_{MSY}[2] and discard options. Their intention was to develop contrasting options to illuminate trade-offs. However, discussions with policy makers and NSAC members narrowed the focus. Policy-makers looked for options that adhered to the legal framework of the CFP's basic regulation but made allowance for the flexibilities that might exist. The NSAC representatives quickly identified those options that would not be plausible from an industry perspective when seen in the light of the practical implementation of the CFP.

At the time of writing this chapter in 2015, work continues and we are not fully certain who will take up and apply the work in shaping management and policy with regard to a North Sea multi-annual multi-species plan. This is partly because uncertainty remains in the roles and responsibilities of management agencies. Moreover, the channels for putting science into regionalised management plans are still in a state of flux and the implications unknown. While we might have been disappointed with the speed of our scientific technical development in the early years, the impact of the delay may not have mattered. In fact, the contrary might be true; the delay being beneficial to alignment with progress on a management plan.

As the structures and processes of the CFP reform start to consolidate, the opportunity for us to make a useful and timely contribution is improving. Despite remaining uncertainty in the roles of the Commission, Member States (MS) and Advisory Councils, a process is emerging, and through this process roles are being defined. In 2014, the Commission initiated two meetings on the development of multi-annual plan(s) in the North Sea, which included input from MS policy-makers, Advisory Councils, ICES, the Scientific, Technical and Economic Committee for Fisheries (STECF) and expert scientists. The Commission took a 'back to square one' process, avoiding being bogged down in the wealth of existing knowledge in favour of clarifying the scope. While the process has a strong spirit of collaboration, it is unclear who is leading it. This is because there are no precedents for guidance.

Technically, it is the Commission who guides the process. They instigated it. But they see themselves as only facilitating it. Legally, the MS's Scheveningen group is meant to lead it. The NSAC is being proactive, drafting a position paper on multi-annual plans that demands their active engagement in the whole process. Even though who leads this remains unclear, the need for scientific support is clear, and

[2]The level of fishing mortality that results in the maximum yield sustained (MSY) over the long-term.

in March 2015 the Commission requested STECF to evaluate options for a North Sea multi-annual plan. The modelling work we have been engaged in will be used in that evaluation, i.e. it has become both relevant and timely – both of which are essential to application and uptake.

11.4 Summary and Perspectives

The new CFP and MSFD provide a solid footing to demand that multi-species and ecosystem considerations are taken into account in fisheries management. But because much of the procedure contained in the reformed CFP is new and the introduction of the landing obligation is likely to change the context in which management plans are implemented, knowledge from various sources will need to be utilised to succeed. While these new policies and the research that supports them are (slowly) leading to a transition towards inclusive approaches aimed at solving such complex, shared problems, their practical realisation hinges on the development of instruments (e.g. management plans; ecosystem models) and arenas which break top-down, linear decision-making structures and allow for interaction among stakeholders, scientists and managers (Röckmann et al. 2015). This does not require abandoning existing decision-making structures but rather is about utilising these structures to enable collaborative research and co-management initiatives to emerge and succeed. Regionalisation (EU 2013: Article 18) of the CFP provides that opportunity.

Reflecting on the progress of our case study, it is probably best described as a 'slow-burner'. The changing policy priorities together with an underestimation of the complexity of the modelling work resulted in scientists spending extended periods working alone, and difficulties in the case study getting the attention it was given to start with. It was never planned or intended that way but progress was dictated by changing priorities, changing needs, changing resources, changing skills and changing lives, all very real day-to-day happenings. Succeeding required persistence, being sensitive to the changing circumstances and taking every opportunity we could to maintain a sense of continuity.

The GAP project strived to establish partnerships where equality of knowledge is respected and responsibility for work and outcomes is shared. Did we achieve this? Did (or will) the collaboration make a difference? What could have been done better? Below we share some of our personal perspectives of this experience of participatory research, reflecting on the success of the case study both in terms of the process and the value of its outcomes.

The Scientist

Being passionate about wanting to make it work, it's been frustrating for me that we were not able to get the dedicated time and continuity I had so hoped for. Nevertheless, the commitment showed that one way or another it was possible to maintain the trust and relationships necessary to co-create knowledge for tackling a shared problem. But while (in 2015) we have not technically got as far as I expected or hoped, neither has the policy that forms a concrete need for the work.

One benefit of the slow burn has been that it has allowed us to establish linkages to the scientific channels most likely to lead to uptake of work (i.e. ICES and STECF). So delays in the timing might not be a detriment in the long run if it leads to better impact. The main risk here is if we are not able to continue the work and realize the benefits beyond the lifetime of this project. We are planning to make sure that does not happen, but doing so might not be straightforward. I admire the professionalism of the NSAC and its members, and this experience has allowed me to understand better how they work and how to work with them. I think that two important challenges for building an effective working collaboration with the NSAC are its (i) low agility – the formal structure and processes can make it difficult to get a quick decision appropriate to the technical interaction needed and (ii) focus on urgent rather than strategic issues – the demanding priority list makes it difficult for the NSAC members to engage well in advance with things they know to be important.

Finally, I regret that we did not do a better job in engaging policy makers with the technical side of the work early on. This oversight was partly due to a complacency that came from working in a science institution connected to a management agency. I thought it would be straightforward. But like the NSAC, their priorities during CFP reform (and implementation) did not resonate perfectly well with the scientific priorities of the case study. Fortunately, we are managing to a better job of this now.

The NSAC Member

The report catches the essence of the difficulties and traditional cultures at play in the EU political system. The fishing industry is only now emerging from years of what can only be described as poor governance where fishing representatives were tolerated rather than treated as stakeholders. This reflection on the 'slow burn' approach provides a clearer picture I believe - a short film as opposed to a snap shot - and has recorded and interpreted some of that change.

Fishermen will still be required to manage expectations with regard to creating law that remains the preserve of politicians. That said, there is a need for business to explain why a certain course should be followed over others. A clear and knowledgeable understanding of the impacts of various actions is a luxury that until recently has been missing from the toolkit of Advisory Councils, and notwithstanding the good work of this project, it still remains a requirement as yet undelivered; certainly untested.

The interaction throughout has been both interesting and challenging, certainly with respect to some of the finer details contained within the model. That said, it has been a learning curve for many in the NSAC; having personally visited Lowestoft to have the workings of the model explained, I know this certainly enlightened me as to what is possible with regards to mapping trophic interactions and the impacts and effects of anthropogenic actions.

A negative is the potential that may have been lost due to a pull-on resources from a scientific perspective, but also from participation from members of the NSACs who themselves have a broad range of roles to fill. It's difficult to assess whether or not the slow burn approach compensated in any way.

We all see the potential of NSACs, but as yet we have largely failed to deliver that potential. This is in part a result of pressure to respond to what can be classed as 'here and now' situations, such as the landing obligation, but also because of a failure to act and think strategically. Many of the tensions are not between NSACs and managers but between the various competing actors within NSACs. So, whereas we have to learn to engage more fully with the system, we also have to come to terms with our different backgrounds and priorities and learn the art of compromise."

References

ACs (2012) Common position on the role of the RACs in the future CFP reform. European Parliament Public Hearing on Regionalisation. Brussels, 21 March 2012. http://www.nwwac.org/_fileupload/Image/InterRAC_Common_Position_Paper_Regionalisation_21March2012_EN.pdf

EU (2013) Regulation (EU) No 1380/2013 of the European Parliament and of the Council of 11 December 2013 on the Common Fisheries Policy

FAO (2011–2018) EAF-Net. What is EAF. FI Institutional Websites. [online]. Rome. Updated 27 May 2011. [Cited 6 March 2018]. http://www.fao.org/fishery/

ICES (2013) Interim report of the working group on maritime systems (WGMARS), 4–8 November 2013, Stockholm, Sweden. ICES CM 2013/SSGSUE, vol 06, p 30

ICES (2014) Interim Report of the working group on multispecies assessment methods (WGSAM), 20–24 October 2014, London, UK. ICES CM 2014/SSGSUE, vol 11, p 104

Johnson TR, van Densen WLT (2007) The benefits and organization of cooperative research for fisheries management. ICES J Mar Sci 64:834–840

Mackinson S, Pastoors M (2014) Putting the science into regionalisation consensus from policy makers, stakeholders and scientists. GAP2 Policy Brief no. 2. June 2014. http://gap2.eu/gap2general/gap2s-policy-briefs/

Mackinson S, Deas B, Beveridge D, Casey J (2009) Mixed-fishery or ecosystem conundrum? Multi-species considerations inform thinking on long-term management of North Sea demersal stocks. Canadian Journal of Fisheries and Aquatic Science 66:1107–1129

Mackinson S, Platts M, Garcia C, Lynam C (2018) Evaluating the fishery and ecological consequences of the proposed North Sea multi-annual plan. PLoS One 13(1):e0190015. https://doi.org/10.1371/journal.pone.0190015

NSAC (2012) North Sea RAC position paper on the evaluation of the cod management plan. http://www.nsrac.org/wp-content/uploads/2012/03/NSRAC-1112-2-2012-03-13-Cod-Management-Plan-Position-Paper-FINAL1.pdf

Pinnegar JK, Trenkel VM, Blanchard JL (2008) 80 years of multispecies fisheries modelling. In: Payne A, Cotter J, Potter T (eds) Significant advances and continuing challenges, in advances in fisheries science: 50 years on from Beverton and Holt. Blackwell Publishing Ltd, Oxford, pp 325–357. https://doi.org/10.1002/9781444302653.ch14

Pope J, Hawkins AD, Tingley D, Mardle S, Cattermoul N (2006) Long-term management of North Sea fisheries. A report to DEFRA and the North Sea regional advisory council. Seafish. 46 pp. http://www.nsrac.org/wp-content/uploads/2011/01/Long-Term-Management-of-N-S-Fisheries-Mar-061.pdf

Röckmann C, van Leeuwen J, Goldsborough D, Kraan M, Piet G (2015) The interaction triangle as a tool for understanding stakeholder interactions in marine ecosystem based management. Mar Policy 52:155–162

Ulrich C, Reeves SA, Vermard Y, Holmes SJ, Vanhee W (2011) Reconciling single-species TACs in the North Sea demersal fisheries using the Fcube mixed-fisheries advice framework. ICES J Mar Sci 68:1535–1547

Chapter 12
Action Research in Tropical Tuna Purse Seine Fisheries: Thoughts and Perspectives

Manon Airaud, Laurent Tezenas, Gala Moreno, Laurent Dagorn, and Jefferson Murua

Abstract The French-Spanish GAP case study is part of a more general movement that aims to link science and technology with democratic and participatory fisheries governance. The stated objective is to 'bring stakeholders involved in tuna fishing together so that they can jointly formulate proposals for the sustainable management of FADs'. In order to promote such collective functioning, a research project was initiated. As part of it, multi-profession meetings were held and a survey conducted. Taking various points of view into account resulted in reflections on how best to design and carry out activities, as well as how to go about doing action research. From the field, it became clear that a greater flow of knowledge and understanding between different actors is important and necessary. When this happens, action research will bring forth more updated information about actors' stakes and interests, as well as their positions in the systems of relations. Collective formulations can be produced and reflective stances taken. This participatory process has raised questions regarding the relationship between actors, how they function, and the implications for knowledge mobilisation. It is difficult for an action research project to be timebound like other projects. Rather, action research projects need to be tried and tested at variable densities and rhythms.

Keywords Fish aggregating devices (FADs) · Participatory governance · Collaborative research · Tuna fisheries · Action research

---

M. Airaud · L. Dagorn
IRD UMR MARBEC, Sète cedex, France

L. Tezenas
Montpellier SupAgro, Montpellier, France

G. Moreno (✉)
International Seafood Sustainability Foundation (ISSF), Washington, DC, USA
e-mail: gmoreno@iss-foundation.org

J. Murua
AZTI, Txatxarramendi Ugartea z/g, Sukarrieta, Bizkaia, Spain

P. Holm et al. (eds.), *Collaborative Research in Fisheries*, MARE Publication Series 22, https://doi.org/10.1007/978-3-030-26784-1_12

12.1 Introduction

The fishery discussed here is that of industrial fishing characterised by the significant use of Fish Aggregating Devices (FADs) by European fleets catching tropical tuna. Schools of tuna are attracted by the presence of objects floating on the surface of the sea and aggregate beneath these objects. Fishing on 'floating object sets' was started by tuna fishers in the early 1980s. These floating objects are known as FADs (see Box 12.1) Purse seine fishing with FADs increases the probability of catching tuna and allows producers to obtain better yield. In comparison with free-swimming school fishing, which is more selective because the tuna is caught mainly as it swims, purse seine fishing with FADs is more likely to catch undersized fish and have negative impacts on the biology of the tuna and biodiversity. The question of managing FADs used by fleets of tropical tuna seiners is of interest to a heterogeneous group of actors, including scientists, fishermen, ship owners, producer organisations, suppliers, canners, food retailers, political authorities, NGOs and, more broadly speaking, civil society. The use of FADs is highly controversial and raises a number of social, economic and environmental issues. This defines the tropical tuna purse seine fisheries arena, in terms of social actors and stakes.

For instance, Greenpeace France denounced tuna fishing with FADs as part of a new campaign in 2014 because of the effects of FADs on bycatch and small-sized tuna. Its regulation, therefore, is currently an important issue for regional fisheries management organisations in the Indian Ocean (Indian Ocean Tuna Commission – IOTC) and the Atlantic (International Commission for the Conservation of Atlantic Tuna – ICCAT). FADs are also of concern to French and Spanish interests; the perceptions and practical deployment of this fishing method differs between French and Spanish fleet.

In this chapter, we tell the story of an action research project on FADs and the processes involved, organized under the GAP project. First, we provide a short reminder of what the GAP is, as it formed the starting point of our work in the FAD case study. By focusing on the actors involved, we outline the paths and actions that led these actors towards better recognition of each other. We discuss the progress made in the research in terms of action taken and discuss whether it led to the establishment of new configurations among the actors. We try to identify the effects of the action research project and what conclusions can be drawn about the so-called participatory approach that informed the case study.

Throughout this story, the attention could be focused on three issues: the politic, the knowledge and the time dimension. Knowledge can be framed by two characteristics: the 'registers' (scientific, professional, administrative, legal, etc.) and the way it is understood and applied – a combination of cognitive and normative dimensions. Thus, a key challenge in this story is the tricky issue of enabling participatory processes in the field of regulatory science. This hybrid of knowledge is an important stake that influences how participants can be involved and who is 'entitled' to decide who participates. It exposes the tension between knowledge as common good carried by participatory ideology and information perceived as a strategic resource.

Box 12.1: What Are FADs?
Floating objects that aggregate marine animals can be natural floating objects (e.g. logs, drifting algae, branches, jellyfish, etc.) and anthropogenic (e.g. marine debris, ropes, oil drums). Fishers built their own floating objects fitted with underwater structures to increase tuna catchability. The latter are called Fish Aggregating Devices (FADs). European purse-seine fishers use drifting FADs monitored by means of satellite buoys. Drifting FADs are deployed and left drifting for a certain period of time before they are harvested (Fukofuka et al. 2004; Moreno et al. 2007).

Global catch of tropical tunas accounted for ~4.5million tons (t) per year in 2012; 60% was recorded by purse-seiners, of which nearly 65% was obtained using FADs. The sharp global increase in the use of drifting FADs in purse-seine fisheries is principally due to a combination of three advantages. First, FADs are often the only way to exploit tropical tunas in off-shore areas when free-swimming schools of tunas are not detected. Second, the success rates for FAD-associated sets are both high and stable (> 90%). Third, the average catch per set is often higher for FAD-associated sets than for free-swimming school sets (Fonteneau et al. 2013).

The use of FADs by purse-seine fisheries have three potential impacts: (i) a reduction in yield per recruit of certain target species; (ii) increased bycatch, or the capture of non-target species associated with floating objects (Hall and Roman 2013); and (iii) modification of the surface habitat (Marsac et al. 2000; Hallier and Gaertner 2008; Dagorn 2012).

FADs are difficult to access for the purpose of collecting data. This makes it difficult to study their impacts on pelagic ecosystems (Dempster and Taquet 2004). The increase in the use of FADs has intensified environmental concerns related to their potential impacts. Scientists, environmental NGOs, tuna RFMOs and other actors of the tuna fishery, such as shipowners and canneries, have called attention for the need to define specific conservation measures to manage FADs. Designing specific conservation measures has been complicated by conflicting pressures and the need for a compromise between achieving a reduction in these potential impacts and allowing the sustainable exploitation of healthy tuna stocks.

12.2 The Participatory Approach

The GAP project is part of a more general movement meant to link science and technology with the democratic and participatory order where scientific communities have assumed the role of pilot figures.[1]

[1] For further explanation, see Pestre (2011), Audoux and Gillet (2011), Blondiaux and Sintomer (2002) and Callon et al. (2001).

12.2.1 The Context from Which the Project Emerged

The declared aim of the GAP project was 'to bridge the gap between science, stakeholders and policy-makers' and 'bring fishermen, scientists and policy-makers together to work towards sustainable fisheries for the benefit of society' (http://gap2.eu/). The first phase of the programme started in 2008 and brought together different stakeholders and their respective knowledge so as to decide upon necessary actions to improve the management of fisheries (see Chap. 1 and 2).

The second phase of the programme started in 2011 by initiating the case studies, one of which – the French and Spanish one described in this chapter –focussed on the sustainable management of FADs in tuna purse seine fishing. This case study brought together four partners. On the French side, the case study included the Institute for Research and Development (IRD), a state-funded scientific and technical institution, as well as the only French professional organisation for producers of frozen and quick-frozen tuna, Orthongel. On the Spanish side, the case study included AZTI (a private Spanish Basque research institute) and one of the two Basque professional organisations for tropical tuna producers, ANABAC.

IRD and AZTI, by being part of the scientific committees of IOTC and ICCAT, are actively involved in the process of managing FADs. IOTC and ICCAT play an important role in governing tuna fishing. They set catch limits, restrict fishing effort, define technical and collection measures for data, and can control the application of obligations. France and Spain are represented as part of the EU delegation. France is also directly represented as its territories are part of the region in question. In addition, these organisations include other coastal states with tuna fishing interests in these regions, like Seychelles, Mauritius, Madagascar, Japan, Sri Lanka, China, Australia, Ivory Coast, Guinea, Vanuatu, Russia, etc.[2] Here, the 'high proximity between research and decision-making which has marked the institutional history of fishing' (Catanzano and Rey 1997) makes it possible to understand the importance given to issues of management within fishery sciences.

The rules governing tuna fisheries have been formulated taking scientific advice into account. At the international level, such advice is formulated within the scientific committees of the tuna commissions. At the supranational level, the European Parliament's fisheries commission promotes these commissions and requires them to ensure that any obligations individual Member States might have should be followed. At the national level, coastal states consult with teams of scientists to design measures applicable in individual sovereign states' waters. In addition, the key economic players are also allowed to impose rules to themselves. For example, French fishing vessel companies have voluntarily imposed a limit of 200 buoys per vessel per year for tuna seiners belonging to the French producers' organisations. No similar restrictions are imposed on Spanish vessels. Nowadays, the limit of active FADs at sea per vessel is 350 for all the vessels operating in the Indian Ocean.

[2]There are 27 and 48 contracting parties to the IOTC and ICCAT, respectively.

Box 12.2: The Regulations Applied to European Tuna Seiners in 2014
In the Atlantic and for all tuna fishing vessels (seiners, longliners, canners), the total allowable catch (TAC) for bigeye was 85,000 t. There was a moratorium on FADs in January and February in the zone between the African coast, 10°S, 5°E and 5°W. In the Indian Ocean and for all tuna fishing vessels there was a ban on tuna discards. The fishing agreements with third party countries that allow tuna seiners to fish within an Exclusive Economic Zone (EEZ) are most often signed between the European Union and the country in question. When there is no agreement between the EU and the country, the fishing vessel companies sign private agreements.

Vessel monitoring systems (VMS) produce data that can be accessed by the Community Control Agency and the third party country with whom a fishing agreement has been signed when the vessel is in the EEZ of this country. From 2015 in the Indian Ocean, it will be obligatory to mark the vessel's registration number on the buoys it owns. Collecting fishing data is carried out by means of logbooks (electronic since 2013).

The ICCAT and IOTC have already set up management measures concerning the monitoring of FADs and the need to eliminate their entangling aspect with regard to the associated species (non-entangling FADs – see photos above – prevent what is known as "ghost fishing", that is, the mortality of individuals from drifting FADs). The IOTC plans to examine at least one proposal for managing FADs (already discussed in 2014) at the 2015 commission, and in December its scientific committee recommended setting up a work group on FADs. The ICCAT has decided to set up a work group from 2015 on FADs, at which an opinion will be requested regarding the management measures for limiting the number of FADs.

The interplay between the status of science and knowledge concerned in the regulations, the decision-making processes and the endogenous processes of rule-making within key economic players is an intriguing facet of the governance of tuna fisheries and the way it is perceived by the various stakeholders. For example, many fishers see science as having control over the process, but scientists would say that their role mainly is to advise, as a complementary part of the policy-making (Box 12.2).

The aim of the FAD case study under GAP was to invite French and Spanish fisheries scientists to test a participatory research approach so as to draw up proposals for the management of FADs. The scientists had two goals:

- To produce scientific knowledge on FADs by integrating more non-scientific knowledge provided by fishing professionals and other stakeholders.
- To produce knowledge that is useful for the sustainable management of fishing.

The highly generic term participatory research thus designates the desire for more collective mobilisation in order to produce knowledge on the use of FADs in the tropical tuna sector.

12.2.2 Participatory Research: Multiple Motivations

The practical rendering of the participatory principle refers to specific interests and diverse intentions being expressed through mutual exchange. In other words, it is necessary to understand the specific interests and goals of different actors and the relations between actors in order to understand how they connect to each other.

For scientists, in general it has been a question of 'getting the actors in the sector to cooperate fully by providing them with data. [...] It is necessary to have data on the number of drifting FADs per zone, per fleet and per season, all information that it is easy to obtain with the support of key players in the sector' (Taquet 2011: 45). In the words of one scientist, getting stakeholders to participate is a means of 'reinforcing the climate of trust and cooperation'. Given that scientists influence and directly frame the work of industry partners, we hypothesize that the objectives of the research will appear more legitimate in the eyes of fishermen and fishing vessel companies if they themselves play a part in their formulation.

For the industrial partners, they participate so as to establish a 'platform for dialogue between scientists and fishermen as a means of bridging the gap between the knowledge of fishermen and scientific knowledge. Entering into a dialogue with the scientific community may motivate it to adopt better argued positions within the RFMO' (an employee of Orthongel). Industry, in other words, hopes that scientific advisors take a firmer stand with management organisations. Industry also opens up for a better understanding about which actors influence more the decision-making process of fishing governance. The initial perception of a 'gap' between scientific knowledge and practical knowledge can be seen as a result of a lack of communication between scientists and fishermen. So, the face-to-face meetings between fishermen and scientists start to provide the means to understand and explore the complexity of this arena.

12.2.3 'Encouraging Participation': The First Steps

Initially, fisheries scientists tried to translate the non-scientific knowledge into their own language with the help of their vocabulary and grammar. The idea was to 'transform qualitative data collected from fishermen into quantitative data' so as to 'integrate empirical knowledge into the production of scientific knowledge' (a scientist). The questions asked to skippers in an attempt to estimate the number of drifting FADs in the Indian Ocean included: What are the conditions necessary for deploying the purse seine? What elements are important when launching FADs and buoys? What observations do fishermen make of the drift and trajectory of wrecks at sea?[3] Ultimately, the results of the research are supposed to emerge as a collaborative product of the know-how of scientists and fishermen who have willingly shared their knowledge.

[3] Questions taken from the work of Alexandra Maufroy, a doctoral student in fisheries science at EME212 IRD Sète, France, to estimate the number of FADs in the Indian Ocean.

At the start of the second phase of GAP, fishing vessel companies refused to give researchers data from the echo-sounder buoys with which FADs are equipped. These data hold high strategic value for the fishing vessels and are strictly guarded. As one scientist involved in GAP explained: 'The problem is that there is also a strong hierarchy, and more people involved than just the fishermen and scientists.' To figure out a solution, the scientists responsible for the project sought the advice from a scientific communication and participatory management consultant. The consultant proposed to bring together scientists, employees of fishing vessels companies, skippers, employees of NGOs and national political authorities responsible for fishing matters to discuss the problem. This led to meetings, one in Quimper, France and another in Sukarrieta, Spain.

The official focus of these meetings was the Ecosystem Approach to Fisheries (EAF), which is one of the leading approaches to fisheries governance. Scientists conceptualised and theorised sustainable management of FADs in line with their own professional training.[4] By arguing and structuring the meeting on the EAF in the way they sought fit, the team of scientists forgot that mobilising other key players could not take place unless they went beyond their own forms of professional logic and interests. The skippers and fishing vessel companies present showed a pronounced lack of interest towards doing so however. When they took to the floor, they immediately refocused the discussion on the management of FADs. Ironically, by doing so, they returned to the initial issue of discussion. As one employee from Orthongel said this first meeting was the real starting point of the GAP project.

The participatory imperative brought into the open the hierarchy of different professional logics and the need for broader circulation of knowledge and understanding. The need for greater awareness of the interests and perspectives of other actors in the sector became apparent. It highlighted the difference between the ideal of horizontal integration implied by participatory actions and the vertical hierarchies structuring some organisations in the sector.

12.3 New Pathways for Getting to Know each Other (and Oneself) Better

In the face of the uncertainties and social stakes raised vis-à-vis FADs and sustainable fishing, the action research approach advocated a dialogue between researchers and other actors. It is a question of 'producing knowledge of and through the concerted and participatory management of resources' (Barnaud 2013: 5). This approach raises a series of questions: What are the epistemological objectives of researchers? What legitimacy can be granted to participation? How can we assume that research is intertwined with change? How can we maintain a healthy balance between distance and proximity in research (Parrini-Alemanno 2005)? What was clear was that continuous reflection was needed.

[4]Via professional socialisation, individuals acquire certain codes and conventions specific to the culture of an organisation or a professional group, allowing them master roles in the professional milieu (Dubar 1992).

12.3.1 Evolution in the Position with the Arrival of New Skills

> The relationship between fishermen and scientists worked, but only up to a certain point. Furthermore, there are many more actors than just fishermen and researchers. One of the most complicated relationships is the one between researchers and ship-owners. We saw that we really needed new skills to facilitate the process and understood that certain aspects were beyond our comprehension (a scientist).

Once the project was confronted by the realities of the field more interest was given to the regulatory mechanisms of the social world of the tuna fishing sector. By including others with new skills (the consultant in scientific communication and participatory management, and then a sociologist), what initially had been taken as a fixed assumption was transformed into an open question, namely how can actors in tuna purse seine fishing be collectively part of the formulation of proposals for sustainable management? Three approaches were identified to address these questions: (a) actors could play a part in the exchange of knowledge in making the idea of co-constructing knowledge in terms of FADs the subject of dynamic and collective reflection; (b) actors could work on sharing knowledge and experiences of current practices specific to each profession; and (c) actors of different professions could increase their exchanges.

A new reading of the official documents of the GAP project, with particular focus on the ideal underpinning the participatory approach, led to a reorientation of the case study. The new reading redefined the main purpose as identifying and generating other questions as opposed to answering questions or validating hypotheses. The participatory approach announced a process that deals with diverse interests and conflicts. Each actor can be invited to participate. While designing action research, it is important to go beyond meeting deadlines and also envisage future evolving effects. Participants are free to define what their commitment will be. As participation cannot be imposed, what remains to be seen is whether actors with different professional logics can be made to see the advantages of exchange amongst each other.

In action research, 'researchers are not on the "outside" but instead "inside" the world that they study' (Petit 2008). A so-called participatory approach implies that knowledge and reality are actively connected: this is how the equivalence of the expression action research should be understood. Adopting a comprehensive approach[5] has made it possible to confront the standards, interests and points of

[5] By raising questions about what we say, what we do and what we think about what we say and do, the comprehensive approach is characterised by the reconstruction of meaning. It sets up an area of comprehension within which people involved in it confront their own questions and hypotheses, as well as those of other participants, the facilitator or the sociologist. The comprehensive approach has little direction, thus leaving room for unexpected information to arise, which then enters into the field of analysis and interpretation. Each participant is able to play an active role in the process, and regardless of his position in this new area-forum, he himself defines its contours one way or another. By taking part in the interactions, he works on himself, which plays a part in the (re)construction and renewal of his identity.

view of all actors without exception. Scientists, who were initially isolated from stakeholders targeted by the project, became stakeholders themselves. Scientists, by occupying a position inside the fishing social area, composed of interests and stakes, and by being themselves committed to the process, must now be considered as actors in their own right as well.

12.3.2 From Knowledge to Recognition

Two operations were set up: the organisation of inter-professional meetings and a survey on the preoccupations and perceptions of actors concerning the sustainable management and use of FADs. These operations focused on three different objectives: (a) understanding the situation, (b) producing new knowledge on the relationships between actors and (c) accompanying the action processes (Faugère 2010). The distinction between these three objectives is a theoretical manner; in practice, they intertwine throughout the process and only acquire meaning in their mutuality.

However, it is important to note retrospectively that the participatory process was very restrictive since the beginning: not including all types of actors limited the scope of the work and the potential for participation. Through a reflexive posture and without detour, we can say that it was an attempt to build on the existing long-established between fishermen, shipowners and scientists, which by its foundation excluded other groups in the sector. Another reason for incomplete coverage is that the geographical and organisational vastness and complexity of tropical tuna purse seine fisheries represents a considerable challenge – particularly within the limited means available for research. Hence, except for two meetings, the work focussed deliberately on only scientists, shipowners and skippers, leaving out public administrations, politicians, NGOs, suppliers and crew. In reflection, we have seen that participatory processes can themselves define their own target groups. Later in our dialogue, the participants themselves questioned the first and restrictive definition of who should be involved.

Interviews undertaken helped the participatory process at three levels. First, interviews were a vector for recognising and promoting everyone's thoughts and experiences. People from outside with few or no interests in the sector acted as intermediaries which helped in ensuring that all views were taken into account. People also felt freer to speak out during the inter-professional meetings. Second, the interviews helped the interviewee be reflexive about his practices, roles and, possibly, his way of communicating. Finally, the interviews promoted the participatory process and group engagement. The interviews thus encouraged a new position of the actors with regard to the group, as well as the adoption of a reflective position. This played a part in enrolment in the participatory process.

12.3.3 Conducting a Survey

As we did not include a sociologist in our Spanish team, the survey concentrated on the French case. Nevertheless, regular exchanges with Spanish fisheries scientists who, at their end, carried out a few interviews with scientists, skippers and employees of the fishing vessel companies, made it possible to cross-reference certain types of information.

The mobilisation of tools from the field of social anthropology made it possible to describe, based on what the actors said, the diversity of experiences and points of view, stakes and interests raised by the issue of the sustainable management of FADs. We focussed on the relationships between actors in tropical tuna purse seine fisheries, on the links that they have or do not have with each other, on the advantages that arise if they collaborate with each other and on the stakes involved in the circulation and mastery of information, etc. The use of technical devices such as FADs is grounded in arguments and processes for complex choices involving multiple motives, be it for fishermen, shipowners or scientists. These uses hence reveal a complex social organisation, including relations with the environment and the context, technical gestures, economic, political, social and environmental stakes, and controversies.

The survey was carried out on the basis of partially directed individual and group interviews, as well as with the use of participant observation. Between May and October 2013, a total of 34 interviews were carried out with scientists, employees of fishing vessel companies, skippers, employees of the French producers' organisation and employees from public administrations in France. Interviews lasted on average 2–5 h and were carried out at the interviewees' place of work, at their homes or in cafés. Notes were taken and then the subject matter was qualitatively processed. Two central questions provided the framework for interpreting the discursive information collected, namely, what meanings do the actors give to their technical, social and economic actions, and to what extent do these meanings shape their relationships with the other players?

In addition, we undertook participant observation of the various meetings with the fisheries scientists involved in GAP and Orthongel, as well as the second interprofessional meeting in Concarneau (Brittany, France). We observed what the actors did and how they talked, and how they constructed common definitions of the situation among themselves. Group meetings, at which identities of and negotiations between actors came into play, provided a good opportunity to understand various networks and the roles played by each person.

Field trips were also privileged moments for reflection on issues, including the processes set in motion by the survey, the identities that were revealed and conflicts that existed. Designing and carrying out the survey were directly connected to the participatory process. By choosing a comprehensive approach, we were able to ensure that the field produced unexpected events and threw up new issues that reoriented the project.

12.3.4 Inter-Professional Meetings

Inter-professional meetings took place around the same time in Sukarrieta, Spain (in June 2012 and November 2013), and in Quimper and Concarneau, France (in June 2012 and February 2014, respectively). The activities proposed and designed had the same structure for the two meetings and therefore made it easier to pool information. Group meetings brought together actors involved at different levels, including scientists, ship-owners, skippers and employees from the producers' organisations Orthongel and Anabac. The aim was to encourage direct exchange to understand the practices and expectations of each participant. The meetings authorised the creation of links in a context exclusively dedicated to sharing experiences, points of view and preoccupations. According to an employee of Orthongel, the meetings made possible an awareness of 'clichés, which aren't as clear-cut as all that in reality'. They also helped facilitate a 'process that aims to bring together scientists and fishing professionals at both the human level and with regard to their respective preoccupations' (Fig. 12.1).

The interviews helped to identify subjects that deserved attention at the inter-professional meetings. The two main topics that emerged were:

- Management of tuna purse seine fishing.
- Sustainable management of FADs.

The sessions were composed of information input followed by collective discussions within profession groups, inter-profession groups, or full plenaries. The task given to these groups was to reflect on issues of conservation and management of tuna fishing with FADs and to develop collective reflections in a well-meaning and respectful atmosphere. The meetings were hosted by a facilitator from outside the

Fig. 12.1 Group photo at the inter-professional meeting in Concarneau, Brittany – February 2014. © IRD/GAP2/Airaud

sector. The facilitator ensured that everyone had the chance to speak, even those who were not used to doing so.

The inter-professional meetings were a means of bringing to life and mobilising key players from various professional milieus. Every work environment is characterised by conventions, codes and constraints internalised by those who work there, as well as specific practices liable to vary within each professional group. Professional socialisation encourages mastery of the specific language of a work environment and makes it possible to recognise and get to know each other. It is through this professional socialisation that modes of reasoning become logical, so logical that it is easy to forget that what is interesting or important for one is not necessarily the case for another.

12.4 Research Through Action

The back and forth between research and action generated circulation of information about and attention to the configuration of interests and conflicts in the tuna sector. It also reconfigured the participatory arena or, at the very least, the representation of which groups could legitimately invite themselves into it. Satellite data, information on the price of fish, and even the sharing of what happens at sea between the skipper and the fishing vessel company are all issues that were tabled as they provided the starting point for the economic activities of ship-owners. At another level, when a skipper decides to replace someone else's buoy with one of his own on a FAD, he claims a property right to the FAD, even if only for a short time. It is a means of mastering, even if only occasionally, the information on the life of the FAD. Finally, between the search for data from fishing vessel companies and skippers and attempts to come up with management advice requested from scientific committees, researchers too are under pressure with regard to the acquisition and promotion of information. Fishermen and ship-owners cooperate but do not divulge everything in order to maintain a certain informational advantage.

The production of knowledge and the use and mastery of information are thus crucial for all actors in the sense that knowledge is power. The sociological survey and mobilisation of actors played a part in producing knowledge related to the management of FADs in tuna purse seine fishing, thus serving the participatory research process well.

12.4.1 Sustainability and Management of Tuna Resources

Two concepts came very clearly to the forefront during the participatory process: sustainability and management of common natural resources.

The term 'sustainable'has different meanings according to how it is used in different contexts. Sustainability is situated at the crossroads of economic, social and environmental concerns, and regardless of the actors involved, refers to a social construct composed of practices, value judgements and scientific or practical knowledge. In the context of tuna purse seine fishing, the management of FADs is central to the management of fish resources. In the last two decades tuna fishing, which traditionally has been considered as ‘hunting’, has come to be regarded as a ‘gathering’ activity. Today, debates on fishing focus on the ‘excessive’ use of FADs. According to estimates for the Indian Ocean, the total number of FADs has risen from 1250 (250 French and 1000 Spanish) in 2007 to 5700 (1100 French and 4600 Spanish) in 2013. In addition, the Spanish fleet uses supply vessels also known as buoy tenders. These are smaller fishing vessels converted to distribute FADs according to the demands of tuna seiner skippers. The uneven use of FADs within the European sector generates technical and informational inequalities between fleets. Hence, the preoccupation with regard to managing FADs is to ensure everyone access to the resource and control of the area and that resources are not over-appropriated. Introducing FADs into tuna purse seine fishing has thus, since the 1990s, changed the social dynamics of the sector. Otherwise, FADs also increase environmental concerns and sustainability issues (Fig. 12.2).

Fishing companies have different ways to manage FADs and consequently manage exploitation, as well as different business strategies. They prefer certain fishing methods over others. For example, French fishing companies encourage free school fishing rather than FAD fishing. There is a cap on the number of FADs available for each vessel. In addition, only fish weighing over 1.5 kg are included in the share-wage system. In contrast, Spanish vessels do not have any limit to the number of FADs available. Here, wages are calculated on the basis of the total tonnage caught by the vessel, including small fish. Strategies thus have an influence on practices, which themselves have an influence on approaches to professional recognition.

The survey revealed that all actors encountered were aware that economic efficiency and its technical means are confronted with the issue of the sustainable development of fisheries. This latter point is of interest to all actors, be it in terms of

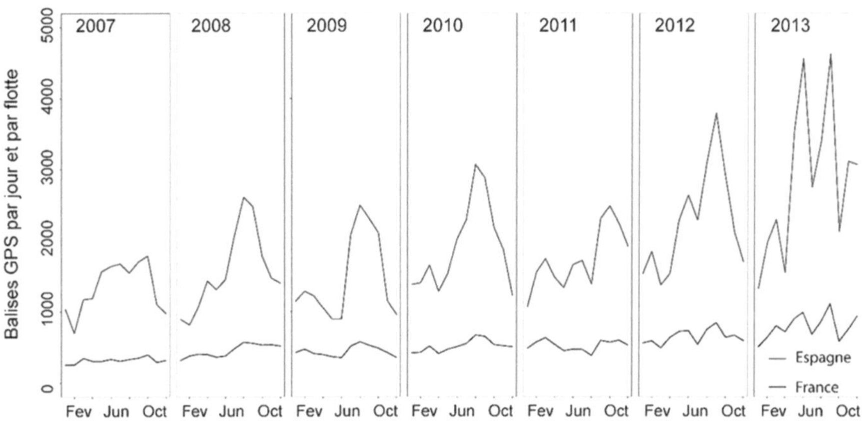

Fig. 12.2 Estimate of the total number of French and Spanish FADs in the Indian Ocean, per day, at the end of each month (2007–2013). (Source Maufroy 2014: 16)

sustainable production of a food-industry resource, maintaining an economic activity, worrying about future generations, maintaining the diversity of marine ecosystems, or professional ethos.

12.4.2 The Social Hierarchy of Knowledge

The knowledge of fishermen and researchers are of different kinds. Researchers acquire their knowledge through a selective set of institutional academic stages. A scientist is seen as a 'dispenser of scientific knowledge that the profession is supposed to apply' (Delbos and Jorion 1984). The legitimacy and authority of scientific knowledge are the result of the social status of such knowledge, giving science a privileged place in the collective decision-making process and in particular in the management systems of natural resources.

Although the skippers encountered clearly supported the idea that 'this isn't a profession that comes from book learning; it's a job you learn through experience' (a skipper), so-called empirical observations alone are not considered as being adequate for generalization by scientists. A low level of education and lack of confidence with the spoken word were regularly cited during interviews with skippers. The verve of scientists to become spokespersons of tuna fisheries[6] is thus often criticised by tuna fishermen.

The social hierarchy of knowledge and the symbolic force given to the number of years of education may explain certain imbalances in terms of speaking out. Despite knowledge exchanges between tuna skippers and scientists, this hierarchy of knowledge was implicit. Two problematic elements stood out in terms of knowledge exchange: (1) the manner in which knowledge is produced and (2) the manner by which it becomes legitimate and valid in the eyes of others. In other words, there is a divide between seamen (skippers) and landmen (scientists, even shipowners), almost as if they were in two different worlds with a social hierarchy between them. There is a conflict of legitimacy over the function of spokesperson within the tuna fishery arena, which is linked to a controversy about which knowledge is legitimate and by whom.

12.4.3 For the Construction of New Relationships Between Actors

The initial intention was to unite the French and Spanish[7],'so that they collectively would be in a position to bolster the influence they are able to exert on European processes to qualify maritime areas and their resources' (Lazuech 2014). Scientists

[6] For further elaboration on this point, we recommend a reading of Callon (1986).

[7] Actor categories include professional producers' organisations, tuna fishing vessel companies, scientists and tuna skippers, French on one hand and Spanish on the other hand.

thus wanted to link French and Spanish producers' organisations. Although scientists have cooperated beyond national boundaries for many years, fishing vessel companies that are members of producers' organisations have not done so. Their operating strategies, financial strategies and political orientations are not the same. As for the skippers, they encounter each other at sea, in a competitive context of catching fish and with differential access to instruments, tools and information. The heterogeneity of practices of shipowners and skippers, at both the national and European levels, are the result of differences in organisational work structures and cultural and political contexts. This heterogeneity makes it difficult to formulate and defend a European position when it comes to adopting measures for managing FADs. One scientist said that within the international tuna commissions and the European Commission, 'no approach has been implemented for "management plans". The only requirement to which these declarations of intent refer is obtaining data and information on how FADs function'. Orthongel, in reaction to this, has since 2012 asked that the sector 'not be limited to data collection projects, but to real management plans that include supervision of the use of FADs' (see paragraph 9). Finally, collective reflection on the harmonisation of French and Spanish national strategies is yet to occur.

Direct meetings between actors, along with the possibility for non-consensual exchanges, played a part in reaffirming identities, opening up the potential for actors to change their positions. For example, a meeting between Orthongel, French shipowners and researchers from IRD revealed two very different perspectives. On the one hand, the commercial actor, operating within a competitive economic environment where it exerts a high level of control on the information that it divulges, believes that the circulation of certain information could jeopardise its activities. On the other hand, researchers, whose aim is to improve knowledge, hold that the circulation of information and data is essential.

At the meeting in Concarneau, skippers said that they expected researchers to 'limit FAD'. The small group of scientists present tried hard to explain to the group of fishermen what they perceive themselves about their role as scientists, namely that 'scientists make proposals, but it is politicians that make decisions' (a scientist). By expressing their opinions, skippers highlighted the confusion that exists when it comes to the role of scientists within the management systems. They also point out the classical researcher's ideology, which is also a controversial one within the scientist community itself.

12.5 Discussion: Critical Input and the Limitations of Action Research

It was possible to identify three effects of the action research: (a) those on the production of knowledge, (b) those of the position of actors and (c) those on the process itself.

(a) Three types of knowledge were produced during the action research process. On the one hand, knowledge was produced that 'came from a dynamic of collective learning' where researchers and other actors collaborated (Barnaud 2013), often highlighting the different expectations of various actors such as scientists and fishing professionals. On the other hand, there was knowledge produced that was the result of a reflective analysis of the participatory process of action research. Interviews, for example, resulted in new positions emerging. Finally, as knowledge was reproduced through the participatory process, knowledge moved away from being exclusively scientific so as to 'consider multiple and subtle relations, describing chains of inter-dependence, which are constantly getting longer' and which 'take into account normative recommendations, as much as production costs or identity expressions' (Petit 2008).
(b) The participatory research process of the GAP project resulted in scientists changing their positions. Initially, they had two hypotheses pertaining to their objective of obtaining data from the fishing industry, based on the belief that science is perceived as for the common good and scientists are the unbiased face to knowledge production. The first hypothesis was that scientists' interests were shared by other actors; the second hypothesis was that other actors would cooperate based on the objectives of researchers. Action research helped diffuse tensions that existed regarding the request for data.
(c) By creating a few shared references, the action research process also made possible a significant shift from a process that was decision-making centered to one that was more consultation centred. Here, consultation centred is defined as a 'collective constructive process for questions, visions, objectives and common projects relating to an object' (Beuret 2006). The co-construction invoked by the action research process pre-supposed the 'implementation of aptitudes, as well as an adequate methodology that recognises the role of everyone'. Aptitudes mentioned by Sebillote (2007) remind us of a certain number of elements outlined in this paper: 'recognising that the points of views of others are worthy, agreeing to adjust language to ensure comprehension, having compatible visions of the present and future, and clarifying each person's conception of the roles of research in society'. What is at stake here is a reinforcement of the process itself through contributing to the researcher's epistemology to act.

12.5.1 From the Ideal to the Illusion of 'Participatory'

The participatory process thus highlighted the decisive role played by the enunciation. Enunciation can be the subject of debates and negotiations of different kinds. The process allowed researchers to create the stage during which the objects of the research were formulated and reaffirmed at the same time researchers' idea of scientific knowledge working on behalf of a collective purpose. In this sense, enunciation can support the establishment of new meanings and rationalities for practices. Enunciation is no longer a constraint imposed from the outside but becomes a basis for the strategies of actors themselves. Multiplicity of exchanges are necessary to create a climate of trust. Just as power plays a role in the control of information and the production of knowledge, so too do contradictory and autonomous forms of logic as opposed to expected compromises. Moreover, it is critical to recognise that actors involved in the process might themselves react. Those most affected by decisions must be able to protest, participate, propose solutions or even be absent. The current injunction to 'go down the participatory road' so as to produce frameworks, standards and codes is liable to deprive the process of 'the continual challenge needed during the implementation of the approach' (Barnaud 2013).

So-called 'participatory' research approaches satisfy a demand to establish a link between research and action, theory and practice, and the logic of the researcher and the researched (Anadòn 2007). But why claim that research is also action? This claim reminds us of a belief: 'the idea' that science is not an entity that responds autonomously to its own ends. Trying to reconnect science with 'society', by breaking out of the framework of 'confined research' (Callon 2001) brings meaning to science. By recognising that the knowledge produced by scientific research plays a part in but does not define a situation, (Stengers 2013) makes it possible to imagine that its objects and results can enter a relationship with the 's' of the social world.

12.5.2 Conclusion, Questions and Perspectives

The GAP project has made meetings possible between skippers, scientists and ship owners of European fleets. Meeting occasions were an opportunity to raise different concerns and points of view and resulted in questioning predetermined views regarding the use and management of FADs. Such meetings were an opportunity to confront what the actors said they were, with what others thought they were. Different actors' views were challenged by other actors. For instance, scientists could explain that even if they play a role in the management process by giving their scientific advice, they are also far off the political decision-making. Management authorities do listen to scientific advice, but they do not necessarily follow them. Skippers expressed their fears with regard to the growing administrative burden of monitoring. Ship-owners gained awareness that they might not really represent fishermen's interests. In other words, key to the participatory process is spending time on deconstructing and reconstructing views that actors have of their counterparts.

Generally, research projects are framed by the prerequisites of administrative and funding requirements. In order to draw up a framework that guides a project, forecasts and budgets are required at the stage of conception itself. Defining objectives, participants, expected results and deliverables are a prerequisite for a project to be deemed coherent and legitimate for funding. This leaves little room for learning anything other than what was planned at the outset. The context of the European GAP project was a little different. A certain freedom was given, as well as time for reflection, once the intermediate deliverables had been submitted. It is interesting to reflect on how the project evolved. From being a project dominated by natural science, a sociological dimension was included. Social scientists are interested in understanding what constitutes the diversity and multiplicity of points of view, interests, forms of compromise and processes that help social groups evolve towards compromises. They focus on understanding how socio-economic institutions and organisations construct knowledge and understandings of phenomena.

Analysing the problem in its complexity ultimately served objectives whose utility were not necessarily clearly understood. What emerged was that the circulation and control of information was a source of power. A participatory approach enabled actors to discuss issues that were controversial. It nevertheless raises the question of the purpose of a participatory approach if the actors have not decided together to discuss the subjects that cause tensions. The difficulties of mobilising around the project showed that the exchanges on what is, what we do and what we say, are not always a priority. There is still a long way to go before actors assert their new positions in actual action.

The French and Spanish groups of scientists, skippers and ship-owners, who met occasionally, are eager to consolidate the participatory process by trying to translate the debates that took place and decisions that were made into durable operational tools. It is up to the French industrial partner to articulate its interest 'to reinforce the exchanges between scientists and skippers in an organised format'. If they do, the inter-professional meetings could become a biennial or even annual event. Also under discussion is a project to start a newsletter. When the French industrial partner announces that 'the wheels must be kept in motion' and that 'it is a project and a process, or a process project', it also raises the question of the viability of the group. It is necessary that viability emerges from an evolving process and that it is based on cohesive work that is still to be carried out. Creating a newsletter, an idea that was suggested by participants at an inter-professional meeting, is now proving to be difficult because of the small number of people involved. 'It takes time, and this action has no immediate, direct impact for anyone. That is, it does not provide more fish or more publications. Everyone thus makes their choice in relation to their priorities, which means in relation to the time they have and what is urgent', concluded one scientist. Although the fruit of the work articulated in this case study is limited, it nevertheless has resulted in the production of operational tools that provide 'a test whose success can play a part in broadening the circle of people interested, and in going beyond the limited lifespan' of the project (Petit 2008).

In conclusion, the choice of restricting the actors in GAP to just scientists, fishing vessel companies and skippers must be questioned. One of the ship-owners expressed interest in understanding the work of NGOs and how they perceived tuna fishing. The action research process also highlighted the fact that most French skippers demanded regulations that would limit the number of FADs. Nevertheless, scientists and fishing vessel companies have not been able to respond to these skipper's concerns. This illustrates the current omerta (agreed silence) that surrounds the management of FADs. Environmental issues are often claimed as a priority, while economic and political constraints have not been decoded. Certain actors have been engaged in the process, but not all. To understand the limits of the project, it is also necessary to understand who are engaged by the process. Including a larger number of stakeholders would certainly have increased the number of views. Whether this added diversity would have resulted in more effective participation in the GAP approach remains uncertain, however.

Acknowledgements To all the persons that contributed towards this contribution, closely or remotely. From scientists to skippers, shipowners, Orthongel, the reviewer and editors.

References

Airaud M (2014) Une recherche-action auprès d'acteurs de la pêcherie thonière tropicale à la senne : méthodologie et perspectives. Communication at the séminaire jeudi du CRH, Sète

Anadon M (2007) La recherche participative, multiples regards. Presse de l'Université du Québec, Québec

Audoux C, Gillet A (2011) Recherche partenariale et co-construction de savoirs entre chercheurs et acteurs: l'épreuve de la traduction [Electronic version]. Revue Interventions économiques 43

Barnaud C (2013) La participation, une légitimité en question. Natures Sciences Sociétés 21:24–34

Beuret JE, Pennanguer S, Tartarin F (2006) D'une scène àl'autre, la concertation commeitinéraire. Natures Sciences Sociétés 14:30–42

Blondiaux L, Sintomer Y (2002) L'impératif délibératif. Politix 15:17–35

Callon M (1986) Eléments pour une sociologie de la traduction. La domestication des coquilles Saint-Jacques et des marins-pêcheurs dans la baie de Saint-Brieuc. L'Année Sociologique 36:169–208

Callon M, Lascoumes P, Barthe Y (2001) Agir dans un monde incertain, essai sur la démocratie technique. Seuil, Paris

Catanzano J, Rey H (1997) La recherché halieutique entre science et action :réflexions sur fond de crise. Natures Sciences Sociétés 5(2):19–30

Dagorn L, Holland KN, Restrepo V, Moreno G (2012) Is it good or bad to fish with FADs? What are the real impacts of the use of drifting FADs on pelagic marine ecosystems? Fish Fish 14:391–415

Davies TK, Mees CC, Milner-Gulland EJ (2014) The past, present and future use of drifting fish aggregating devices (FADs) in the Indian Ocean. Mar Policy 45:163–170

Delbos G, Jorion P (1984) La transmission des savoirs. Editions de la Maison des sciences de l'homme, Paris

Dempster T, Taquet M (2004) Fish aggregating device (FAD) research: gaps in current knowledge and future directions for ecological studies. Rev Fish Biol Fish 14(1):21–42

Dubar C (1992) Formes identitaires et socialisation professionnell. Rev Fr Sociol 33(4):505–529

Faugère E, Navarrete M, Charles M, Etienne M, Fauriel J, Lasseur J, Lécrivain E, Napoléone M, Paratte R (2010) Des connaissances scientifiques en quête de connaissances d'acteurs. Natures Sciences Sociétés 18:395–403

Fonteneau A, Chassot E, Bodin N (2013) Global spatio-temporal patterns in tropical tuna purse seine fisheries on drifting fish aggregating devices (DFADs): taking a historical perspective to inform current challenges. Aquat Living Resour 26:37–48

Fukofuka S, Brogan D, Itano D (2004) The development, design and current status of anchored and drifting FADs in the WCPO. In: 17th standing committee on Tuna and Billfish. Majuro, Republic of the Marshall Islands, 11–18 August 2004, INF-FTWG-3

Hall M, Roman M (2013) Bycatch and non-tuna catch in the tropical tuna purse seine fisheries of the world. FAO Fisheries and Aquaculture Technical Paper No. 568

Hallier J, Gaertner D (2008) Drifting fish aggregation devices could act as an ecological trap for tropical tuna species. Mar Ecol Prog Ser 353:255–264

Lascoumes P (2001) La productivité sociale des controverses. Dissertation, séminaire Penser les sciences, les techniques et l'expertise aujourd'hui, Paris
Lazuech G (2014) Comment gérer les ressources marines? Les jeux d'acteurs autour de la PCP. Revue économie rurale, Nov–Dec, 2014
Mackinson S, Neville S, Raicevich S, Worsøe Clausen L (eds) (2008) Good practice guide to participatory research between fisheries stakeholders and scientists. In: GAP project deliverable, vol 1, p 23
Marsac F, Fonteneau A, Menard F (2000) Drifting FADs used in tuna fisheries: an ecological trap? In: Le Gall JY, Cayré P, Taquet M (eds) Pêche thonière et dispositifs de concentration de poissons, vol 28. Ifremer, Actes de colloques, pp 537–552
Maufroy A, Bez N, Kaplan D, Delgado de Molina A, Murua H, Chassot E (2014) How many fish aggregating devices are currently drifting in the Indian Ocean ? Combining sources of information to provide a reliable estimation. Indian Ocean Tuna Commisssion, IOTC-2014-WPTT16–21. 2014
Moreno G, Dagorn L, Sancho G, Itano D (2007) Fish behaviour from fishers knowledge: the case study of tropical tuna around drifting fish aggregating devices (DFADs). Can J Fish Aquat Sci 64:1517–1528. https://doi.org/10.1139/F07-113
Olivier de Sardan JP (1993). Le développement comme champ politique loca, Bulletin de l'APAD [Electronic version], 6
Parrini-Alemanno S (2005) La recherche-action en communication des organisations est-elle une méthode qualitative constructiviste?, Recherche qualitative et production de savoirs, Hors-Série n°1
Pestre D (2011/1). Des sciences, des techniques et de l'ordre démocratique et participatif», Participations, 1, 210–238
Petit S, Fleury P, Michel V, Mougenot C (2008) Raconter la recherche-intervention. Retour sur trois opérations de gestion de la biodiversité. Natures Sciences Sociétés 16:326–336
Sebillote M (2007) L'analyse des pratiques :réflexions épistémologiques pour l'agir du chercheur. In: Anadon M (ed) La recherche participative, multiples regards. Presse de l'Universitédu Québec, Québec
Stengers I (2013) Cultiver une déloyauté envers ceux qui nous gouvernent, entretien réalisé par Michail Maiatsky, from www.contretemps.eu
Taquet M, Blanc M, Dagorn L, Filmalter JD, Fonteneau A, Forget F, Gaertner JC, Galzin R, Gervain P, Goujon M, Guillotreau P, Guyader O, Hall M, Holland K, Itano D, Monteagudo JP, Morales-Nin B, Reynal L, Sharp M, Sokimi W, Tanetoa M, Yen Kai Sun S (2011) DCP employés par les pêcheries artisanales et industrielles : une question d'échelle. L'utilisation et le développement technique des DCP au coeur de la conférence de Tahiti sur les DCP / Artisanal and industrial FADs: A question of scale. Tahiti conference reviews current FAD use and technology. Fish Newsl 136:35–45

Chapter 13
From Planning for Society to Planning with Society. Integration of Coastal Fisheries into the Maritime Spatial Planning

Robert Aps, Mihhail Fetissov, and Madli Kopti

Abstract The aim the GAP project is to promote and enable processes for open and effective participation of stakeholders in research and management. This case study builds on mutual learning as a basic principle of transdisciplinarity that incorporates the processes, methodologies, knowledge and goals of stakeholders from science, industry and politics. This paper addresses mutual learning as an important element of the Maritime Spatial Planning processes. A step-by-step approach towards collaboration is based on understanding the other side's thinking, focusing on shared interests and looking for solutions to common problems. It is technically supported by the participatory Geographic Information System (GIS)-based mutual learning methodology. The results of integrating fisheries into the real process of MSP in Estonia are presented and discussed.

Keywords Maritime spatial planning · Mutual learning · Baltic Sea · Stakeholder participation · Mapping

13.1 Introduction

Maritime spatial planning (MSP) is a tool that has been gaining importance for supporting the management of activities at sea. The European Union's (EU) road map for MSP (EC 2008a) aims at improving decision-making and providing a framework for arbitrating between competing human activities and managing their impact

---

R. Aps (✉) · M. Fetissov
Estonian Marine Institute, University of Tartu, Tallinn, Estonia
e-mail: robert.aps@ut.ee

M. Kopti
Estonian Maritime Academy, Tallinn, Estonia

P. Holm et al. (eds.), *Collaborative Research in Fisheries*, MARE Publication Series 22, https://doi.org/10.1007/978-3-030-26784-1_13

on the marine environment. The ultimate objective relates to balancing sectoral interests to achieve sustainable use of marine resources.

The roadmap was made concrete by the EU Directive establishing a framework for MSP (EC 2014) and defining objectives of maritime spatial planning (MSP) as follows: 'When establishing and implementing maritime spatial planning, Member States shall consider economic, social and environmental aspects to support sustainable development and growth in the maritime sector, applying an ecosystem-based approach, and to promote the coexistence of relevant activities and uses. Through their maritime spatial plans, Member States shall aim to contribute to the sustainable development of energy sectors at sea, of maritime transport, and of the fisheries and aquaculture sectors, and to the preservation, protection and improvement of the environment, including resilience to climate change impacts'. MSP is also seen as one of the central elements to the smart management of the Baltic Sea fisheries (Aps and Lassen 2012).

According to Ehler and Douvere (2007), MSP is first of all '… a public process of analysing and allocating the spatial and temporal distribution of human activities in marine areas to achieve ecological, economic, and social objectives that are usually specified through a political process'. In the same paper, the authors talk about the concept of 'ecosystem based sea use management', referring to the management of human uses of marine resources, including the use of marine space in such a way that ecological, social and economic objectives are achievable.

This case study applies mutual learning as a basic principle of transdisciplinarity, which incorporates the processes, methodologies, knowledge and goals of stakeholders from science, industry, and politics. Hirsch-Hadorn et al. (2008) state that transdisciplinary research is necessary when knowledge about a societally relevant problem is uncertain, when the concrete nature of problems is disputed, and when there is a great deal at stake for those concerned and involved in investigating them. Scholz (2000) argues that transdisciplinarity aspires to make the change from research for society to research with society, while the mutual learning can be conceived of as the adaptation process inherent in the interaction and joint problem-solving between science and society.

The first pilot of MSP for the Pärnu Bay (Gulf of Riga, Baltic Sea) was developed by the Baltic Sea Regional Programme 2007–2013 BaltSeaPlan Project (Martin et al. 2012). The outcomes of BaltSeaPlan were fed into the initial stages of GAP as a way to find possible solutions to the problem of integrating fisheries into the process of MSP with regard to place-based management of the spring spawning Baltic herring (Clupea harengus) fishery in the Pärnu Bay (Kopti et al. 2011). The outcome of BaltSeaPlan was also used to evaluate the role of MSP in promoting sustainable economic development including tourism and recreation in the Pärnu Bay sea area (Martin et al. 2013).

The question of importance to us here is how to increase collaboration among MSP negotiating stakeholders. According to our working hypothesis, it was expected that a possible move from a primarily positional negotiation format to a more collaborative interest-based negotiation format, underpinned by mutual learning, would more likely enable MSP stakeholders to reach acceptable and consensual agreements on marine space allocation.

13.2 Case Study Area and Context

Pärnu County's sea area (Fig. 13.1) is characterised by intensive human use as well as sensitive environmental conditions. In addition to traditional uses such as shipping, fishing and recreation activities, there are also plans to use the sea for offshore wind park development. The local small-scale fishery lands up to 80% of Estonian total small-scale fishery catches. In 2014, Pärnu County's small-scale fishery was the source of income to 396 professional fishermen who landed 7472.4 tonnes of fish. The main fish landed was spring spawning Baltic herring (Clupea harengus) which accounted for 5716 tonnes of the total catch landed and a first sale value of approximately 1 million euros.

The GAP Project's Baltic Sea MSP case study was initiated by the project's science partner, the Estonian Marine Institute, University of Tartu, and stakeholder partners, the Estonian Fishermen's Association and the Association of Fisheries of the Gulf of Liivi. The main objective of the case study was:

1. To collaboratively identify and map Pärnu County's small-scale fisheries and its problems
2. To further develop salient (relevant and timely), credible (authoritative, believable and trusted) and legitimate (developed in a process that considers the values

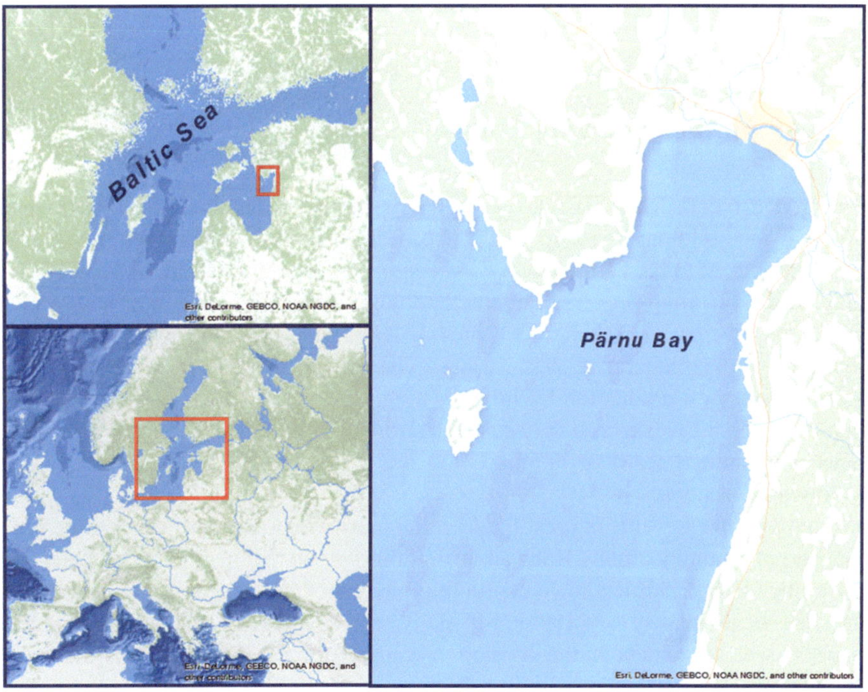

Fig. 13.1 Geographical location of Pärnu Bay and Pärnu County's sea area

and interests of all relevant stakeholders) arguments to promote the small-scale sector as a means to balance environmental, economic and social interests in the MSP process

3. To build the capacity of the small-scale fishermen and other stakeholders for informed interest-based and collaborative participation in the process of MSP

The Government of Estonia initiated the MSP process for Pärnu County's sea area in October 2012. The GAP Project's Baltic Sea case study was actively contributing to mutual learning events carried out by the Pärnu County government and the planner (contracted private company Hendrikson & Ko) in Pärnu town and local communities concerned. These events involved representatives of the Estonian Ministry of Interior (responsible for planning in Estonia at that time), the Ministry of Rural Affairs, the Ministry of the Environment, the Ministry of Economic Affairs and Communications, representatives of local communities and environmental NGOs, as well as important stakeholder groups from (1) shipping and ports, (2) tourism (marinas, sand beaches), (3) recreation (surfing, jetting, yachting), (4) fisheries (commercial, recreational) and (5) wind park developers. The Association of Fisheries of Gulf of Liivi and the Estonian Fishermen's Association actively participated in and contributed to the mutual learning process. The results derived from this collaborative mutual learning events contributed directly to further development of consensual planning solutions at the Pärnu County Government level.

13.3 Methodology

13.3.1 Mutual Learning Process

The GAP Project's Baltic Sea case study has developed and actively used the mutual learning methodology – step-by-step approach – towards collaboration that has been introduced to the Pärnu County's sea area MSP stakeholders at the very beginning of the planning process. This methodology is actually used to structure and guide the process of stakeholder's mutual learning events and the MSP-related interest-based collaborative negotiations. The aim is to raise the quality of the science-policy co-production by strengthening networking between science, stakeholders and policymakers so they can learn from each other's experiences and practices. The mutual learning approach towards collaboration is comprised of the following five steps.

Step 1. Move towards interest-based collaborative negotiations. Negotiation is at the core of mutual learning. Rahwan et al. (2003) argue that negotiation is essential in settings where partners have conflicting interests, but yet a desire to cooperate, and that the process of negotiation is usually seen as a type of interaction in which partners seek agreement on the division of scarce resources, while each partner tries to maximise his/her share or utility. Rahwan et al. (2003) define the negotiation position of a partner in terms of the resource(s) that a partner wants to acquire from his/her negotiation counterpart, while a partner's negotiation interests reflect his/her underlying goals in terms of use of the resources. The drawback of positional negotiation is that the dialogue between participants is focused on their negotiating positions. On the other hand, the interest-based negotiation format allows negotiators to exchange additional information and correct misconceptions during interaction (partners may argue about each other's beliefs and other mental attitudes in order to justify and influence their negotiation positions).

Fisher and Ury (1991), authors of the classic 'Getting to Yes', argue that 'the basic problem in a negotiation lies not in conflicting positions, but in the conflict between each side's needs, desires, concerns, and fears…. Such desires and concerns are interests'. However, Panzarasa and Jennings (2001) suggest that a key problem in most real-world negotiations is the uncertainty and ambiguity of the information needed to reach an agreement. They further argue that real agents not only have limited cognitive ability, but they also have their own private interests, which are rarely perfectly aligned with the interests of the other agents with whom they need to interact.

Step 2. Understand the other side's thinking. Fisher and Ury (1991) point out that understanding the other side's thinking is not simply a useful activity that will help you solve your problem – their thinking is the problem. Furthermore, they argue that the '… conflict lies not in objective reality, but in people's heads. Truth is simply one more argument — perhaps a good one, perhaps not — for dealing with the difference. The difference itself exists because it exists in their thinking. Fears, even if ill-founded, are real fears and need to be dealt with. Hopes, even if unrealistic, may cause a war. Facts, even if established, may do nothing to solve the problem'.

Step 3. Focus on shared interests. Fisher and Ury (1991) go on to argue that in theory it is obvious that shared interests help produce agreement, while by definition, inventing an idea which meets shared interests is good for you and good for them. Three points are highlighted about shared interests that are worth remembering: (1) shared interests lie latent in every negotiation and they may not be immediately obvious, (2) shared interests are opportunities, not godsends – to be of use, you need to make something out of them – and (3) stressing your shared interests can make the negotiation smoother and more amicable.

Step 4. Look for solutions to common problems. Panzarasa et al. (2002) suggest that negotiation can be seen as a mechanism for finding a solution to a common problem in a collaborative manner. Panzarasa and Jennings (2001) argue that '… negotiation can be regarded as a transformation of commitments. When successful, negotiation generates an agreement on a joint plan. An agreement is a composite concept that reflects what practical judgments the agents have brought about via practical reasoning and to what extent the agents have compromised with one another over their own views and preferences'. Furthermore, an agreement reflects a joint commitment to acting in accordance with the agreed-upon plan – coming to an agreement thus transforms a joint commitment towards a state into a joint commitment to performing a plan for achieving that state.

Step 5. Apply the Participatory GIS-based Mutual Learning. Today both knowledge and reasoning tend to be physically and organisationally distributed on the

Web. Knowledge sources required for tasks reside in distributed subsystems while reasoning may also be distributed. Knowledge-rich systems today are distributed, have many users with different degrees of expertise and integrate many shared knowledge sources of varying quality (Gil 2011). According to Harvey (2009), the boundary objects play an important mediating role in interactions across information, geographic and social spaces and are subject to change. As a result, these objects are more stable in interactions because of their unique characteristics that connect as well as separate different interest groups in a network. Furthermore, it is stated (Harvey and Chrisman 1998) that GIS technology connects different social groups in the construction of new localised social arrangements, while the negotiation of differences between different groups is fundamental to the construction of GIS technology-based boundary objects.

Rinner (2001) provides the theoretical foundations for participatory GIS tools by introducing argumentation maps as an object-based model for geographically referenced discussions. Argumentation elements and geographic reference objects are defined by an argumentation map model (Rinner 2006) as independent entities distinguishing between reference objects that are part of the map and reference objects that are created by users, e.g., to mark a point location or highlight an area.

13.3.2 The MSP Meetings

The first MSP meeting was held in Pärnu on 7 March 2013. Twenty-three people participated from the Estonian Fishermen Association, Estonian Fund for Nature, Fisheries Information Centre, Pärnu County Government, Estonian Ministry of Agriculture, Estonian Environmental Board, Estonian Maritime Administration, Tahkuranna Local Community and Eesti Energia, Hendrikson & Ko (Pärnu County Marine Spatial Plan developer). The GAP project and the concepts of mutual learning and collaborative mutual-interest-based negotiations were introduced. The second MSP meeting was organised by the Pärnu County Government at the Saulepa Civic Centre close to Pärnu on 28 March 2013. The meeting was attended by fishermen representatives, local communities, local administrations and the Estonian Ministry of Interior. The floor was open to all stakeholders to express their views, concerns and interests. Afterwards the work continued in three workshops, namely discussing MSP visions and the interests of local fishermen. In addition to these multi-stakeholder meetings, a further eight specific fishery stakeholder meetings were conducted by the Pärnu County Government in cooperation with the GAP Project: four in Pärnu (5 June 2013, 6 December 2013, 28 April 2014 and 27 October 2014), one in Häädemeeste (5 June 2013), one in Kihnu Island (6 June 2013), one in Audru (7 June 2013) and one in Tõstamaa (7 June 2013).

The mutual learning approach to MSP was also discussed amongst the wider stakeholder community of the Baltic Sea region at the Baltic Sea Regional Advisory Committee (BS RAC) stakeholder meetings in Riga (5 May 2011) and Copenhagen (11–12 January 2012, 4 July 2013 and 29 October 2014).

13.4 Outcomes of the Mutual Learning Events

The main collective interests that fishermen identified and highlighted during the MSP mutual learning events conducted by Parnu County Government in 2013–2014 were:

1. Maintaining a good environmental status of the concerned marine area
2. Maintaining essential fish habitats so that fish could spawn, breed, feed and grow to maturity
3. Minimising the wind farm's possible impact on fish resources and the fishery
4. Balancing the use of marine space related to potential aquaculture and mussel farming
5. Maintaining and developing traditional coastal culture
6. Taking into account transboundary fisheries-related issues (e.g. the internationally regulated Gulf of Riga spring spawning Baltic herring stock shared by Estonia and Latvia)
7. Focusing on the Blue Growth objectives – environmental, economic and social sustainability of the local small-scale fishery

The following is a summary of small-scale fishermen's views and outcomes of mutual learning events and interest-based negotiations that were structured and guided according to suggested by GAP project – step-by-step approach towards collaboration. At that, the format of interest-based collaborative negotiations (mutual learning step 1) was introduced and accepted by all negotiating partners from the very beginning.

13.4.1 Good Environmental Status

It is important to point out that high-level political commitments are in place that recognise the importance of the fisheries. For example, reference is made to qualitative descriptors for determining good environmental status of the EC Marine Strategy Framework Directive (EC 2008b). In a course of dedicated mutual learning events, the focus was made on understanding the stakeholder thinking (mutual learning step 2) and on the shared interests (mutual learning step 3).

Accordingly, it is expected by fishermen that Pärnu County's marine area MSP process should ensure that the planned collective pressure of human activities is kept within levels compatible with the achievement of good environmental status. What is also expected is that the capacity of the marine ecosystem to respond to planned human-induced changes is not compromised and that it is possible to ensure sustainable use of marine goods and services, including the sustainable use of fishery resources.

It was reiterated, in a course on interest-based collaborative negotiations, that Pärnu County's small-scale fisheries are environmentally friendly. The general shared vision and conceptualisation of 'Pärnu Bay as a cradle of marine life including fish' were proposed, discussed and accepted. In a course of looking for solutions to common problems (mutual learning step 4), such a vision supported by participatory GIS-based mutual learning sessions (mutual learning step 5) was also used efficiently by fishermen in MSP-related stakeholder interest-based negotiations. As a result, fishermen's interests in maintaining the good environmental status of Pärnu County's marine area was taken into account and supported by all concerned MSP stakeholders.

13.4.2 Essential Fish Habitats

One of the ways of approaching fish stocks as a spatial resource could be to treat it as an Essential Fish Habitat (EFH) which is described as a subset of all habitats occupied by a species and is defined as those waters and substrate necessary for fish to spawn, breed, feed and grow to maturity (NOAA 1998). In the context of Baltic Sea fisheries management, the implementation of EFH in general and the assessment and protection of related spawning and nursery grounds/habitats of important commercial fishes in particular have attracted little attention so far.

The main concern of fishermen is to ensure through the MSP process that traditional fishing grounds and associated fishing opportunities are maintained. Further, fishermen suggest the need to protect fish spawning grounds in Pärnu County's marine area but they also said that doing so was not enough. In addition, they stressed it was necessary to protect fish larvae and fish during their early life stages, e.g. against increased mortality because of recreational jetting in the Pärnu Bay during the fish spawning season.

During mutual learning meetings, fishermen's and other stakeholder's knowledge and relevant scientific evidence on habitats occupied by fish (mutual learning steps 2 and 3) were drawn on a Pärnu County sea area map for further analysis and inclusion into the MSP. This exchange of stakeholder's knowledge is a real manifestation of the kind of shared learning that creates a belief that both fishermen and environmental NGOs should join efforts to protect the Baltic Sea EFH (Fig. 13.2). The outcome of these contributions to MSP (mutual learning steps 4 and 5) was concrete planning options for temporal and spatial restrictions related to jetting. In addition, due to the need to protect EFHs in Parnu County's marine area, temporal and spatial constraints have also been imposed on yachting and surfing which are major recreation sports in the summer.

13.4.3 Wind Farm Development

The importance of the spring and autumn spawning herring fishery for the local small-scale fishing community was discussed in relation to possible wind parks being established near the spawning grounds. The potential grounds for autumn

Fig. 13.2 Mutual learning and interest-based negotiation meeting. (Photo by Robert Aps)

herring spawning were mapped based on fishermen's knowledge and existing scientific evidence (Fig. 13.3). During one of the mutual learning meetings for Parnu County's MSP, several fishermen raised concerns with regard to the potential damage to fish and fish stocks likely to be caused by the construction, operation and decommissioning of wind farms (mutual learning steps 2 and 3). These concerns were discussed in necessary detail involving specialists in underwater noise and wind farm cabling-related electromagnetic fields. The extensive mutual learning involving public authorities and stakeholders resulted in constructive and collaborative dialogue between authorities, fishermen, scientists and wind farm developers.

What was also discussed were safety limits necessary for underwater noise and electromagnetic fields as set by the EC Marine Strategy Framework Directive qualitative descriptors for determining good environmental status. These were to be observed with reference to descriptor 11 – introduction of energy, including underwater noise – so that the marine environment was not adversely affected.

So, to summarise, the mutual learning meetings facilitated collaborative and productive discussions between authorities, fishermen, scientists and wind farm developers (mutual learning steps 4 and 5). All agreed to take time to study the possible impact and disturbance of a planned wind farm areas during the construction, operation and decommissioning phases and to secure funding for studies needed to understand the adverse impact of wind farms on fish spawning grounds and fish migrations. However, despite such initiatives, this has not stopped the possibility of allocating of potential sea area for wind farm development during the MSP process.

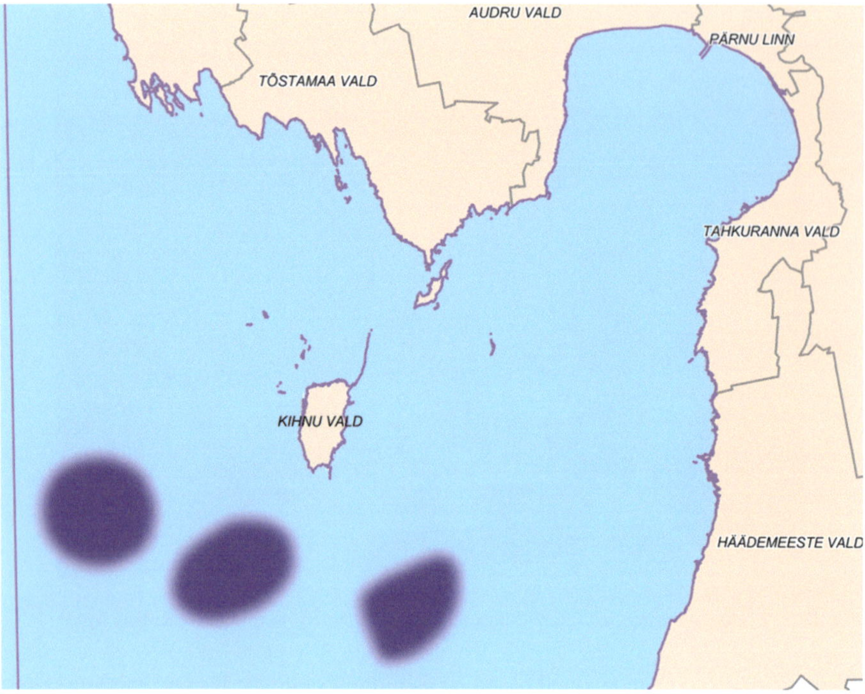

Fig. 13.3 Spatial visualisation of the herring spawning grounds (autumn spawners) in Pärnu Buy based on fishermen's and scientist's knowledge

13.4.4 Aquaculture and Mussel Farming

The issue of allocating sea space for aquaculture and future mussel farming ventures was also discussed. Although at present no business interests exist in these areas, a lively argument took place (mutual learning step 2). Allocating marine space for mussel-feed farming so as to clean the water and produce high-quality mussel biomass for feed production (in line with the EU Blue Growth aspirations) is an emerging issue within Pärnu County's marine area MSP process. Mussel-feed farming, by counteracting eutrophication, contributes to a process of closing the nutrient loop. However, mussel farms would compete for space with coastal fisheries, recreation and tourism (mutual learning step 2). Due to the low salinity of Pärnu County's marine area waters and the voracious filter feeding ability, the zebra mussel (*Dreissena polymorpha*) was chosen as the most suitable candidate for the mussel-feed farming pilot. Discussions were supported by the latest scientific research. Oganjan and Lauringson (2014) recently studied the relationship between the grazing rates of an invasive bivalve *Dreissena polymorpha* and the ambient environmental factors in the turbid eutrophic Pärnu County's marine area waters.

The possible allocation of marine space for mussel farming is based on the principle of spatial efficiency and co-use of marine space (mutual learning steps 3 and

4). Such co-use of space already exists in marine areas characterised by high standardised abundance of *Dreissena polymorpha* and where seasonal coastal spring spawning and autumn spawning herring trap net fisheries exist simultaneously (Fig. 13.4). Fishing, as is the case with mussel-feed farming, counteracts eutrophication by removing considerable amounts of nutrients through fish landings – e.g. the content of N and P in the Baltic Sea herring is 2.4% and 0.43% of the wet biomass (Hjerne and Hansson 2002).

The potential wind park sea area could also be co-used for future mussel-feed farm installations (Aps et al. 2014). However, the technical suitability of this option still needs to be studied in further detail (mutual learning steps 4 and 5). If necessary, there is sufficient sea area available that can be used even for large-scale mussel-feed farming in Pärnu County's marine area. Moreover, discussions on actual planning solutions for spatial and temporal allocation of sea area for aquaculture and mussel farming have been postponed because of insufficient interest expressed by potential mussel farming developers.

13.4.5 Traditional Coastal Culture

One of the most appealing ideas to stakeholders was to establish the Kihnu Strait Marine Park with the aim of protecting sensitive nature and fish spawning grounds and preserving the unique coastal culture and history (mutual learning steps 2 and 3). Kihnu is the largest island in the Gulf of Riga and the seventh largest of Estonia's islands. First mention of Kihnu Island was made in 1386 (as Kyne) and of its people in 1518. Evidence exists, however, that that Kihnu was inhabited by seal hunters and fishermen 3000 years ago.

Today, Kihnu Island is home to a community of 600 people. UNESCO proclaimed Kihnu's culture and traditions as a Masterpiece of the Oral and Intangible Heritage of Humanity on 7 November 2003. Kihnu museum hosts exhibits about the history of the island and the life and times of the famous local captain, Kihnu Jõnn (Fig. 13.5). However, concern has been expressed that establishing Kihnu Strait Marine Park may result in overly stringent constraints being imposed on local fishermen – the reason the idea of a marine park has not been sufficiently supported up to now. It was decided (mutual learning step 4) that discussions on the declaration of a marine park would continue with the aim of finding a consensual solution.

13.4.6 Transboundary Issues

Estonia as a Contracting Party to the Convention on Environmental Impact Assessment in a Transboundary Context, the Espoo Convention (UN 1991), is either individually or jointly taking all appropriate measures to prevent, reduce and control adverse transboundary environmental impacts of proposed or planned activities.

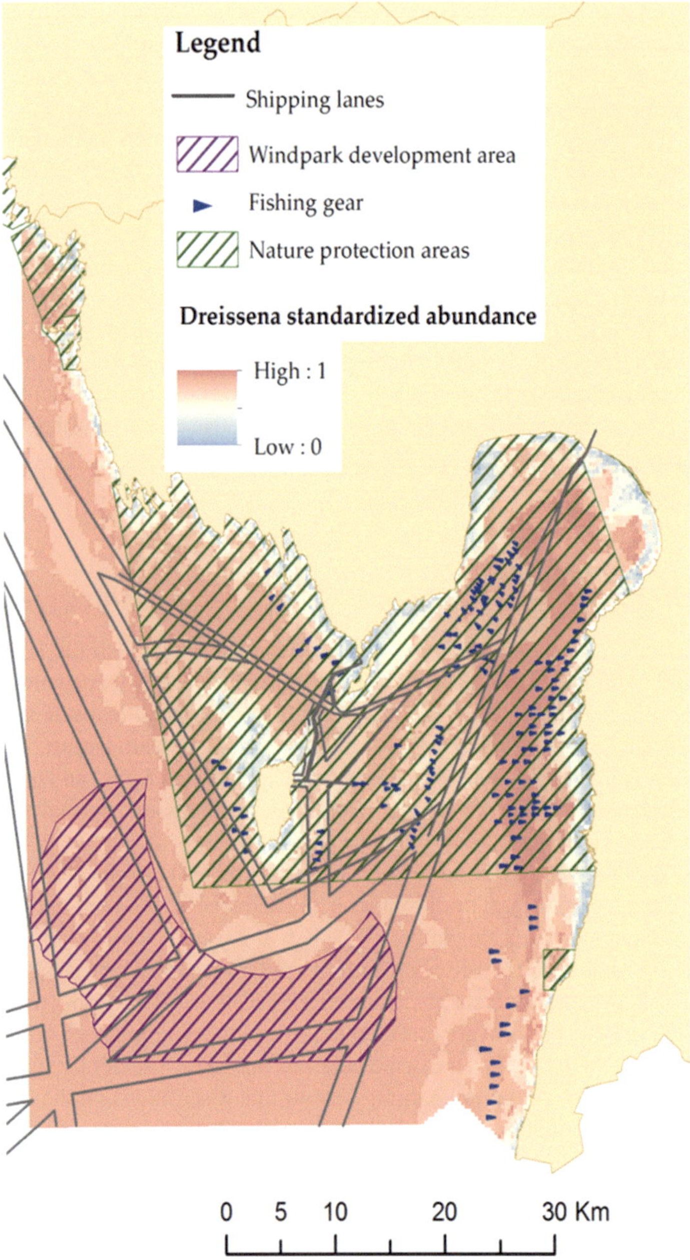

Fig. 13.4 Spatial distribution of Dreissena standardised abundance against the background of shipping lines, planned wind park development areas, fishing gear positions and nature protection areas. (Aps et al. 2014)

Fig. 13.5 Left: Unique coastal culture and history of Kihnu island's inhabitants. Right: Famous local captain Kihnu Jõnn. (Photos by Robert Aps)

Gulf of Riga herring is a separate herring population and part of the international stock management unit of Baltic Sea herring (Clupea harengus). According to the International Council for the Exploration of the Sea (ICES 2014), the Gulf of Riga herring fishery is exploited by both Estonia and Latvia, using trawls and trap nets. In recent years trap net fishing has accounted for about 30% of all fishing. Therefore, when planning and allocating the herring fishery in Pärnu County's marine area, relevant transboundary issues and regulations should be taken into account.

In 2011–2013, total Gulf of Riga herring landings were, respectively, 35,024 tonnes, 31,733 tonnes and 30,360 tonnes. The total herring landings, on the other hand, from Pärnu County's marine area fisheries were 7281 tonnes, 5925 tonnes and 5675 tonnes respectively. Based on latest information from ICES (2014), total Gulf of Riga herring landings are below the Maximum Sustainable Yield (MSY) level. Moreover, from a precautionary approach point of view, the herring stock is harvested sustainably. Herring catches in the Gulf of Riga include the local Gulf herring and the open-sea herring, which enters the Gulf of Riga for spawning.

During a course of mutual learning events carried out between 2013 and 2014, scientists, fishermen and concerned authorities discussed the status of the Gulf of Riga herring stock and related fishing opportunities (mutual learning steps 2 and 3). Fishermen's demands to integrate protection of the internationally regulated Gulf of Riga herring stock's Essential Fish Habitats into maritime space planning solutions has been well received and taken into account (step 4) in the second MSP draft presented in 2014 (Fig. 13.6).

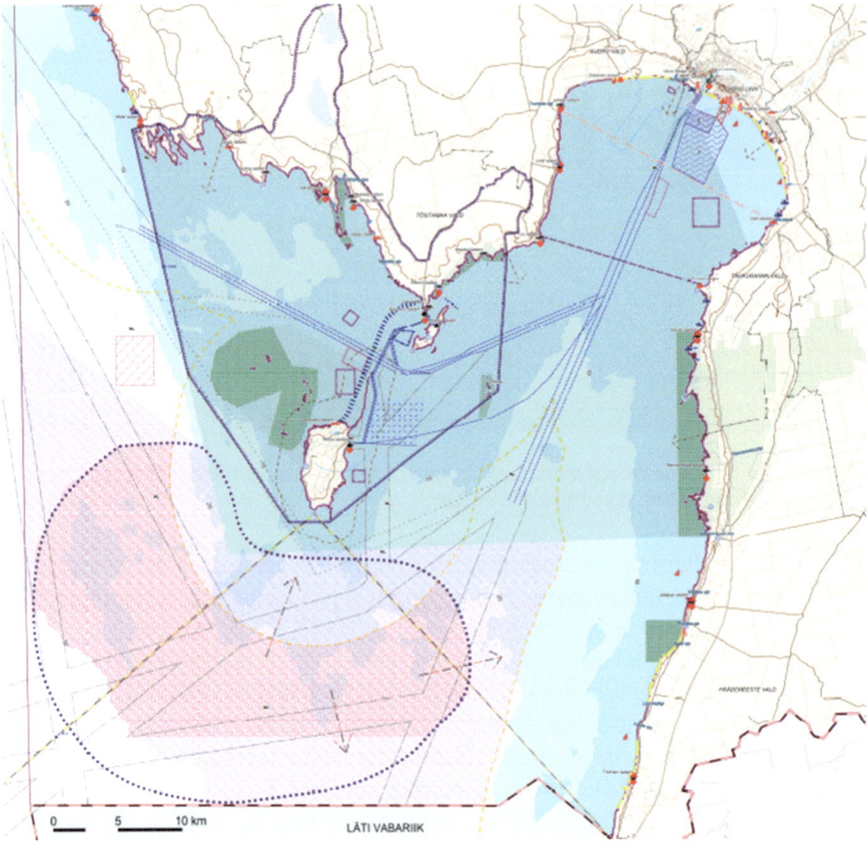

Fig. 13.6 General view of the draft MSP for Pärnu County's marine area, autumn 2014. (Pärnu County Government)

13.4.7 *Blue Growth*

An EU Directive establishing a framework for maritime spatial planning (EC 2014) defines the objectives of maritime spatial planning as follows: 'When establishing and implementing maritime spatial planning, Member States shall consider economic, social and environmental aspects to support sustainable development and growth in the maritime sector, applying an ecosystem-based approach, and to promote the coexistence of relevant activities and uses'. It is underlined that through their maritime spatial plans, Member States shall aim to contribute to the sustainable development of energy sectors at sea, of maritime transport, and of the fisheries and aquaculture sectors, and to the preservation, protection and improvement of the environment, including resilience to climate change impacts. Importantly, the EU Directive establishing a framework for maritime spatial planning (EC 2014) is providing the platform for Blue Growth (EC 2012) realisation.

As a first practical step towards realisation of Blue Growth objectives, Pärnu County's marine area draft spatial plan has already linked the fishing grounds and fishery to the network of small-scale fish landing ports where the first sale of fish takes place (mutual learning steps 2 and 3). It is expected (mutual learning step 4) that linking fishing, fish landing and fish markets in the ongoing process of MSP would contribute to the economic efficiency of fisheries through improved mutually beneficial collaboration between fishermen and fish markets representatives.

13.5 Conclusions

The mutual learning methodology, a step-by-step approach towards collaboration developed by the GAP case study, has been successfully used to structure and guide a genuine application of multi-stakeholder collaboration (and negotiation) on MSP. In particular, it provided a mechanism to formulate fishermen's collective interests in an applied context that was meaningful to them.

The draft copy of the MSP for Pärnu County's marine area was presented and discussed at the end of 2014. It included collaboratively developed planning solutions for most fishery-related conflicts identified at the beginning of the MSP process including (1) spatial and temporal allocation of the yachting sea area, (2) exit corridors for surfers to access the open area of Pärnu County's marine space and (3) spatial and temporal allocation/prohibition of jetting sea areas to protect fish larvae during the spawning period.

One of the decisions arising from the process was that more scientific evidence on fish migration and spawning grounds is needed before developing final decisions for the allocation of wind parks. Acceptable technical parameters of wind parks were also deliberated upon. Discussions on actual planning solutions for spatial and temporal allocation of the sea area for aquaculture and mussel farming were postponed because of insufficient interest expressed by potential fish and mussel farming developers.

The fisheries science and stakeholder partnership created by the GAP project's Baltic case study will continue to facilitate the integration of fisheries into the process of MSP in Estonia. This integration will include addressing transboundary issues pertaining to the internationally regulated Baltic herring, sprat, cod and salmon fisheries. Further collaborative work in support of the integrated planning of spatial development of the Estonian coastal zone and marine waters will be largely focused on the development and practical implementation of the Quality Management System for Estonian MSP processes.

References

Aps R, Lassen H (2012) Towards smart management of the Baltic Sea fishery. In: Intilli JS (ed) Fishery management. Nova Science Publisher Inc, New York, pp 97–107
Aps R, Kotta J, Lauringson V, Oganjan K, Pärnoja M, Kartau K (2014) Planning marine space for mussel farming with focus on use of scientific knowledge – Estonian Case Study. ICES CM 2014/3654 K:02
EC (2008a) Communication from the commission: roadmap for maritime spatial planning: achieving common principles in the EU. COM (2008) 791 final, Brussels
EC (2008b) Directive 2008/56/EC of the European Parliament and of the council of 17 June 2008 establishing a framework for community action in the field of marine environmental policy (Marine Strategy Framework Directive). Off J Eur Union 164:19–40
EC (2012) Communication from the commission to the European Parliament, the Council, the European Economic and Social Committee and the Committee of the Regions. Blue Growth – opportunities for marine and maritime sustainable growth. COM (2012) 494 final, Brussels
EC (2014) Directive 2014/89/EU of the European parliament and of the council of 23 July 2014 establishing a framework for maritime spatial planning. Brussels
Ehler C, Douvere F (2007) Visions for a sea change. Report of the First International Workshop on Marine Spatial Planning. Intergovernmental Oceanographic Commission and Man and the Biosphere Programme. IOC manual and guides. UNESCO, Paris
Fisher R, Ury W (1991) Getting to yes: negotiating agreement without giving in. Penguin Books, New York
Gil Y (2011) Interactive knowledge capture in the new millennium: how the Semantic Web changed everything. Knowl Eng Rev 26(1):45–51
Harvey F (2009) Of boundary objects and boundaries: local stabilization of the polish cadastral infrastructure. Inf Soc 25(5):315–327
Harvey F, Chrisman N (1998) Boundary objects and the social construction of GIS Technology. Environ Plan 30:1683–1694
Hirsch-Hadorn G, Biber-Klemm S, Grossenbacher-Mansuy W, Hoffmann-Riem H, Joye D, Pohl C, Wiesmann U, Zemp E (2008) The emergence of transdisciplinarity as a form of research. In: Hadorn GH, Hoffmann-Riem H, Biber-Klemm S, Grossenbacher-Mansuy W, Joye D, Pohl C, Wiesmann U, Zemp E (eds) Handbook of transdisciplinary research. Springer Verlag, New York, pp 19–39
Hjerne O, Hansson S (2002) The role of fish and fisheries in Baltic Sea nutrient dynamics. Limnol Oceanogr 47(4):1023–1032
ICES (2014) Baltic Sea, herring in subdivision 28.1 (Gulf of Riga), advice for 2015. ICES Advice 2014, Book 8
Kopti M, Aps R, Fetissov M, Suursaar Ü (2011) Integration of fishery management into the process of Maritime Spatial Planning. In: Brebbia CA, Zubir SS (eds) Management of natural resources, sustainable development and ecological hazards III. Southampton, WIT Press, pp 183–194
Martin G, Aps R, Kopti M, Kotta J, Remmelgas L, Kuris M (2012) Towards a pilot maritime spatial plan for the Pärnu Bay. BaltSeaPlan report no 13
Martin G, Kotta J, Aps R, Kopti M, Martin K, Remmelgas L, Kuris M (2013) Marine spatial planning as a tool for promoting sustainable economic development including tourism and recreation in a coastal area: an example from Pärnu Bay, NE Baltic Sea. In: Brebbia CA, Beriatos E (eds) Sustainable development and planning VI. Southampton, WIT Press, pp 393–404
NOAA (1998) A primer for federal agencies, essential fish habitat: new marine fish habitat conservation mandate for federal agencies. National Marine Fisheries Service, Southwest Regional Office, Long Beach
Oganjan K, Lauringson V (2014) Grazing rate of zebra mussel in a shallow eutrophicated bay of the Baltic Sea. Mar Environ Res 102:43–50

Panzarasa P, Jennings NR (2001) Negotiation and joint commitments in multi-agent systems. Sozionikaktuell 3:65–81
Panzarasa P, Jennings NR, Norman TJ (2002) Formalizing collaborative decision-making and practical reasoning in multi-agent systems. J Logic Comput 12(1):55–117
Rahwan I, Sonenberg L, Dignum F (2003) Towards interest-based negotiation. In: Proceedings of the 2nd International Conference on Autonomous Agents and Multi-Agent Systems (AAMAS), Melbourne, Australia. ACM Press, New York, pp 773–780
Rinner C (2001) Argumentation maps: GIS-based discussion support for on-line planning. Environ Plan B Plan Des 28(6):847–863
Rinner C (2006) Argumentation mapping in collaborative spatial decision making. In: Dragicevic S, Balram S (eds) Collaborative GIS. Idea Group Publishing, Hershey, pp 85–102
Scholz RW (2000) Mutual learning as a basic principle of transdisciplinarity. In: Scholz RW, Häberli R, Bill A, Welti M (eds) Transdisciplinarity: joint problem-solving among science, technology and society. Workbook II: mutual learning sessions. Haffmans Sachbuch Verlag, Zurich, pp 13–19
UN (1991) UN Convention on Environmental Impact Assessment in a Transboundary Context. United Nations, New York

Chapter 14
Implementing the Landing Obligation. An Analysis of the Gap Between Fishers and Policy Makers in the Netherlands

Marloes Kraan and Marieke Verweij

Abstract The introduction of the landing obligation is a radical change of the European fisheries policy that has widened the gap between fishers and policy officers in the Netherlands. Especially the mixed demersal trawl fisheries have to adjust to this new measure, which requires a concerted effort between ministry, fleet and research institutions. This chapter describes the implementation process of the landing obligation in the Netherlands (between 2013–2015) and how it has been met with strong opposition by Dutch fishers. This chapter argues that such opposition stems not only from interests, but also from strongly held perceptions about the (ecological) consequences of the measure, as was clear from observing meetings and interviewing key actors. Although several meetings had been organised by the ministry to discuss the landing obligation with fishers, perception differences were not discussed explicitly. Discussions rather seemed to be parallel monologues where parties failed to meet a shared understanding. This situation jeopardises real cooperation in the preparation for full implementation of the landing obligation. The chapter ends with a discussion on possible ways to bridge the gap between fishers and policy officers.

Keywords Perceptions · Landing obligation · Discards · Trust · Collaborative research

M. Kraan (✉)
Wageningen Marine Research, IJmuiden, The Netherlands & Environmental Policy Group, Wageningen University, Wageningen, The Netherlands
e-mail: Marloes.kraan@wur.nl

M. Verweij
Wing, Wageningen, The Netherlands

P. Holm et al. (eds.), *Collaborative Research in Fisheries*, MARE Publication Series 22, https://doi.org/10.1007/978-3-030-26784-1_14

14.1 Introduction

The European Commission (EC) introduced a landing obligation in European fisheries in 2013. As a result, fisheries policy will not only focus on landed commercial catches, which have been at the centre of fisheries research and management for decades, but also on undersized catches and bycatch. This is a revolutionary change that has consequences for how fisheries are managed and studied and how fishing is practised. In particular, mixed demersal fisheries that have in the past discarded significant parts of their catch will be impacted by this measure.

In the Netherlands, the landing obligation is a serious challenge for the fishing industry because mixed demersal fisheries comprise a very large part of the Dutch fishing fleet (Catchpole et al. 2008). The precise social and economic impact of the landing obligation will differ for the various gear-species combinations in the fleet, and will depend on the details of how it is implemented. But clearly the most important Dutch demersal fishery, the 80 mm sole fishery in the North Sea (Centrum voor Visserijonderzoek 2019), of which the Netherlands holds 75% of the European Union (EU) total allowable catch, is confronted with a major challenge to solve the bycatch of undersized plaice and dab which is high and difficult to solve.

The GAP case study in the Netherlands first studied research cooperation between researchers of Wageningen Marine Research[1] and demersal fishers sampling their catches. Halfway through the project, the discard ban was introduced and had a direct impact on the dynamics of the case study.[2] All of a sudden discards assumed centre stage, becoming as it were a 'political commodity' for policy makers and fisher representatives by changing the technical approach to discards (getting good estimations on the amount of discards to improve stock assessments) to an ethical and political discussion (discarding as wasteful practice). The implementation of the discard ban (which was later relabelled as the 'landing obligation') at the national level was soon set in motion. Implementation was a fascinating process to study from a GAP perspective (see Chaps. 1 and 2), as it put existing (good) relations under pressure. The gap between industry, research and policy in the Netherlands was previously not so wide as a result of years of successful research cooperation. However, the landing obligation unveiled fundamental differences between the fishing industry and the ministry regarding the goal and consequences of the landing obligation and ideas about cooperation in policy implementation. The gap between these parties started widening.

Central to this chapter is an in-depth analysis of two meetings that were held in 2013 between the minister and fishers to discuss the landing obligation, which were studied within the context of a longer period of participant observation of meetings between 2013 and 2015 (see the methods section below). These meetings provided a great opportunity to observe the gap between the two parties. The context of the landing obligation and the different goals it aimed to serve are described in the third section. The analysis of the different perceptions of both parties is made in the

[1] At the time of the GAP project, Wageningen Marine Research was called IMARES.

[2] For a presentation of the case study see GAP (2019).

following section by making use of a theoretical lens on where perception differences are rooted. This chapter explains why the widened gap is problematic and ends with a discussion on possible ways to bridge the gap. The main contribution of our analysis is how relations between fishers and policy makers can be conducted in the future, leading to more effective fisheries management.

14.2 Methods

Two meetings in 2013 were observed, which can be seen as a form of interface ethnography – a particular kind of fieldwork in boundary areas where a relatively closed community or organisation presents itself to the public (Ortner 2010 in De Koning 2015:133). Permission to observe was requested from the organisers of the meetings and granted prior to the meetings. The observer's role was not made known to the wider audience to prevent influencing the nature of interactions. All statements made by the minister and the audience (mostly fishers, but also consultants, policy officers, local politicians, fisher representatives and scientists) were noted. The notes were then entered in a computer the same day, coded and analysed soon thereafter. The analysis of the perceptions and lines of causal reasoning was based on observations and notes, and was directed at what was said, which topics were addressed and how the discussion had taken place. In particular, attention was given to:

- what the atmosphere had been like;
- how the actors had engaged;
- how the minister defended the policy;
- how the minister tried to convince the fishers to cooperate with her and her ministry, and,
- how the fishers had responded.

The two reports of the meetings were analysed by using the computer program Atlas TI. In total 43 codes were used in the first round of coding (free coding), which were later clustered into 10 'families': metaphor, management, emotion (beleving, emotie), effects landing obligation, knowledge and perspective, fishers, gap, research, political strategy, principles and room to manoeuvre (rek en ruimte). After that, the number of words used per family-code in each meeting was used as a proxy for topics discussed most per meeting. Then the number of quotes and amount of words per code per meeting were assessed. Lastly, the code-family 'effects landing obligation' was assessed by looking at all codes within the family to see which effects were discussed most in both meetings. These queries of Atlas TI were used as tools to analyse the content of both meetings, in addition to a more qualitative analysis.

In addition to analysing these two key meetings, information was derived from (participatory) observation between 2013 and 2015 of other meetings between policy makers and the fleet, and of the project group 'Implementation agenda' dealing with the Landing Obligation. This project group was set up in 2013 and contained policy officers of the ministry, fisher representatives of the two main organisations (Visned and the Vissersbond), representatives of non-governmental organisations (NGOs) and scientists. Their goal was to discuss progress of the research projects undertaken by industry and the implementation process at the regional[3] and EU level. Meetings of the project group were held on a monthly basis. In addition, a number of key actors (fishers and policy makers) were interviewed (structured and semi-structured). People quoted in this chapter remain anonymous unless what they said has been quoted in published articles.

14.3 The Goal of the Landing Obligation

In June 2011, the EC put forward a proposal to introduce a discard ban. The main reason for this was that discards were seen as a wasteful practice (European Commission 2019a) and hence the discard ban was introduced as a measure to encourage fishers to avoid catching discards. The framing of the landing obligation has been very much influenced by a powerful media campaign, which started in the United Kingdom in 2010 by celebrity chef and food campaigner Hugh Fearnley-Whittingstall called the 'fish fight'(River Cottage 2019). The framing of his campaign was that discards are a 'senseless waste': fish is caught and thrown back (dead) in the sea 'because of crazy EU laws'. Since then, the European Council of Ministers and the European Parliament have been negotiating how best to reduce or even eliminate discards. In February 2013, the European Parliament and the European Council voted in favour of a landing obligation (Wageningen Marine Research 2019).

Discards have in fact for long been a 'collateral damage' of fishing operations. Fishers land what they can sell at the market given existing landing quotas. Whatever cannot be sold (no market) or may not be landed (by law) is discarded (see Box 14.1).

Box 14.1: The Main Reasons for Discarding

1. The catch is of little or no value. For example: we do not eat starfish; there is only a limited market for dab, which affects the price (supply/demand); there's no European market for jellyfish;
2. The catch is damaged;
3. The catch may not be landed because:
   (a) The fish are too small (there are legal minimum sizes);
   (b) Individual fishers have exhausted their quota for that species;
   (c) There are rules about the daily catch composition;
   (d) The fish is a protected species (e.g. sturgeon or certain species of shark);

(continued)

[3] The Scheveningen Group is the regionalisation platform of the group of member states bordering the North Sea: the Netherlands, Belgium, France, the United Kingdom, Denmark and Germany.

Box 14.1 (continued)

4. Fishers sometimes throw fish of a certain size back because other sizes are more profitable (particularly important if the quota are tight). This practice is known as high grading and is illegal (whereas reasons named under 1–3 were legal, at least until the enforcement of the landing obligation).

Source: Wageningen Marine Research (2019).

There thus are several reasons for discarding, all requiring different solutions. In the societal debate, there was not much attention for such a nuanced perspective. Rather the image of tonnes of 'good fish' being wasted by discarding (such as cod caught without having the quota to land it) became a symbol of the failure of fisheries management in the EU. The landing obligation was seen to be key in the reform of the common fisheries policy (CFP). The discourse of a failed CFP has been commonplace in scientific articles (i.e. Raakjaer 2011) as well as in the communiqués of NGOs (Ocean 2012, 2019) and politicians (i.e. European Commission 2009:8). The upcoming reform was the new window of opportunity for change (Salomon et al. 2014).

At the same time, also in the Netherlands, the landing obligation is presented as a 'tool' to increase selective fishing practices (Rijksoverheid 2014). By obliging fishers to bring discards to the shore, and by counting these as catch under their quota, fishers are 'stimulated to increase their selectivity' and catch less unwanted bycatch (Salomon et al. 2014:79).

For the implementation of the landing obligation, it became important to be more specific about the goal (is it to end wasteful practices or to increase selectivity?) and how best that goal can be achieved (causal reasoning about the effect of the landing obligation). Before we assess how this plays out in the Dutch context, we will first discuss some theoretical insights on perceptions.

14.4 Theorising on Different Perspectives

Fisheries management is a multi-stakeholder arena. Measures are introduced by policy makers while NGOs and fisher representatives try to influence decisions and scientists monitor catches, assess fish stocks and evaluate the effectiveness of measures (Verweij and Van Densen 2010:1145). Input given by different stakeholders often stems from different (ecological) perceptions towards the topic. These differences are often attributed to different interests. Simply put, fishers do not want to comply with the landing obligation because it will imply more work and less marketable catch, thus higher costs. This is also a legitimate concern and contributes to the perception and causal reasoning about the landing obligation, but it hides part of the story. Differences in perceptions between stakeholder groups and/or policy makers are rooted in differences in information capturing and processing that are structured by factors other than interests and functional positions alone (Verweij et al. 2010: 523). Interests do influence the capturing and processing of information (Hommes et al. 2009; van Densen 2001), but differences in perceptions between groups also originate from other characteristics that have been shaped by '(1) differences in their information environments and (2) through the individual and collective life histories of the group they are part of (education and experience)' (Verweij et al. 2010:523). Analysing the perceptions and lines of causal reasoning can be fruitful as it sometimes leads to unexpected breakthroughs in issues that appear to be unresolvable. Underlying principles or images, if they are not accounted for, can have a significant influence on processes of participation (Kooiman et al. 2005). Verweij et al. (2010) and Verweij and van Densen (2010) unravel different perceptions in North Sea fisheries. Perceptions are built up over the years and are based on and shaped by information from the information environment and knowledge (see Fig. 14.1).

The information environment is the environment in which people are confronted with data (facts that may or may not be useful) and information (data that has a form and content that is appropriate for a particular use), the latter which develops (through education and experience) to knowledge. The information environment can be visualised as a physical surrounding, which plays a role in the way information is presented. The information environment of a policy officer at the ministry, in a shared office, is completely different than that of a fisher on his vessel at sea; they thus have access to different sources of information. Discard information for a fisher

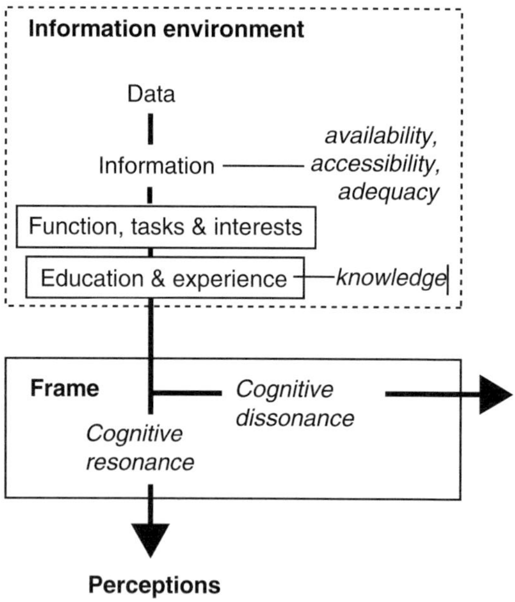

Fig. 14.1 The capturing and processing of information, leading to perceptions. (Verweij et al. 2010:523)

is what he sees every day on his vessel and which moves through his hands, whereas for a policy officer, it is a graph or table in a scientific report. There are also differences between groups and individuals in the way information is captured and processed, depending on education and life experience, leading to knowledge. This means that even individuals presented with the same type of information may process and understand this differently, which can also result in perception differences. Finally, how new information is dealt with also depends on whether it fits into a frame as a set of current beliefs or not. If it does, this leads to cognitive resonance, and the information then shapes (existing) perceptions. However, if the new information contradicts one's belief system, it results in tension (cognitive dissonance), which can be reduced by minimising, devaluing or even completely disregarding the information (Verweij et al. 2010: 523).

Based on the above theory, perception differences about discards and the effects of a landing obligation are not only because of interests, but also because of the different information environments and how this information is 'treated' internally. Where discards are concerned, this leads to different perspectives regarding the aim of a landing obligation and why this is important, and what the ecological and socio-economic consequences will be of the measure taken.

14.5 The Implementation of the Landing Obligation

During 2013 it soon became clear that the landing obligation would not be a total and absolute 'ban' on discarding (including benthos); it would only apply to commercial fish stocks subject to catch limits (i.e. quotas) or minimum landing sizes; it would be introduced gradually and there would be a number of exceptions, giving some room to manoeuvre (Salomon 2014:78). Despite it not being an absolute ban, the fishing industry generally opposed the measure and questioned its goal (EAPO 2013) (see further on).

It was decided that the landing obligation would be introduced gradually between 2015 and 2019 for all commercial fisheries in EU waters (European Commission 2019b). The pelagic fisheries were the first to introduce the obligation in 2015 and the other fisheries were to follow shortly after. In what order the landing obligation would be implemented was unclear from the outset. Would it be gradually introduced according to the type of fisheries (and if so, how would a fishery be defined) or by species? Moreover, how would the conditions for exemptions be decided upon?

Implementation of the landing obligation takes place at multiple geographical levels, with the final decision taken by the EU. Member states prepare the details of specific implementation measures in so-called discard plans while keeping in mind national, regional and international research and preparation trajectories. However, consideration of the regional level cooperation among member states has become increasingly important because the reformed CFP introduced 'regionalisation' as a new approach to improve fisheries management (European Commission 2019c). Regionalisation is the process through which all the member states that border a particular EU water body (such as the North Sea) may together formulate joint recommendations in terms of managing that particular water body.

14.5.1 *The Implementation Process in the Netherlands*

As soon as the landing obligation was decided upon in 2013, the Netherlands started preparing for implementation. The ministry created a project-group with representatives from fisheries, NGOs, research institutions and the ministry. It also announced a subsidy for research, allowing fishing companies and fisher organisations to bid for funding by submitting research proposals. The project-group (see 'Methods') would guide the research and monitor the progress of projects and would also discuss the implementation process at the regional level (Scheveningen group and the North Western Waters group, the two groups dealing with the two main seas in which Dutch vessels are active) and at the EU level.

The fishing industry opposed the landing obligation right from the start, partly for ecological reasons (later explained in Fig. 14.3). Before the landing obligation was decided upon, the joint fishing industry representative organisations in the EU (the European Association of Fish Producers Organisations, Europeche and Cogeca) had written an explanatory memorandum in which they stated the following: 'simply imposing a simplistic solution of an overall discard ban will not solve the real issue behind discarding which are complex and multi-faceted'. They also presented an alternative way they felt the discard ban could be implemented, with a focus on stock sustainability (EAPO 2013).

In addition, one of the two demersal fisher organisations, the Vissersbond, launched a campaign 'Landing Obligation NO!' in May 2013 (Visserijnieuws 2013). They published a manifesto stating why they were against the landing obligation for ecological and economic reasons (Vissersbond 2013). Fishers posted pictures on twitter of a banner with the slogan 'You should not bring young fish (to the shore), they should remain in the sea – Support us!' tied to their vessel (Fig. 14.2).

The other demersal fisher organisation, Visned, who did not run such a campaign, nonetheless agreed with the general message: 'we can't and will not comply' (Roos 2013).

Once the landing obligation was officially adopted, the fisher organisations continued to express their concerns to the Dutch government, but they also pragmatically joined the project group and applied for funding in order to carry out research to prepare for the landing obligation. They developed a cooperative research plan (named the houtskoolschets or charcoal sketch) via their cooperative platform the CVO.[4]

[4]CVO stands for Cooperatieve Visserij Organisatie (cooperative fisheries organisation). It was established in 2011 and is meant to have a platform for cooperation for the two Dutch fishers' organisations, as in many cases (starting with MSC certification processes), it makes a lot of sense to work together. Throughout the years the two fisher organisations have tried to become 1 fisher

Fig. 14.2 Vissersbond campaign banner tied to a fishing vessel. (Source: Nederlandse Vissersbond)

The project group decided to present the sketch to the fishers in a so-called harbour tour in the summer of 2013 in order to inform them about the landing obligation and invite them to share ideas as to how the Dutch fishing industry could best deal with the landing obligation. Three meetings were planned, in Urk (a prominent fishing village in the centre of the Netherlands), Den Helder (in the north of the Netherlands) and Stellendam (in the south). The meetings were organised by the Blueports[5] and prepared by the project group. In the first meeting held in Urk on the 31st of August, the fishers present (about 70) responded very emotionally, expressing their disbelief and despair:

> Why don't they just buy us out of business, then they are rid of us altogether!? (Fishtrend 2013:10)
>
> The discard ban is the end. We don't have anything to lose. Please buy us out of business if we aren't allowed to fish anymore. (Visserijnieuws 5-9-2013)

In the week following the first harbour meeting, the rest of the meetings were cancelled by the fisher organisations. Their decision to cancel these meetings was based both on the emotional response of many fishers at the first meeting as well as on a letter from the ministry that was received the day before the meeting (30-8-2013). The letter was a response to a letter from the industry in which the industry had

representative organisation, but the latest attempt failed in 2014. The CVO has a board in which board members (fisher representatives and (mostly non-active) fishers) from both Visned as well as de Vissersbond take place.

[5]The Blueports were erected in 2011 with financial support of the European Fisheries Fund (EFF). There are 5 regional Blueports and one national platform. The Blueports are the follow up of the national Fisheries Innovation Platform and are meant to stimulate bottom up ideas for innovation. They in fact aim to bring different parties together at the regional level.

expressed their concern with the landing obligation (4-7-2013). In the letter the minister addressed the points made by the fishing industry, but did not agree with the core assessment of the fishing industry (in a mixed fishery, discards can't be avoided and reductions come at a high cost, perhaps even so high that the fleet needs to be decommissioned). In the newsletter of 6 September, Visned wrote to the fleet:

> The minister does not seem to recognise the main (problematic) issues fishers have pointed out nor does she acknowledge the impossibility of the task that she has set to the fishers. (Visned 2013a).

The reaction of the ministry disappointed them a lot and hence they demanded a face-to-face meeting with the minister. That meeting took place, and it was decided that instead of having planned harbour tours regarding cooperative research plans, they would have two meetings in which the minister herself would explain the policy and listen to the concerns of the fishers.

14.5.2 Meetings Between the Minister and the Fleet

The goal of the two meetings organised on 27 September 2013 in Scheveningen and 4 October 2013 in Urk was, according to the minister:

> to explain to you (my) vision and explain how I foresee how we can achieve a workable implementation [of the landing obligation] together. I want to, together with you, achieve an implementation agenda […]. (Source: Visned 2013b)

Both meetings had the same set-up. In one meeting the minister was introduced by a fisher representative acting as chairman of the meeting (Scheveningen) and in the other (Urk) by the mayor. In both meetings the minister explained her vision and emphasised how she wanted to achieve this vision together with the fishers. Then the minister responded to questions and comments. The meetings lasted about an hour and a half.

The general message of the minister in both meetings was the following: the decision to have a landing obligation has been taken and cannot be reversed. Therefore, there is no sense in discussing the principles and goals of the landing obligation. Saying 'no' to Brussels will weaken the position of the Dutch fleet. In other words, that would only be counterproductive because the room to manoeuvre in Brussels would disappear. She stated that the Netherlands have agreed to the ban, although it was not her personally but her predecessor, and that meant that the ministry has a shared responsibility which she has to defend. Nevertheless, she reiterated that not everything had been decided upon yet and that there was room to manoeuvre with regard to a number of issues. These issues were still open to discussion and needed to be addressed through research. She promised to negotiate in Brussels to ensure that the economic concerns of fishers were addressed as well as the issue of technological innovation for increasing sustainability. She wanted to combine the practical knowledge of fishers and her knowledge of the political arena. The ministry and the fishers, she believed, needed each other. She also believed that there was an important role for industry and research institutes to play in terms of gathering evidence and building arguments as to what is and isn't possible. All objections to the ban were welcome, as long as they were sustained with arguments and evidence and did not outrightly question the principle of the landing obligation. During the discussion with fishers, the minister emphasised that she wished to discuss the research agenda by posing relevant questions ($N = 8$) herself and emphasising regularly that she wanted to cooperate with the fishers ($N = 13$).

The fishers came to the meetings with different goals. First they wanted to hear the minister defend her policy, as they did not understand the logic of it (Holm et al. 2013). Second, many fishers hoped that they could convince the minister that the ban was a bad idea and should be reversed. Asking questions and presenting examples functioned for them as a way to illustrate how senseless and detrimental the ban was for the fleet. The fishers emphasised how the ban would limit their economic manoeuvrability:

> You say that you are looking for the maximum flexibility, but you know what happens if you overstretch a rubber band, it snaps! (Fisher, meeting observation in Urk, 4-10-2013).

Although the minister said she did not want to discuss the principles of the landing obligation, the fishers did try to discuss these. They focused on ecological arguments against the obligation, namely, on the impacts on small fish (that when landed will be 100% dead), the effects for fish stocks and the ecosystem and the need to have room to manoeuvre in the 'survivability' paragraph. They were convinced that the fish were not dead when thrown overboard, and thereby landing them would mean increased mortality of fish. And even though some fish would be dead, the fishers reasoned that these dead fish still had a function in the ecosystem – as food for foragers. They also reasoned that, whatever they do to fish more sustainably, it would never be enough and that the discard ban undermined the fundamentals of fishing policy (protecting small fish) and the accomplishments made by the fleet in earlier years:

> We have compromised for 25 years, we have adjusted our gears to reduce impact on the ecosystem, we have scrapped 50% of the fleet – but if this continues our fishery will collapse. We have done enough. To land young fish instead of throwing it back into the sea does not make sense to us. We can't understand this! (Fisher, meeting observation in Scheveningen, 27-9-2013)

14.5.3 Parallel Monologues

What the observations of the meetings illustrate is that there were two parallel monologues going on. The first one was about the goal and consequences of the landing obligation. Where the minister emphasised that the landing obligation was an irreversible 'fact', the fishers emphasised that the obligation was 'impossible' to implement. The fishers wanted to discuss the fundamentals of the landing obligation (what is the goal), while the minister had no interest to do so:

Fisher 1: I have fished now for 12 years, which isn't as long as others here present, but still it is quite a long time. For 12 years I have spent 100 hours at sea per week and I have observed that the presence of all commercial species are increasing, not only plaice and sole, but also red mullet and greater weaver fish. This is the case in the north and south. We spend much more time at sea than the biologists.

Fisher 2: When discussing evidence that fish survives discarding; I am a shrimp fisher, 40 years ago there was a factory to process undersized fish [puf fabriek – in Dutch]; but that was bad. In the sixties a shrimp-sorting machine was introduced resulting in fish being discarded alive instead of being landed to the factory, and the fishery improved greatly. So the proof is there! We did not know what happened! We suffocated in our own success. We made sacrifices, but now all is fine. And now we are being asked to land all small fish again? You see, the 3000 HP vessels, that was too much, but we have reduced those type of vessels... How long will this last...?

Minister: How can we reach out to each other? Again the principle debate has been held. The question is what can we do to prevent the fleet from drowning? (Fisher 1 and 2 and minister, meeting observation in Scheveningen, 27-9-2013)

What these quotes highlight is that there are significant differences in perceptions with regard to the (ecological) consequences of the discard ban (see Fig. 14.3); and these differences have never been discussed explicitly in the Netherlands by ministry and fleet. The landing obligation had widened the gap between policy makers and fishers. In Fig. 14.3, the line of reasoning about discards and the effects of a discard ban on policy makers and fishers are positioned next to each other for comparative purposes.

Policy makers reason that many of the fish that are caught and discarded are dead because of the process of catching and sorting. They reason that the discard ban is a tool to trigger more selective fishing behaviour from fishers. This will result in less catches of undersized fish (discards in the current system) and thus in a lower fishing pressure. The end result is less waste of protein (e.g. less fish that are not commercially used, are caught), a further fish stock recovery (as less unwanted and unaccounted removals take place) and increased profits for fishers (as the stocks improve in the long term).

Fishers (especially Dutch demersal trawl fishers) have an almost opposite reasoning; first of all, they think that many discarded fish do not die. They refer to the 'aliveness' of the fish on board that they witness during the sorting process. They also argue, however, that dead fish still have a role to play as food in the ecosystem, both for birds and for scavengers on the seabed. They reason that the discard ban cannot be achieved, as there are many technical, economic and social constraints to increasing selectivity (without significant loss of marketable fish). Therefore, landing all undersized fish will result in increased mortality of fish and thus higher fishing pressure which will be detrimental to fish stock recovery and result in increased costs for fishers (as sorting the catch will take much longer and fish stocks will deteriorate). Moreover, the fact that discards now need to be landed means that food is lost from the ecosystem, resulting in increased mortality for

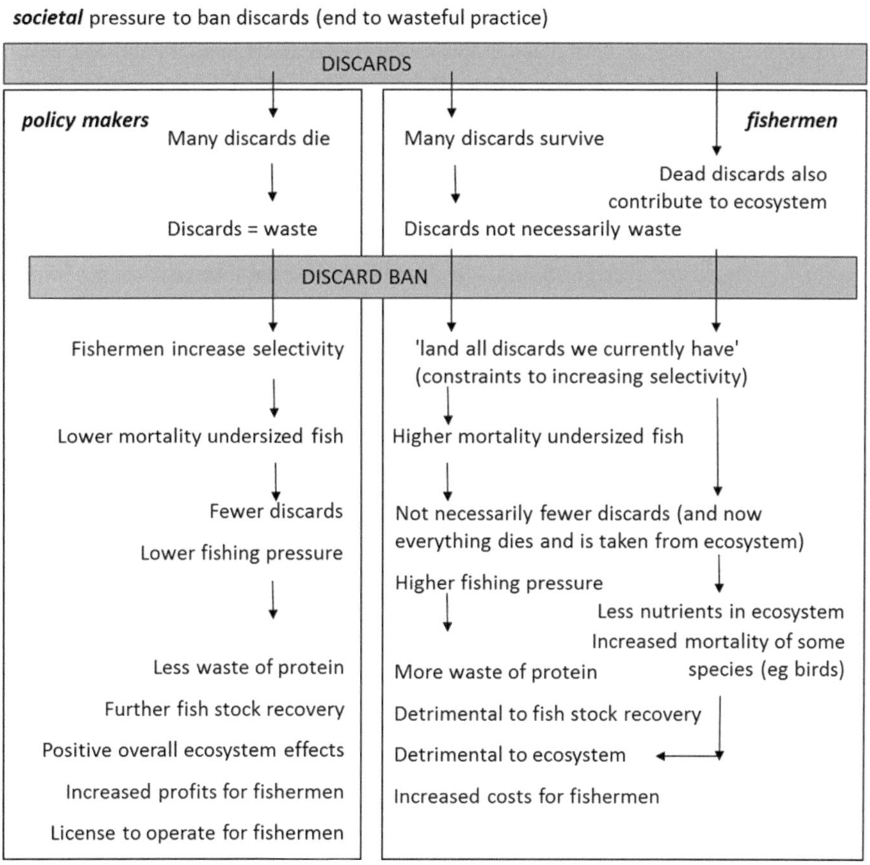

Fig. 14.3 Different perceptions of fishers and policy makers of (ecological) consequences of the landing obligation

some species (including birds) which will consequently have detrimental effects on the ecosystem as a whole.

The fact that fishers' reasoning is not discussed in relation to the line of reasoning of policy makers leads to misunderstanding and misinterpretation. The fishers' reasoning is not only related to their own 'interests' but is very much linked to their understanding of the ecosystem, their belief system (frame) and to their experience (fish looks alive when it is thrown overboard, birds feed on their discards). Moreover, many fishers have a fundamental ethical problem with landing undersized fish. They compare it to 'child murder'. Furthermore, they can't understand why on the one hand they have been required to use bigger meshes and to fish further away from the coast to avoid catching small fish (i.e. the Plaice box), but on the other hand, they are now required to land all small fish they catch. As a young fisher in Urk said:

> Young fish with a chance to live should be thrown overboard so that they can grow! (Fisher, meeting observation in Urk, 4-10-2013)

Given their strong views, it is likely that fishers will not ‘hear’ arguments that counter their fundamental belief that small fish either survive being caught and discarded, or fuel the ecosystem when they do die. Policy makers have the same sense of conviction. They are convinced that fishers can increase their selectivity a lot. They are also convinced that fishers don’t want to discuss changes to their practices because of their economic interests. Leaving these images and perspectives undiscussed will not bridge the gap.

A way forward in finding common ground between the two sets of actors is the ‘room to manoeuvre’ (rek en ruimte) – as highlighted by the minister – in the implementation of the discard ban:

> But what we will try (commit ourselves to) is to look for maximum space to maneuver, within certain margins. (Minister, meeting observation in Scheveningen, 27-9-2013)

Most fishers were eager to tell the minister why the ban would not work, why it is pointless and the minister wished to set up a proper research agenda in order to prepare for implementation. Fishers mentioned that there is no space in the vessel to store undersized fish (as in a good fishing week in the summer their load was full). They argued that relative reductions in discards have already taken place, for instance, when they fished with the pulse trawl instead of the beam trawl and because of the huge reduction in fishing effort over the last few decades. Only some fishers expressed faith in innovation as a way out. The minister, on the other hand, wanted to use the landing obligation in such a way that it improves the selectivity of the demersal fishery, without jeopardising the fleet’s economic performance. Despite fishers’ continued attempt to get the policy cancelled, the minister insisted this was not possible:

> I am not deceiving you, I am not saying: ‘I will go to Brussels and get the policy cancelled for you’. (Minister, meeting observation in Scheveningen, 27-9-2013)

The second parallel monologue was about cooperation. At one level, the minister emphasised the need for her ministry to cooperate with the fishing industry; however, fishers questioned what cooperation meant. They expressed the following:

1. Why cooperate with the implementation of a rule you can’t relate to (we don’t see why it is necessary and we are apprehensive that it will result in us being put out of business)?
2. Policy has no understanding of fishing in practice. We’d like to share knowledge if we have the impression that it has a positive effect. Evidence from the past, however, shows us that there is no point in sharing knowledge and experiences:
   - Why should we prove that something is wrong and doesn’t work? Rather, why shouldn’t policy makers prove that a measure will work?
   - Policy is never evaluated (not even when there is scientific evidence that policy measures were ineffective).

The fishers had the feeling that the minister had no idea what the ban would mean in practice. She asked them to explain it to her and to support their explanation as much as possible with (scientific) evidence. Fishers felt that such a line of reasoning was unfair. Why did they have to do that, whilst policies are not evaluated by their (possible detrimental) ecological effects? It is their experience that existing 'good' policies have not been acknowledged to be effective (such as the improved fish stocks in the North Sea); on the contrary Brussels has tended to emphasise that fisheries policies have failed. On the other hand, if policy measures are proven to be ineffective, they are never changed (such as the Plaice box). Thus fishers are sceptical about the potential benefits of cooperation. In fact, the fishers said the following to the minister: if you want us to cooperate, it should be two ways. There is hardly any direct interaction between the fishers and policy makers. As a result, fishers mistrust policy makers and hence are hesitant to share knowledge and cooperate – when asked.

The minister tried to bridge the gap personally at the meeting in Scheveningen, by being very open to comments, showing she listens and cares. The difficulty, however, is that she stands for a system, a system which often is not characterised by cooperation, but by a top-down management and control. The latter characteristic was evident in Urk when the minister was directly challenged by fishers who asked her 'what will happen if we don't adhere to the landing policy'. The minister reverted back to the 'typical' governmental response:

> We act upon illegal activities, as you have seen today' (referring to police action in the south of the Netherlands that same day (which was the 'talk of the town' in the Dutch fishing community), in response to alleged illegal trade in sole. (Minister, meeting observation in Urk, 4-10-2013)

14.6 Discussion and Conclusions

In this section we will conclude on two main issues that impair understanding between policy makers and fishers and discuss the implications of the (widened) gap for the implementation of the landing obligation in the Netherlands as well as how this gap might be bridged.

The first conclusion is that two important lines of reasoning in relation to the landing obligation differed between the fishers and the ministry; the first on the consequences of the landing obligation and the second on cooperation. Where an explicit discussion of the differences would have been useful, in fact, a parallel monologue was held. The result of this is a widened gap between fishers and the ministry in the Netherlands.

The decision to not discuss the goal and consequences of the landing obligation was partly based on the assumption that fishers would simply oppose the ban based on their interests. Yet different perceptions also stem from different information environments and experiences (see Sect. 14.4). While a discussion might not result in a common position, it would at least avoid parallel monologues (one could agree to disagree). Although the minister's strong resistance to discuss the fundamental principles of the policy measure might be understandable from the perspective that some discussions never come to an end, it did leave the goal of the landing obligation remaining in limbo – at least in the fishers' minds. If the goal is to end wasteful practices, why then land all undersized fish to be destroyed on land? This adds to the confusion and frustration of fishers (and to widening the gap between policy makers and fishers).

Parallel monologue also prevailed on the issues of cooperation. While the minister stressed the need for cooperation, fishers expressed their experience that cooperation in the past has not always resulted in a lot of good (to them). Therefore, they find it difficult to trust the minister and her officers and feel that sharing knowledge will not help their cause. They also expressed concern with lack of proper evaluation of policy measures in the past and were not too hopeful therefore that this would change in the future. The lack of clarity on the goal of the landing obligation would also continue to undermine cooperation (you need a shared goal) and effectiveness (how can you evaluate?).

The second conclusion is that the fishing sector is not a uniform body, but that there are differences between the different fishers in terms of how they view policy measures. Although this might seem an open door, often, this heterogeneity is not explicitly considered, which obscures how policy affects different fishers differently. Some fishers expressed despair and said that they see no future in fishing if such measures are adopted. Others were seeking ways to deal with the measures and believed in innovation and a pragmatic implementation of the landing obligation. Such pragmatism can be linked to personality traits and business interests, but is also linked to differences in perceived effects of the landing obligation (Fig. 14.3) and the impacts upon them. For example, fishers with large plaice quotas (fishing with 120 mm + mesh sizes) will be much less affected than fishers with large sole quotas (fishing with 80 mm mesh sizes).

Also the two fisher representative organisations in the Netherlands have developed two different strategies in dealing with the landing obligation. The Vissersbond is more clear in opposing the measure, whereas Visned while also opposing the ban does so in a more strategic way (articulating less their opposition). These differences between fishers, as well as the different strategies that the fisher organisations have developed in communicating their concerns about the landing obligation, impact the 'voice' of the fishing industry in the wider process (Gezelius and Raakjaer 2008). Such fragmentation also works to paralyse the process in two ways: (1) in terms of fishers' coming to a unanimous position vis-a-vis the ban (if one of the two organisations is 'too cooperative' in the eyes of some fishers, that organisation can be called back by pointing to the other organisation not making that move) and (2) in terms of them not giving a collective message to the ministry, allowing the latter room to manoeuvre.

By observing the process and listening to the reasoning of fishers, the (widened) gap between policy and fishery became manifest. The fact that fishers and policy officers reason differently, but don't discuss that, and the lack of fishers' trust in the ministry, will make the implementation of the landing obligation more difficult. Compliance of the fishers with regulations is often the best guarantee to successful policy implementation. In the current situation, it is not likely that fishers will (voluntarily) comply with the new set of rules; thus, there remains a gap still to be bridged. That bridge can perhaps be found when taking time to jointly discuss the foreseen economic and ecological effects of the landing obligation. The minister needs this information for her negotiations at regional and EU level, and the fishers wish to make clear that the ban is not the best measure to achieve ecological and economic sustainable fisheries. It takes considerable time and effort to convince parties to talk about these practical and fundamental expectations.

In the project group meeting, an attempt was made to build bridges between policy officers and fishers' representatives. The policy officers explained to the fishers' representatives what the (international) political reality is and how much space the Netherlands has (or does not have) to tweak the landing obligation. Fishers' representatives explained the (im)possibilities of going through with the landing obligation. In the project group, the room to manoeuvre was carefully discussed, by talking about how survivability of discards can be studied, what is needed for an exemption, when each fishery would be phased in and what works best for fishers.

Considering the gap that continues to exist between the ministry and fishers, the fact that the project group is still working together, building trust and understanding each other's roles is really positive. However, the ministry remains opposed to discussing the principles of the landing obligation. And it is important to emphasise that whatever space is sought and found in this group, this needs to be renegotiated at the international level and needs to be 'sold' to the fishers. It remains to be seen what the end results will be. Cooperation of the fishers' representatives and those who implement policy might turn out to be a temporary bridge if the foundations turn out not to be so solid.

As the landing obligation will be phased in over a number of years, it might be worthwhile to spend time discussing fundamental lines of reasoning (i.e. what are the ecological consequences of the landing obligation?), fundamental processes of cooperation (i.e. how can governance processes be made more suitable for stakeholder participation than they currently are?) as well as fundamental questions about policy effectiveness and evaluation (i.e. how can policy measures that don't deliver expected results be reverted?). This will most likely result in more buy-in of fishers in the implementation of the landing obligation and result in a more solid bridging of the gap between the different actors.

References

Catchpole T, Keeken van O, Gray T, Piet G (2008) The discard problem – a comparative analysis of two fisheries: the English Nephrops fishery and the Dutch beam trawl fishery. Ocean Coast Manag 51(11):772–778

Centrum voor Visserijonderzoek (2019) Noordzeetong. Retrieved from https://www.wur.nl/nl/Onderzoek-Resultaten/Wettelijke-Onderzoekstaken/Centrum-voor-Visserijonderzoek-1/Onderzoek/Beleidsadviezen/Zeevisbestanden/Noordzeetong.htm. Access date: 30-1-2019

EAPO (2013) EAPO alternative proposal discards approach by EU Industry. Retrieved from: http://nsrac.org/wp-content/uploads/2013/01/Paper-5.1-EAPO-Alternative-proposal-discards-approach-by-EU-Industry.pdf. Access date: 30-1-2019

EC (2009) Green paper. Reform of the common fisheries policy. EU, Brussels

EC (2019a) CFP reform – the discard ban. Retrieved from: https://ec.europa.eu/fisheries/sites/fisheries/files/docs/body/discards_en.pdf. Access date: 30-1-2019

EC (2019b) Discarding and the landing obligation. Retrieved from: http://ec.europa.eu/fisheries/cfp/fishing_rules/discards/index_en.htm. Access date: 30-1-2019

EC (2019c) Regionalisation. Managing fisheries. Retrieved from: http://ec.europa.eu/fisheries/cfp/fishing_rules/regionalisation/index_en.htm. Access date: 30-1-2019

GAP (2019) The Netherlands discards. Research cooperatio, discards and self-sampling. Retreived from: http://gap2.eu/gap2wordpress/wp-content/uploads/2015/03/Netherlands-CS-12.pdf. Access date: 30-1-2019

Gezelius S, Raakjaer J (2008) Making fisheries management work. Implementation of policies for sustainable fishing. Springer, Amsterdam

Holm P, Dubois M, Tveiteras K, Bjorkan M, Wilson DC, Kraan M (2013) Collaborative research and participant attitudes. GAP2, WP4 report. Deliverable D4.2.1

Hommes S, Hulscher S, Mulder J, Otter H, Bressers H (2009) Role of perceptions and knowledge in the impact assessment for the extension of Mainport Rotterdam. Mar Policy 33:146–155

Koning de A (2015) 'Het veld' in de antropologie van steden en van beleid. In: Driessen (ed) Het hart van de wetenschap. Over de waarde van veldwerk. Wereldbibliotheek, Amsterdam

Kooiman J, Bavinck M, Svein J, Pullin R (2005) Fish for life: interactive governance for fisheries (No. 3). Amsterdam University Press, Amsterdam

Ocean 2012 (2019) Transforming European Fisheries. Retrieved from: https://www.youtube.com/watch?v=xc2dDQnie980. Access date: 30-1-2019

Raakjær J (2011) The EU common fisheries policy: a fisheries management system that has failed! Acta Agric Scand Sect C Food Econ 8(2):105–113

Rijksoverheid (2014) Vragen en antwoorden aanlandplicht. Retrieved from: https://www.rijksoverheid.nl/documenten/publicaties/2014/11/26/vragen-en-antwoorden-aanlandplicht-basisset. Access date: 30-1-2019

River Cottage (2019) Hughs fish fight. Retrieved from: https://www.rivercottage.net/campaigns/hughs-fish-fight. Access date: 30-1-2019

Roos (2013) Europese aanlandplicht voor vis lijkt pyrrusoverwinning van milieulobby. In: Reformatorisch Dagblad 1-6-2013. Retrieved from: https://www.rd.nl/vandaag/economie/europese-aanlandplicht-voor-vis-lijkt-pyrrusoverwinning-van-milieulobby-1.313851. Access date: 31-1-2019

Salomon M, Markus T, Dross M (2014) Masterstroke or paper Tiger – the reform of the EU's common fisheries policy. Mar Policy 47:76–84

Van Densen W (2001) On the perception of time trends in resource outcome. Its importance in fisheries co-management, agriculture and whaling. PhD thesis, Twente University, Enschede

Verweij M, Van Densen W (2010) Differences in causal reasoning about resource dynamics and consequences for the participatory debate on North Sea fisheries. Mar Policy 34:1144–1155

Verweij M, van Densen W, Mol A (2010) The tower of babel: different perceptions and controversies on change and status of North Sea stocks in multi-stakeholder settings. Mar Policy 34:522–533

Visned (2013a) Nieuwsbrief. 6 September 2013

Visned (2013b) Nieuwsbrief. 27 September 2013

Visserijnieuws (2013) Aanlandplicht voldongen feit. Retrieved from: https://www.visserijnieuws.nl/nieuws/aanlandplicht-voldongen-feit. Access date: 30-1-2019

Vissersbond (2013) Manifest Aanlandplicht NEE. Retrieved from: https://www.facebook.com/permalink.php?story_fbid=283181665152725&id=283175375153354&__tn__=K-R. Access date: 30-1-2019

Wageningen Marine Research (2019) Discards unwanted catch. Retrieved from: https://www.wageningenur.nl/en/Dossiers/file/Discards-Unwanted-catch.htm. Access date: 30-1-2019

Chapter 15
Taking the Initiative on Maltese Trawl Industry Management. Industry and Science Collaboration on Identifying Nursery and Spawning Areas for Trawl Fisheries Target Species

Nicholas Flores Martin

Abstract Malta has managed a Fisheries Management Zone (FMZ), which extends to 25 NM from the baseline of the Maltese Islands, since 1971. The key aim of the Malta FMZ is to protect the fisheries resources of Malta's marine area and the ecosystems on which they depend. While bottom trawling is limited in terms of the number of vessels, it is one of the major contributors to landings. As of the start of the GAP project (April 2011), 12 bottom trawlers were licensed to trawl within the FMZ. The study was originally the initiative of fishers, motivated by the need to have data that could be used as a basis to advise on the management of the trawling fleet working within the FMZ. Throughout the sampling design stage, the methodology was discussed between fishers and scientists with the aim of using fishers' knowledge to determine sampling locations while at the same time obtaining sound results. A 13-month study was conducted, using modified versions of "mazara" type nets traditionally used by Maltese bottom trawlers. The nets had two square mesh cod-ends with mesh sizes of 40 mm and 20 mm. Data were collected on seven target species and three non-target species which will be used to address data gaps with respect to nursery and spawning areas of local populations of targeted stocks. This chapter provides an initial discussion on the potential contribution of the information collected to provide management advice for Malta's trawl fisheries management plan, the main focus of which is the control of fishing effort.

Keywords Co-management · Mediterranean · Stock assessment · Fishers' knowledge · Malta

N. Flores Martin (✉)
Fisheries Resource Unit, Department of Fisheries and Aquaculture, Marsa, Malta
e-mail: nicholas.floresmartin@soton.ac.uk

P. Holm et al. (eds.), *Collaborative Research in Fisheries*, MARE Publication Series 22, https://doi.org/10.1007/978-3-030-26784-1_15

15.1 Introduction

Spatial ecology of fishery target species is a major issue of current fisheries research (Ciannelli et al. 2008). In order to ensure long-term sustainability of the Maltese trawling industry, an integrated approach to the development of future management strategies is clearly needed. Reducing adverse impacts, primarily fishing, on habitats essential for the completion of a full life cycle of a species (such as sites used for spawning and growth to maturity), is one aspect of an ecosystem approach to fisheries management (FAO 2003). While improved biological data is imperative to evaluate the status of demersal stocks, a participatory approach to achieving these aims is invaluable to devise management plans that are more readily acceptable to fishers than the present scheme.

Knowledge on spatio-temporal distribution of critical habitats of key commercial species in the General Fisheries Commission for the Mediterranean (GFCM) Geographical Sub-Area (GSA) 15, and particularly within Malta's 25 NM Fisheries Management Zone (FMZ), is limited and lacks seasonal and smaller-scale resolution. Previous studies, often based on data obtained in the annual Mediterranean International Trawl Survey (MEDITS), have been carried out on hake (*Merluccius merluccius*), pink shrimp (*Parapenaeus longirostris*), red mullet (*Mullus barbatus*) and common octopus (*Octopus vulgaris*) (Fiorentino et al. 2003a, b, 2004; Garofalo et al. 2008, 2010, 2011; MedSudMed 2007; Camilleri et al. 2008; Abella et al. 2008; Fortibuoni et al. 2010), no similar studies have been conducted for other species and data from the 25 NM FMZ has not always been included.

In informal conversations, fishers had previously voiced concern and lack of trust towards scientists and policy-makers due to what fishers perceived as lack of biological data on commercially important demersal species such as hake and red shrimp within the 25 NM FMZ. It was partly awareness of this knowledge gap that had been a major contributor to the lack of trust between fishers, policy-makers and scientists. Increasing knowledge about the state of fisheries in the FMZ was relevant to them because management decisions would directly affect them, and hence it was vital that such decisions were made based on high-resolution data. Collaborating with the fishers directly on the project proposal and agreeing on what parameters ought to be studied was necessary to ensure that the project deliverables were both furnished with traditional ecological knowledge, as well as to bridge the divide that in the past has hindered communication between fishers and the scientists.

The collaborative approach allowed use of fishers' ecological knowledge of the species studied and their in-depth knowledge of the sea bottom. It was also necessary to consult with fishers to obtain exact locations for sampling based on commercial trawling lanes. The fact that fishers are the main stakeholders meant that it was imperative that they be included and were allowed to provide input.

The aims of the Maltese case study were therefore as follows:

- Improve working relations between fishers and scientists by conducting collaborative research and spending time working with one another.

- Collect information on commercially important species targeted by Maltese trawl fishers – particularly with respect to spatial seasonal changes in density and biomass of juvenile and mature fish of the species of interest. These species were as follows: red mullet (*Mullus barbatus*), striped red mullet (*Mullus surmuletus*), hake (*Merluccius merluccius*), common octopus (Octopus vulgaris), giant red shrimp (*Aristeomorpha foliacea*), rose shrimp (*Parapenaeus longirostris*) and Norway lobster (*Nephrops norvegicus*).
- Use this data to improve and update Malta's fisheries management plans using the participatory approach to ensure that these plans are more readily accepted by the fishers.

15.2 Background to the Case Study

The FMZ and GSA 15 geographically form part of the Strait of Sicily (Fig. 15.1). The region is characterised by two layers of water flow. The upper layer, Modified Atlantic Water (MAW), has a relatively low salinity and flows in a West (W) to East (E) direction. The lower layer, Levantine Intermediate Water (LIW), is more saline and flows in the opposite direction like an undercurrent. The MAW enters the strait from the western boundary along Adventure Bank (South West (SW) shelf of Sicily), coming close to the Sicilian shore in the middle of the south coast of Sicily, and separates again when it encounters the Malta bank on the SE shelf of Sicily. It

Fig. 15.1 Map of GSA 15 with the 25 NM zone border

encircles two large cyclonic meanders, one around Adventure bank and another on the Malta bank (to the NE of Malta). In the meso-scale, these meanders have been identified as important areas for nurseries and spawning for the Strait. This is due to the fact that the water circulation favours permanent upwelling to the left of the MAW, reinforced by the predominant westerly winds. The offshore transport associated with this upwelling affects the survival/mortality rates in the early life stages of species of interest, influencing the strength of recruitment (Levi et al. 2003).

In this area, Maltese bottom trawlers target demersal resources. The main species targeted include giant red shrimp, pink shrimp, red mullet, hake and Norway lobster of which the former three form a sizeable proportion of the total catch both in terms of tonnage and catch value of the Maltese fleet. The fleet is small in capacity, representing less than 2% of Malta's Fleet Vessel Registry. At the start of the project (2011), the trawlers consisted of 12 boats between 17 m and 24 m length overall with an overall tonnage of 1056 GT, and a total main engine power of 3700 kW. Trawling in designated areas within the FMZ is allowed, although the total trawling capacity within the 25 NM mile zone will not be allowed to increase from its present level. Furthermore, only trawlers shorter than 24 m are allowed to trawl within the FMZ. As a further restriction, in areas where the depth of the sea floor is less than 200 m, trawlers cannot have an engine capacity that exceeds 185 kW (Cap. 425 Act II of 2001).

Trawling is undertaken both during day and night depending on species targeted. Three different types of trawling activities are undertaken during the year:

- At night in depths of between 50 and 150 m on heterogeneous bottoms targeting fish species such as red mullet, comber, pandora and weaver.
- During the day at depths of between 150 and 200 m on clay and mud bottoms targeting pink shrimp, red mullet, hake, squid, octopus and various species of dogfish.
- During the day at depths of 600 m targeting king prawns, Norway lobster, hake and forkbeard.

The regulation of trawl fisheries within Malta's FMZ has so far been based on limitations on fishing capacity (number of vessels, licenses and engine size), minimum landing sizes and net mesh sizes (Council Regulation (EC) no. 1967/2006). However, the establishment of no-fishing zones, particularly within nursery areas, has been increasingly advocated as a further component of the EU's fishery management strategy (Council Regulation (EC) no. 1967/2006). Malta's management plan outlines a strategy to reduce capacity by 20% and to potentially reduce effort by a further 10% by introducing a closed season between 15 August and 15 September if capacity reduction is not sufficient to reduce the fishing effort. This is in line with plans by the General Fisheries Commission for the Mediterranean (GFCM) to introduce a regional management plan that includes harmonising closed seasons between Tunisia, Libya, Malta and Italy (MedSudMed 2013, 2014, 2015).

In line with Article 19 (Council Regulation (EC) no. 1967/2006), Malta is obliged to implement a trawling management plan. Malta's Fishing Effort Adjustment Plan outlines a strategy to reduce capacity by 20%, and potentially by a further 10%

(MSDEC 2013; MEA 2014). This adjustment plan also takes into consideration Recommendation GFCM/39/2015/2, which set minimum standards for bottom trawling fisheries on stocks in the Strait of Sicily prior to adoption of a multiannual plan in 2016 (GFCM/40/2016/4). Recommendation 40/2016/4 established Fisheries Restricted Areas within the Straits of Sicily for demsersal stocks of pink shrimp and hake, along with a closed season in the Gulf of Gabes between 1 July and 1 September.

Prior to the present study, a single unpublished study had been carried out within GSA 15 by Knittweis and Dimech (Unpublished). This was based on 2003–2008 Mediterranean International Trawl Survey (Medits) data from GSA15 which included data on the following species: *Mullus barbatus, Mullus surmuletus, Merluccius merluccius, Parapenaeus longirostris, Aristaeomorpha foliac*ea and *Nephrops* norvegicus. Critical habitats for *A. foliacea* and *N. norvegicus* showed a scattered distribution, while P. longirostris spawners were mainly found in the northern half of the zone with juveniles showing a more scattered distribution. Other species showed spawner concentrations towards the East (E), North East (NE) and South East (SE) ends of the FMZ, and in many cases outside on the Malta bank. It was also found that spatial distribution of critical habitats, in addition to being affected by biology, is also heavily influenced by fishing pressure. The species most impacted by legal trawl zones are the two species of Mullus. Having said that, despite the fact that critical habitats were impacted by fishing pressures, both species have other remaining nursery and spawning sites which are not subject to such pressures. In the case of spawning and nursery grounds in international waters, an international management approach is important (Knittweis and Dimech Unpublished).

15.3 Results

15.3.1 The Collaborative Process: Fishers and Scientists

The overall aim of this case study was to enhance the scientific knowledge base underpinning management decisions regarding resources of the Maltese 25 NM FMZ through an integrated approach, with a view towards improving the acceptance rate of future management plans by fishers – in the eyes of resource users, co-management can produce legitimacy and more effective regulations (Wilson 2003). Prior to the start of the project, trawl fishers in Malta generally had a negative perception of the management regime employed in the FMZ, in particular with respect to the usefulness of closed areas for trawling inside the FMZ. The negative perception of trawl fishers had been ascertained during the recurrent management-related negotiations and meetings held with fishers, and occasional comments made during informal conversations. Similarly, the notion that trawling effort should be increased in terms of total tonnage as well as vessel power had been expressed by fishers in a number of informal talks. Prior to the start of the project, the specific objectives were as follows:

- To carry out stock assessments of the following species: *M. barbatus, M. surmuletus, A. foliacea, N. norvegicus* and *P. longirostris* with particular focus on comparing the population status within the FMZ and in adjacent waters outside it. Data on *Merluccius merluccius* and *Octopus vulgaris* and discards such as sharks and rays were also to be collected;
- To analyse fishers' knowledge of the selected species;
- To investigate differences in perspectives between fishers and scientists regarding management of the trawl industry within the FMZ.

Expected outcomes included stock assessment of the selected species, collaborative workshops between scientists and stakeholders, reports based on workshop outcomes, management proposals such as seasonal and/or area closures agreed upon by both scientists and stakeholders, reports on project results submitted to Maltese enforcement and control authorities and the GFCM, and peer-reviewed scientific papers. One particular benefit of note to fishers was that results would lead to a sound assessment of demersal stock status within the FMZ which might subsequently be used to uphold exclusive fishing rights for Maltese fishers within it. Moreover, in accordance with the opinions of fishers, if fleets were fishing at lower than maximum sustainable yield, adjustments would be made based on these findings, in turn increasing the catch per unit effort by trawlers and thus the long-term viability of the trawl fishery as well.

A kick-off workshop where project ideas were to be presented was originally planned. This workshop would culminate in the formation of a steering committee of scientists and stakeholders tasked with coming up with a joint work plan. The work carried out by the steering committee would include data obtained from analyses of secondary data such as stock assessments in GFCM GSA 15, and data available through the EU's Data Collection Framework (DCF) (Commission Decision (2010/93/EU)). Knowledge gained from semi-structured interviews with fishers would also have been put to use.

Meetings were held in early 2012 between the Department of Fisheries and Aquaculture (then known as the Fisheries Control Directorate) and representatives of the Ghaqda Kooperattiva tas-Sajd, one of the two local fisheries cooperatives, and one of the partners of the case study. During these meetings the sampling methodology, the species under study and expenses to be incurred by the stakeholders to carry out the study were discussed. Fishers suggested a double-bag study involving the addition of a second cod-end with a mesh size of 25 mm, over and above the normal 50 mm mesh-sized cod end. This would have provided data to conduct a comparative study on the % selectivity of the 50 mm mesh in comparison with the 25 mm mesh cod end. This suggestion came about as a result of fishers insisting that catches of pink shrimp (*P. longirostris*) had been on the decrease as a result of implementing a 50 mm mesh size protocol. Fishers further suggested the inclusion of a study on either the swordfish longline, or dolphinfish fisheries.

Successful trials were carried out in September 2012 and a contract was drawn up with the Fishing Trawler Owners Association (FTOA), a cooperative for Maltese bottom trawler owners. The contract specified the number and location of hauls to be carried out and financial remuneration for trawl fishers to cover the costs of fuel and those of the Department of Fisheries to purchase the catch.

Fishers had a further interest in the project. They pointed out that vessels were dumping marine litter while conducting bunkering activities and that the sea bottom was being destroyed by the anchors of such vessels. As a result, areas where fishers normally carried out their trawling activities were being destroyed. Anecdotal data collected from various trawl fishers had made the Department aware that trawl fishers frequently brought up tins of anti-fouling paint, oil drums and other various miscellaneous litter in their nets. If paint and oil were present in the nets, the entire haul was discarded. As a result, data was also to be collected on the suitability of the sea bottom for trawling and the presence of marine litter.

Following the signing of the service contract, a meeting was held with the trawlers to decide upon how sampling was to be carried out. The initial plan was to use the same nets as those used for the annual MEDITS. Hauling would take place for half an hour for hauls carried out at depths of <200 m (shallow water) and an hour for hauls carried out at depths of >200 m (deep water). Hauls were to be carried out during the day, namely starting half an hour after sunrise and finishing latest half an hour before sunset. Fishers criticised this rule on several counts. Maltese trawl fishers use two types of nets – one designed for shallow waters and one for deeper waters. The fishers argued that using two types of nets as opposed to a single more "general" one would improve sampling. Furthermore, it was pointed out that commercial trawling in shallow water is normally carried out at night – sampling during the day would lead to the reporting of decreased catches. Finally, fishers reasoned that a haul time of half an hour to 1 h was too short. In the words of one fisher, "in half an hour you'll probably only bring up water in your net".

It was finally agreed to extend the haul times for both shallow and deep water hauls by half an hour. Two types of nets would be used, and to this end, two of each would be purchased by the DFA although responsibility for obtaining said nets belonged to the fishers. Hauls in shallow water would, however, still be carried out during the day so as to be able to compare data with that of other trawl surveys – MEDITS in particular for which 15-year time series data exists. One final and major point of disagreement was the location of some of the 36 haul transects selected. Although some of the hauls selected for sampling were those used by the fishers for commercial purposes, to cover a larger area, a number of MEDITS hauls had been included. Fishers were largely reluctant to trawl in sites used for MEDITS due to fears that the nets might be damaged or in some cases lost. It was agreed that prior to carrying out sampling in those areas, fishers would be allowed to "scope" the bathymetry of the haul using their bottom echo sounder and stop the haul if any problem arose.

A training session was subsequently held to explain to owners and skippers of vessels involved the method that would be used during the sampling stage. The session was led by the author.

The first sampling trips were carried out in April 2013 using two trawling vessels: MFA0047 Eolo and MFA0128 Ignazio Padre. The former carried out the hauls in shallow water while the latter in deep water. Although these were the first official sampling trips, they also served as a trial for the hauls chosen. At the end of the sampling trip, a revision exercise, incorporating information and advice from the fishers, was carried out to assess the success rate of the hauls, in order to revise the hauls specifications, mainly the coordinates. It was decided that a final, amended list of 30 hauls would be used from the second month of sampling onwards.

Although the sampling stage ended by May 2014, at the time of writing, data analysis was still ongoing. Plans were being made to extend sampling for another year using national funds to obtain more high-resolution, seasonal data.

15.3.2 Preliminary Results and Observations

While the scope of this chapter is to tell the story of the case study and the results of collaboration, not to present scientific findings, a summary of the results is nevertheless presented below.

Initial results seem to indicate that in autumn and winter juveniles of *Mullus surmuletus* can be found in the shallow heterogeneous bank (Hurd's bank) to the NE of Malta and in deeper continental slopes. During spring and summer, they are mostly found in the shallower waters, including Hurd's bank. The adult spawners, on the other hand, are generally found on Hurd's bank in the first half year, and then move towards deeper waters (of up to 200 m). *Mullus barbatus* shows similar patterns, moving to shallower waters later in the year as compared to *Mullus surmuletus. Catches* of juvenile *Merluccius merluccius* seem to indicate that juveniles of this species prefer deeper waters (approximately 600 m) in the Wand NW of the FMZ. As the year progresses, they move E to the shallower slopes in the SE of the FMZ (up to 200 m depth). Adult spawners of *Merluccius merluccius* are normally found in the same area towards the second half of the year.

With respect to the crustacean species studied, *Nephrop*s norvegicus and *Aristeomorpha foliacea* tend to remain in deeper waters to the NW, except for in the summer when some of the population move to the shallower waters of the continental slope. Unlike *Nephrops*, however, *Aristeomorpha* can also be found to the S and SW during the spring and summer. *Parapenaeus longirostris* is normally found in continental slope waters to the N, NE and SE of the FMZ. In the winter, they are found in deeper waters towards the W and NW.

A key observation made by Maltese fishers is that the fish they target within Malta's FMZ forms part of a stock (or, rather, a sub-stock) separate from that targeted by Sicilian, Libyan and Tunisian fishers. If correct, this would have important implications, as current management measures and plans in the Malta region are based on the assumption that stocks targeted by fishers from these countries are shared stocks. In fact, a study by Fiorentino et al. (2004) on *Mullus barbatus* indicates that at least for this species, there is a possibility that due to differences in growth rates (Levi et al. 1992) and oceanographic and geographic features (Levi et al. 1995) in the Strait of Sicily, at least two sub-stocks of *Mullus barbatus* exist, one each in the Gulf of Gabes and the Sicilian side of the Strait. Parasitological observations (Levi et al. 1993) also provide evidence of the possibility that there exists one sub-stock in Tunisian/Libyan waters and another on the southern coast of Sicily. The Sicilian stock itself may be split further, with one population found on Adventure bank on the south western coast and another in the Malta bank. Whether these populations can be considered separate sub-stocks depends on whether the currents in that area maintain enough gene flow between the two populations via larval dispersion.

In other cases, fishers shared their knowledge of certain aspects pertaining to the biology of species of interest and their (fishers') work more generally. With regard to the former, this included the fact that *Mullus spp.* tends to seek shallower waters in the spring for purposes of spawning and that traditionally fishing for red prawns was only carried out during the day. They also told tales of how trawl fishers would find their bearings before the advent of Global Positioning System (GPS).

Impressions of bottom trawling as a destructive means of fishing were also changed to a certain extent. Trawling in Malta tends to be carried out along a limited number of existing lanes. These lanes are "cleared" of debris and items which may snag the net over a few preliminary trawls, which has the added benefit of re-suspending sediment and consequently providing food for species which the trawl fishers target. Trawl fishers stress the importance of trawling in a given lane because if this is not done, other species of little interest to trawlers (such as *Capros aper*) will move into the area. Fishers look at trawling as a form of tilling a field that may become fallow and full of weeds if allowed to fall into disuse.

15.4 Discussion

The nature of and value of using fishers' knowledge in management have been documented in the literature (Neis et al. 1999; Kaplan and McCay 2004; Johnson and Van Densen 2007) and examples of these include the following: reduction in mortalities of elasmobranch bycatch thanks to better handling techniques (Poisson et al. 2011, 2014; Kneebone et al. 2013), stock assessment (Palsson et al. 1989) to complement fisheries data and overcome data limitations on biodiversity changes (Fortibuoni et al. 2010; Coll et al. 2014), migration patterns (Hutchings 1996; Maurstad and Sundet 1998) and current and historical spawning grounds (Ames 2004; Johnson and Densen 2007). Fishers' knowledge may also provide information about schooling behaviour (Parrish 1999) and insights on habitat preference and gear selectivity (Hall-Arber and Pederson 1999).

Throughout the Maltese CS, good working relationships were established with some of the fishers as they appreciated time spent working face-to-face, and placed a certain value on building a working rapport – gaining a measure of the person they are working with and showing what their work is like, and they regularly voiced their desire that scientists spend even more time aboard. This would have been ideal, as fishers spend most of their time offshore, and boats targeting different species are active at different times of the day which made it a challenge to organise regular meetings. Explaining any scientific findings in a more informal setting, such as while spending time at sea, may have been more suitable than the traditional model of presenting results in a report format. The main opportunity to discuss the project with fishers was during a brief meeting to pass on sampling sheets and revisit instructions, normally either with the skipper or the owner of the vessel, just prior to their assigned sampling period.

In common with other projects (e.g. Kraan et al. 2013), two diverging attitudes were encountered among fishers: that participation may improve management and that data they were collecting may eventually be used against them. By and large fishers with a more sceptical attitude remained relatively sceptical, but cooperative, while fishers who were initially optimistic about potential outcomes were eager for more collaborative work. A number of trawl fishers who participated in Gap become more involved in other aspects of data collection for fisheries management, namely, allowing scientists on board their vessels to carry out data collection for the purposes of the Data Collection Framework (Commission Decision (EC) 2010/93/EU). Good working relationships were established with some of the fishers as they appreciated time spent working face-to-face, and place a certain value on building a working rapport – gaining a measure of the person they are working with. A number of trawl fishers who participated in GAP have become more involved in other aspects of data collection for fisheries management, namely, allowing scientists on board their vessels to carry out data collection for the purposes of the Data Collection Framework (Commission Decision (EC) 2010/93/EU). As also encountered in other projects, fishers viewed their knowledge, particularly details of locations of their trawling runs, as private intellectual property.

Looking back at the data collection period, perhaps unsurprisingly, training the fishers to carry out unsupervised sampling was successful even in the earlier stages of the project. The main shortcoming of the case study was the lack of time spent aboard fishing vessels. This would have led to stronger working ties, made diffusion of results easier, and allowed a much greater sense of trust to be built on both sides. Fishers were somewhat discouraged by the length of time for preliminary results to be communicated – and while somewhat mitigated by a meeting held shortly after the conclusion of the sampling period, this should have been made clear from the start.

The Gap project is based on the premise that the knowledge base of fishers can contribute to fisheries management. It has been suggested that fishers are the "true ecologists of the ocean" (Holm et al. 2013). While the latter moniker might not be completely accurate, since fishers' knowledge is limited by their own experiences and the experience of people whom they work with and talk to, dismissing experiential knowledge as having no value solely because it does not come from people with a scientific background is also misplaced. Non-empirical data plays an important role as it may offer insights not considered by scientists, and given proper checks, self-sampled data can play a role in spatial and temporal cover of fisheries data. While collaboration with scientists might make fishers better ecologists (in the traditional sense), this should not be the main aim. This does not however rule out the fact that fishers might be curious about the scientific aspects of fisheries management – an exchange of knowledge of sorts. People whose livelihoods involve fishing would very likely be interested in seeing what a fisheries scientist thinks of their ideas and what these scientists can then come up with. In view of the recent move towards interdisciplinary research among scientists, it would be incongruent to not consider collaborative research to be a more ambitious version of this.

At the time of writing, final results are being analysed and are yet to be published. This means that the impact of these findings and contribution to management decisions is yet to be determined. Despite the fact that Gap as a project has officially come to an end, fishers are keen to continue similar research activities in the near future – possibly by using national funds. It would be naïve to assume that a single 4-year project can bridge all the differences between fishers, scientists and policy-makers, particularly when one considers that this project focused only on a single gear amongst several. However, it has certainly started the process of doing so. The fact that fishers have remained eager and shown interest in further collaborative work is a sign of that success.

15.5 Conclusion

Although the Maltese case study led to the successful collection of data, as mentioned above, it has not met all of its aims with respect to collaboration. There is scope for further research, particularly with regard to collecting time-series data on seasonal changes and fishers are eager to partake in it. In that sense the case study has been groundbreaking vis-a-vis collaborative research between fishers and the state within the Maltese 25 NM FMZ.

What will be necessary is that future research is of a longer timespan. This is the case because time is needed to foster genuine collaboration between fishers and scientists and fishers and policy-makers. It is necessary that tangible and in-depth analyses of any data collected would first need to be published and explained by scientists and equally importantly that policy-makers while using this data involve fishers and get their input. In addition, any new management plans would need to show clear signs of bearing fruit prior to fishers fully endorsing the concept of collaborative research.

References

Abella A, Fiorentino F, Mannini A, Orsi Relini L (2008) Exploring relationships between recruitment of European Hake (*Merluccius merluccius* L. 1758) and environmental factors in the Ligurian sea and the Strait of Sicily. J Mar Syst 71:279–293

Ames EP (2004) Atlantic cod structure in the Gulf of Maine. Fisheries 29:10–28

Camilleri M, Dimech M, Drago A, Fiorentino F, Fortibuoni T, Garofalo G, Gristina M, Schembri PJ, Massa F, Coppola S, Bahri T, Giacalone V (2008) Spatial distribution of demersal fishery resources, environmental factors and fishing activities in GSA 15 (Malta Island). GCP/RER/010/ITA/MSM-TD-13. MedSudMed Technical Documents 13:97

Ciannelli L, Fauchald P, Chan KS, Agostini VN, Dingsor GE (2008) Spatial fisheries ecology: recent progress and future prospects. J Mar Syst 71:223–236

Coll M, Carreras M, Ciercoles C, Cornax MJ, Gorelli G, Morote E, Saez R (2014) Assessing fishing and marine biodiversity changes using fishers' perceptions: the Spanish Mediterranean and Gulf of Cadiz study. PLoS One 9(1):e85670

Commission Decision (EC) 2010/93/EU Adopting a multiannual Community programme for the collection, management and use of data in the fisheries sector for the period 2011–2013

Council Regulation (EC) no. 1967/2006 concerning management measures for the sustainable exploitation of fishery resources in the Mediterranean Sea, amending Regulation (EEC) No. 2847/93 and repealing Regulation (EC) No. 1626/94

Fiorentino F, Garofalo G, De Santi A, Bono G, Giusto GB, Norrito G (2003a) Spatio-temporal distribution of recruits (0 group) of *Merluccius merluccius* and *Phycis blennoides* in the Strait of Sicily (Central Mediterranean). Hydrobiologia 503:223–226

Fiorentino F, Garofalo G, Gristina G, Levi D (2003b) The ratio between bottom-dwelling and overall fish biomass (BOI) as an indicator of trawling impact on demersal assemblages. Biol Mar Mediterr 10(2):819–823

Fiorentino F, Garofalo G, Gristina G, Giacintano S, Norrito G (2004) Some relevant information on the spatial distribution of demersal resources, benthic biocenoses, and fishing pressure in the Strait of Sicily. In: Levi D, Bahri T, Camilleri M, Jarboui O, Massa F, Ragonese S, Zgozi S (eds) Report of the expert consultation on the spatial distribution of demersal resources in the straits of sicily and the influence of environmental factors and fishery characteristics, pp 50–66. GCP/RER/010/ITA/MSM-TD-02. MedSudMed technical documents no. 2

Fisheries Conservation and Management Act, Chapter 425, ACT II of 2001

Food and Agriculture Organisation of the United Nations – FAO (2003) Fisheries management: 2. The ecosystem approach to fisheries, FAO technical guidelines for responsible fisheries 4(2). FAO, Rome, Italy

Fortibuoni T, Bahri T, Camilleri M, Garofalo G, Gristina M, Fiorentino F (2010) Nursery and spawning areas of deep-water rose shrimp, Parapenaeus longirostris, in the Strait of Sicily. J Crustac Biol 30:167–174

Garofalo G, Bel Hassen M, Jarboui O, Zgozi S, Gristina M, Fiorentino F, Ragonese S, Camilleri M (2008) Preliminary results on spatial distribution of abundance indices, nurseries and spawning areas of *Merluccius merluccius* and *Mullus barbatus* in the central Mediterranean. GCP/RER/010/ITA/MSM-TD 19. MedSudMed technical documents no. 19

Garofalo G, Ceriola L, Gristina M, Fiorentino F, Pace R (2010) Nurseries spawning grounds and recruitment of Octopus vulgaris in the Strait of Sicily, Central Mediterranean Sea. ICES J Mar Sci 67:1363–1371

Garofalo G, Fortibuoni T, Gristina M, Sinopoli M, Fiorentino F (2011) Persistence and co-occurrence of demersal nurseries in the Strait of Sicily: Central Mediterranean: implications for fishery management. J Sea Res 66:29–38

General Fisheries Commission for the Mediterranean – GFCM (2015) Recommendation GFCM/40/2015/4, On the establishment of a set of minimum standards for bottom trawling fisheries of demersal stocks in the Strait of Sicily, pending the development and adoption of a multiannual management plan

General Fisheries Commission for the Mediterranean – GFCM (2016). Recommendation GFCM/39/2015/2, Establishing a multiannual plan for the fisheries exploiting European hake and deep-water rose shrimp in the Strait of Sicily (GSAs 12 to 16)

Government of Malta. Ministry for European Affairs (2014) European maritime and fisheries fund – operational programme for Malta 2014–2023, v 1.2. Ministry for European Affairs, Malta

Government of Malta. Ministry for Sustainable Development the Environment and Climate Change – MSDEC (2013) Fishing effort adjustment plan for bottom otter trawl fishery 2013–2015. Malta, MSDEC

Hall-Arber M, Pederson J (1999) Habitat observed from the decks of fishing vessels. Fisheries 24:6–13

Holm P, Dubois M, Tveiteras K, Bjorkan M, Clyde Wilson D, Kraan M (2013) Collaborative research and participant attitudes. GAP2, WP4, Deliverable D4.2.1

Hutchings JA (1996) Spatial and temporal variation in the density of northern cod and a review of hypotheses for the stock collapse. Can J Fish Aquat Sci 53:943–962

Johnson TR, van Densen WLT (2007) The benefits and organization of cooperative research for fisheries management. ICES J Mar Sci 64:834–840

Kaplan IM, McCay BJ (2004) Cooperative research, co-management and the social dimension of fisheries science and management. Mar Policy 28:257–258

Kneebone J, Chisholm J, Bernal D, Skomal G (2013) The physiological effects of capture stress, recovery, and post-release survivorship of juvenile sand tiers (Carcharias taurus) caught on rod and reel. Fish Res 147:103–114

Knittweis L, Dimech M (Unpublished) Assessment of nursery and spawning areas of commercially important species in GSA 15

Kraan M, Uhlman S, Steenbergen J, Van Helmond ATM, Van Hoof L (2013) The optimal process of self-sampling in fisheries: lessons learned in the Netherlands. J Fish Biol 83:963–973

Levi D, Andreoli MG, Rizzo P (1992) Growth curves from representative samples of fish populations as possible hints for the identification of unit stocks: the case study of red mullet from the Sicilian Channel. Marine eutrophication and population dynamics. Proceedings of the 25th EMBS, Olsen & Olsen, Denmark

Levi D, Andreoli MG, Rizzo P (1993) An analysis based on trawl-survey data of the state of the "Italian" stock of *Mullus barbatus* in the Sicilian channel, including management advice. Fish Res 17:333–341

Levi D, Andreoli MG, Rizzo P (1995) Independence of water masses and independence of stocks. Rapp Comm Int Mer Medit 34:248

Levi D, Andreoli MG, Bonanno A, Fiorentino F, Garofalo G, Mazzola S, Norrito G, Patti B, Pernice G, Ragonese S, Giusto GB, Rizzo P (2003) Embedding Sea surface temperature anomalies into the stock recruitment relationship of red mullet (*Mullus barbatus* L. 1758) in the strait of Sicily. Sci Mar 67(S1):259–268

Maurstad A, Sundet JH (1998) The invisible cod E – fishermen's and scientists' knowledge. In: Jentoft S (ed) Commons in a cold climate. Coastal fisheries and reindeer pastoralism in North Norway: the co-management approach. UNESCO, Paris, pp 167–184

MedSudMed (2007) Report of the MedSudMed expert consultation on marine protected areas and fisheries management. GCP/RER/010/ITA/MSM-TD-03. MedSudMed technical documents, 3, 100 pp

MedSudMed (2013) Report of the eleventh meeting of the Coordination Committee, Sliema, Malta, 23–24 April 2013

MedSudMed (2014) Report of the twelfth meeting of the Coordination Committee, Tunis, Tunisia, 14–16 April 2014

MedSudMed (2015) Report of the twelfth meeting of the coordination committee. FAOMiPAAF assessment and monitoring of the fisheries resources and ecosystems in the straits of Sicily. GCP/RER/010/ITA/MSM-TD-35. MedSudMed Technical Documents 35:37

Neis B, Schneider DC, Felt L, Haedrich RL, Fischer J, Hutchings JA (1999) Fisheries assessment: what can be learned from interviewing resource users? Can J Fish Aquat Sci 56:1949–1963

Palsson OK, Jonsson E, Schopka SA, Stefansson G, Steinarsson BÆ (1989) Icelandic groundfish survey data used to improve precision in stock assessment. J Northwest Atl Fish Sci 9:53–72

Parrish JK (1999) Using behavior and ecology to exploit schooling fishes. Environ Biol Fish 55:157–181

Poisson F, Vernet AL, Filmater JD, Goujon M, Dagorn L (2011) Survival rate of silky sharks (Carcharius falciformis) caught incidentally onboard French tropical purse seiners. IOTC-2011-WPEB07–28, p 6

Poisson F, Seret B, Vernet A-L, Houjon M, Dagorn L (2014) Collaborative research: development of a manual on elasmobranch handling and release best practices in tropical tuna purse-seine fisheries. Mar Policy 44:312–320

Wilson DC (2003) Fisheries co-management and the Knowledge Base for management decisions. In: Wilson DC, Nielsen JR, Degnbol P (eds) The fisheries co-management experience, Fish and Fisheries Series, vol 26. Springer, Dordrecht

Chapter 16
People, Sharks and Science

What Can It Take for Industry-Led Research to Make a Difference to the Management of Elasmobranchs of Conservation Concern in UK Waters?

Stuart J. Hetherington and Victoria A. Bendall

Abstract Drawing on personal experience of industry-led research in the South-west of the UK, this chapter takes you on a journey through what it can take for industry-led research to make a difference to the sustainable management of elasmobranchs (sharks, skates and rays) considered to be of conservation concern. Throughout this journey, the research design, implementation and resulting analysis are a collaborative effort between scientists and fishermen. Between 2009 and 2011, three species of elasmobranch in the North-east Atlantic, spurdog (*Squalus acanthias*), common skate (*Dipturus batis 'complex'*) and porbeagle (*Lamna nasus*), went from being perceived as a valuable, commercially fished quota species (limits on the amount of fish that can be landed for sale by fishermen) to being a conservation species with either a zero total allowable catch (TAC) or with a prohibited species listing. Due to the large sizes and aggregative nature of these shark species, by-catch and discard issues emerged in the offshore commercial net fisheries of the Celtic Sea where there were no previous issues. The resulting high rates of by-catch and discards of what are now considered conservation species had a follow-on negative financial impact on the fishing industry. United through the Cornish Fish Producers Organisation (CFPO), fishermen from the South-west of the UK together with the Shark Trust (an environmental non-government organisation (eNGO)) raised awareness with the UK Government of the need to generate the evidence base required to help find a solution for effective management and conservation of spurdog, common skate and porbeagle. Here we highlight two industry by-catch and biological data collection projects that have explored what is needed to create effective engagement and problem-solving by Government scientists, fishermen and their representatives, Government policy advisors and an eNGO. This experience provides a road-map for current best practice and methods for scientists to collaboratively

S. J. Hetherington (✉) · V. A. Bendall
Centre for Environment, Fisheries and Aquaculture Science (Cefas), Lowestoft, Suffolk, UK
e-mail: stuart.hetherington@cefas.co.uk

P. Holm et al. (eds.), *Collaborative Research in Fisheries*, MARE Publication Series 22, https://doi.org/10.1007/978-3-030-26784-1_16

work with the fishing industry to collect by-catch data and biological information on data-deficient species of conservation concern. It serves as an example of how industry-led research can generate scientifically robust data to feed into policy and influence management.

Keywords Industry research · By-catch · Elasmobranchs · Conservation species · Management

16.1 Introduction

One of the biggest problems with managing elasmobranch fisheries when compared to other commercially valuable species such as cod (*Gadus morhua*) and plaice (*Pleuronectes platessa*) is that relatively little is known about their populations in UK waters. Consequently, in terms of fisheries management, elasmobranchs are referred to as data-limited species. This makes stock assessments difficult, and often results in the adoption of more pronounced precautionary management approaches, for example, Prohibited Species listings, where the species in question cannot be targeted, retained, transhipped or landed. Such measures are often viewed by some sectors of the fishing industry as overly punitive and ineffective, and often do not reflect what the fishermen report observing whilst fishing at sea.

In the Celtic Sea (ICES Div. VIIe-j), offshore commercial gill and trammel net fishermen from the South-west of the UK fish on open ground targeting whitefish such as cod, hake (*Merluccius merluccius*), monkfish (*Lophius spp.*) and turbot (*Scophthalmus maximus*), with resulting by-catch (unintended capture) of a diverse variety of elasmobranch species that supplement commercial landings. Fishermen within this region consider populations of some elasmobranchs such as spurdog, common skate and porbeagle to be locally and/or seasonally common. However, with little scientific information available to support regional fishermen knowledge and conservation concerns resulting in restrictive year on year quotas for these species, between 2009 and 2011 a series of precautionary management measures were put in place to move spurdog, common skate and porbeagle from being commercial species to conservation species with a zero TAC or prohibited species listing. This means that irrespective of the actual levels of abundance of spurdog, common skate and porbeagle, fishermen could no longer land these species, resulting in immediate loss of earnings for South-west UK commercial net fisheries.

Due to their large size and aggregative nature, these elasmobranchs are hard to avoid and are susceptible to by-catch (Fig. 16.1). Precautionary management measures can generally result in high levels of live and dead discarding (throwing unwanted catch back into the sea), as the interaction between fishermen and spurdog, common skate and porbeagle continues regardless. Fishermen experience further negative by-catch-related economic impact due to associated gear damage and extended periods of down-time from clearing nets of these sharks and loss of income from previously saleable catch. Fishermen report that levels of by-catch can be so

Fig. 16.1 (**a**) A porbeagle caught in a gill net; (**b**) study species of interest, spurdog, porbeagle and common skate by-catch on-deck of a fishing vessel before being discarded to sea. (Photos by V. Bendall & S. Hetherington)

substantial that vessels may be displaced from their historic fishing grounds so as to negate these impacts.

To address the by-catch problem, there is a need to address the data gaps that ultimately make these species 'data deficient', finding out more about their life history, movement, distribution, behaviour, by-catch levels and discard survival and to use our improved understanding to support and inform policy, for example, by providing options for possible by-catch mitigation.

16.1.1 Catalyst for Collaboration and Initial Research

With high discarding of spurdog, common skate and porbeagle, and the impact of zero allowable catches on the UK's South-west offshore commercial gill net fisheries, the fishing industry united under the umbrella of the CFPO, highlighted to the UK Government the need to act. Additional pressure was applied through the voice of the Shark Trust, championing a need for greater conservation and sustainable exploitation. Together, these voices raised awareness in Government of the need to establish effective policy, evidenced by the best available scientific information and data.

Government policy advisors sanctioned two preliminary data collection initiatives designed to provide information on the by-catch. The first, in 2009, took a conventional fishery-independent scientific approach to study the movement, distribution and behaviour of spurdog and porbeagle using state-of-the-art electronic tags (Righton et al. 2013). The second, in 2011, was an industry-led initiative that was established under the UK Fisheries Science Partnership (FSP) programme, called 'Spurdog, porbeagle and common skate by-catch and discard reduction'. Proposed by the fishing industry, and led by scientists, this latter one-off project used skipper's knowledge, vessels, and fishing gear to catch elasmobranchs in three, 1-week research surveys. This project involved the gill net fishing industry from the

South-west of the UK for the first time in elasmobranch by-catch research, demonstrating their capacity to be involved in collecting policy-relevant data (Bendall et al. 2012).

In trying to ensure the best available evidence, salient questions of these studies were asked by all parties such as: 'How many sharks must be tagged to generate enough data on which to base sound policy decisions?' 'When you have enough data, how do you get it used to inform national and European policy?' These were the type of questions that shaped the journey of our engagement experiences from 2011 onwards, where efforts were dedicated to empowering industry-led research that could make a real difference in how UK Government policy managed elasmobranch fisheries. Here we chart the journey, identifying significant waypoints along the way.

16.1.2 From Pistols at Dawn to Collaboration and Co-specification

To initiate open engagement with the fishing industry and 'discuss' the importance and current issues of elasmobranch by-catch and discards, our first stakeholder event was held on October 2011 in Newlyn in the South-west of the UK, at the largest port for commercial offshore netters (Box 16.1). And what a meeting it turned out to be! Advertised as 'an opportunity to share your views and make a difference', the fishing industry certainly rose to the challenge. In sharing their views, and unknown to them at the time, they changed the course of direction on elasmobranch by-catch research in the UK. Hindsight is indeed a wonderful thing, as the first engagement event proved equivalent to waving a red flag in front of a bull. Passionate, angry fishermen, who felt that their voice had not been heard, were provided with a platform from which to vocalise issues close to their hearts, with a captive audience of scientists and policy-making officials. The stage was set!

Box 16.1: The Advert for the First Meeting in October 2011
Officials from Department for Environment Food & Rural Affairs (Defra), will be present to provide details of fisheries policy in the UK and Europe and scientists from Centre for Environment, Fisheries and Aquaculture Science (Cefas) will be present to share their knowledge gained from recent efforts to develop a greater understanding of by-catch and discards. The Shark Trust and Marine Management Organisation will also be represented.

The event began with an overview of ICES assessments and management rules for spurdog, common skate and porbeagle. This was the cue for the collective voice of the fishing industry to begin their infantry barrage. No matter the justification provided by the scientist, the barrage was relentless, species after species, for what seemed like hour after hour. At the heart of the problem in practically every case raised was a lack of data. Government policy for elasmobranchs is based on data-limited scientific advice, which leads to precautionary management measures. This frustrates the fishing industry because they perceive their livelihoods as being detrimentally affected by policy decisions based on a lack of scientific data. The fishermen openly expressed their view that the science, fisheries policy, and restrictive management measures did not reflect the actual situation which they were encountering during their fishing trips. From the fishermen's viewpoints, far greater local seasonal abundances occurred within their fisheries than the available scientific evidence portrayed. Furthermore, the information used in stock assessments was perceived to be based on unrepresentative or incomplete fisheries data, and it was felt that there was an argument therefore that the assessments did not reflect the true state of the stocks in question.

The fishermen aired their mounting frustrations that both the methodology of data collection and scientific conclusions drawn from it were poor. They criticised the use of a scientific research vessel as being inherently inappropriate to survey elasmobranch abundance. Their argument was that the research vessel did not survey the same grounds as those fished, and the vessels used different, non-comparable, fishing gears, which were responsible for what they saw as inconclusive data. After heated discussions, a moment of light relief was paramount, and a short break for tea and biscuits was convened. From experience, we cannot emphasise enough the power of tea and biscuits to re-engage and refuel excited stakeholders! A far less controversial presentation on scientific data collected from electronic tagging of porbeagle followed, which pricked the fishermen's interest, with genuine intrigue for a novel scientific approach that revealed insights into fish behaviour which the fishers could directly relate to. Following this respite, the fishermen soon turned their focus back to the initial presentation of scientific data collection and questioned why scientists were not using fishing vessels instead. They argued for the incorporation of skippers' knowledge and for the use of more appropriate fishing gears. As a reference point, the fishermen referred to the previous preliminary FSP project on spurdog, common skate and porbeagle by-catch as an acceptable way forward. Looking back, this was a defining moment for our work, a small moment in history, the significance of which was not apparent at the time.

The chair of the meeting, the Chief Executive of the CFPO, was pivotal in the ultimate success of the meeting, bringing proceedings to a controlled conclusion with a cloud parting moment of clarity where upon all parties, fishermen, scientists, Government policy and the eNGO agreed that fishermen's knowledge and information would help to strengthen the evidence base used in stock assessments, scientific advice and management. Subsequently, many offers followed from fishermen both to share their knowledge and to help improve the data on which scientific advice is based. Policy officials from Defra and the scientists from Cefas had a clear take-home message: to develop in collaboration with the CFPO and the fishing industry, a proposal for an industry-led scientific sampling programme to map species distributions and reduce discard mortality.

16.2 Project Inception

16.2.1 The First Path: The Shark, Skate and Ray Scientific By-Catch Fishery

One year on, December 2012, we were back in Newlyn for a second event which was to become an annual forum for all interested stakeholders (local fishermen, Cefas, Defra, Marine Management Organisation (MMO) and the Shark Trust) to discuss the issues surrounding elasmobranch by-catch and discards. The event had one key aim, to launch the first of the two industry data collection programmes considered in this chapter, a Defra-funded, Cefas-led project, 'The Shark, Skate and Ray Scientific By-catch Fishery', with a purpose to increase the understanding of spurdog, common skate and porbeagle distributions in Celtic Sea fisheries by mapping out the seasonal distribution of by-catch.

True to form, the touch paper was ignited by the opening gambit of ICES advice and fisheries policy for the species of interest. Once again, the fishermen vented their frustration at the perceived lack of progress, and at least one skipper voted with his feet, leaving the event with threats involving a Cornish pasty!

On the flip side, one skipper had decided to come to the meeting having prepared a presentation all of his own accord. He stood up and presented to the assembled group with a passionate account of elasmobranch discards based on his own experience, quite possibly a first at such an event. Next, stakeholders fell silent and united around recent scientific advances on understanding the biology, behavioural movements and discard survival of spurdog, common skate and porbeagle. There was a collective understanding that these scientific data were truly relevant to fisheries management and conservation, and better still, a profound realisation that collaboration would be key to any successful way forward. It was at this same event that another foundation block was laid for industry-led data collection, but again somewhat unknowingly and not entirely with planned intent.

16.2.2 The Second Path: The Spurdog Programme

Likely influenced by the skipper's passionate presentation portraying graphic photographs of the high levels of spurdog by-catch frequently encountered, the attending Defra policy advisor took a key message home; the pressing need to tackle spurdog by-catch and dead discards. Later, a joint press article with Cefas and the CFPO highlighted the outputs of the meeting:

> the discarding of dead fish is quite rightly seen as unacceptable among fishermen and the general public. We share this frustration. Eliminating this wasteful practice is something we have been pushing for reform of the CFP. ...and through projects such as the Shark Skate and Ray Scientific By-catch Fishery that will provide information to help us make future management decisions for species such as ...spurdog. (Hetherington et al. 2013)

Like all good engagement events, the discussion continued long after the formal meeting had finished. The participants lingered, openly discussing what could be done in good time for spurdog. Ultimately, these discussions generated the second project considered here, 'Common Fisheries Policy reform implementation: aligning zero quota species and improving fisheries management – a spurdog case study' (Hetherington et al. 2015b), a project that later led to the Spurdog By-catch Avoidance Programme (Bendall et al. 2014; Hetherington et al. 2016, 2018), herewith referred to as the Spurdog Programme.

16.3 Implementation

16.3.1 Data Collection – Priorities and Challenges

The main challenge at sea for the scientists in both projects was not actually the training of the crew in the scientific methods required but rather first gaining the trust of the crew as to what information was being collected, what it was required for, and ultimately what it would be used for. Until now, those fishermen engaged in the process and in attendance at the stakeholder events were the skippers and vessel owners, not the crew. Once the crew had accepted the scientific reasoning and explanation (with underpinning reassurance from the skipper), only then could the logistical issues of collecting quality assured data be tackled.

At sea for up to 9 days at a time, often in challenging conditions, scientists worked with the gill-net fishermen, training skipper and crew in data and sample collection procedures. For example, for the Shark, Skate and Ray Scientific By-catch Fishery, which involved numerous vessels at sea simultaneously, the crew were trained in species identification, measurement, sexing, recording and landing of scientific samples. The on-board procedures were designed to be incorporated into the day-to-day vessel routine, to make them as easy as possible for the crew while at the same time seeking to ensure scientific rigour. On subsequent trips, the scientists would step back and trust the trained crew, providing support only as and when required. For live elasmobranchs in good condition on-deck, crew were trained to tag and release them.

For the Spurdog Programme, the industry-led data collection programme was far easier to achieve. Only requiring catch data for one species (total by-catch by weight) at each location, these data were to be recorded and reported daily. In order to avoid burdening the fishermen by asking for duplication, biological information (length and sex) and on-deck vitality of spurdog was obtained from the FSP project and the Shark, Skate and Ray Scientific By-catch Fishery. From these combined studies, carried out between August 2011 and October 2013, 444 spurdog were captured and biological data were collected by scientists in partnership with fishermen during four separate week-long trips in the commercial net (gill, trammel & tangle) fisheries across the Celtic Sea. The typical gill net soak time was less than 26 h, reflecting typical commercial practice when targeting hake and other whitefish.

Table 16.1 Summary details of spurdog caught over four scientific field studies (2011–2013), under commercial conditions, to improve knowledge of spurdog by-catch survival (on-deck vitality) within commercial net fisheries in the Celtic Sea

Vessel	Month	Total No. Spurdog	Males		Females		Lively		Sluggish		Dead	
			No.	Length range	No.	Length range	No.	%	No.	%	No.	%
Govenek of Ladram	Aug 2011	149	84	56 – 84	65	57 – 107	73	50%	18	12%	58	38%
Charisma	Sept 2011	182	115	64 – 91	67	60 – 123	135	72%	10	5%	37	23%
Charisma	Mar 2012	53	50	67 – 86	3	69 – 82	40	75%	7	13%	6	12%
	Total	384	249	56 – 91	135	57 – 123	248	65%	35	9%	101	26%
Govenek of Ladram	Oct 2013	60	38	75 – 100[a]	22	50 – 100[a]	14	23%	NR	NR	46	76%
	Overall total	444	287	56 – 91	157	57 – 123	262	59%	35	8%	147	33%

Shaded figures based upon data from the FSP project. Unshaded figures based upon data from the Shark, Skate and Ray Scientific By-catch Fishery

[a]Length range estimate only (individuals not recorded to nearest cm); NR = 'Sluggish' health state not recorded for this study

Survival rates were consistently high for spurdog regardless of gillnet soak time (varying between 11 and 45 h), and 262 (59%) were discarded back to sea in what is described as a 'lively' condition. The remaining 41% of spurdog scored poor health state conditions, with 35 individuals classed as 'sluggish' and 147 classed as 'dead' prior to discarding (Table 16.1).

16.4 Outreach and Outcomes

Outreach resulting from the Spurdog Programme went far beyond the annual stakeholder events in Newlyn, forging robust collaboration between fishermen, their representatives, Government policy advisors (both nationally and European), eNGO's, scientists and the European Commission.

In 2014, the Spurdog Programme proposed an alternative management option to the precautionary approach for spurdog. The alternative was intended to ensure that there was no incentive for targeted fishing while at the same time allowing for incidental dead spurdog by-catch in mixed fisheries to be landed, i.e. a move away from a zero catch and landing policy to a situation where dead spurdog by-catch could be landed.

The means to test and monitor the success of this alternative option was proposed as a UK pilot project: 'Spurdog By-catch Avoidance Programme'. This empowered fishermen to make informed decisions, facilitating avoidance of recent by-catch 'hotspots' within their own fishery, with the potential to reduce incidental by-catch and fishing mortality of spurdog. In May 2014, Defra policy advisors travelled to Brussels to brief the European Commission of the Spurdog By-catch Avoidance Programme proposal.

Back in the UK, an extensive and lengthy consultation process was undertaken with a diverse range of stakeholders (fishermen, Defra, Cefas, Marine Scotland, Scottish Association for Marine Science, School for Marine Science & Technology, MMO, Zoological Society of London, several fishermen's associations and organisations, Client Earth, Marine Conservation Society, Shark Trust and the Dutch Elasmobranch Society) to develop the initial proposal presented to the European Commission. To gain cohesion from the fishing industry, the UK National Federation of Fishermen's Organisations (NFFO) was chosen to lead the consultation process. Two regional meetings were held with broad representation of experienced fishermen to evaluate the strengths and weaknesses of the proposal from scientific and management perspectives. These were followed by a National consultation workshop in October 2014, held in London to facilitate convenient access for national and international attendees and speakers from Advisory Councils, eNGOs, Government and science. The key outcome from this meeting was a mutually agreed, finalised proposal for final submission, first to the Scientific, Technical and Economic Committee for Fisheries (STECF) for expert review in November 2014, before being refined for final consideration at the European December Fisheries Council 2014 (Hetherington et al. 2015a)

Although reviewed positively by STECF (STECF report 2014), a dead by-catch allowance to eliminate dead spurdog discards and offset the burden to fishermen of having to potentially move to less productive fishing grounds was not secured at the December Fisheries Council 2014.Undeterred, the Defra and Cefas team reported back to the fishing industry at the annual stakeholder event in Newlyn in 2015. Despite a frosty and frustrated reception from the fishing industry for having failed to secure a dead by-catch allowance, there was appreciation for the effort invested, and progress was made in highlighting the issue with the European Commission. The fishing industry agreed to remain engaged with a limited 'proof of concept' trial of the Spurdog By-catch Avoidance Programme, but without a dead spurdog by-catch allowance (Hetherington et al. 2016).

With single stock management of spurdog across the North-east Atlantic, the stakeholders were keen to engage more widely with European countries. The proposed programme was therefore presented at, and won the support of, the North Western Waters Advisory Council, who recommended the participation of other Member States in the project.

The limited 'proof of concept' trial allowed ongoing engagement with the European Commission, and the programme was reviewed for a second time by STECF in November 2015 (STECF report 2015). With strong underlying conviction, in 2016, Cefas and Defra continued to engage with the European Commission, meeting Commissioners to discuss the finer details of the proposed programme. Finally, in July 2016, after 6 years of extensive collaborative engagement, a dead spurdog by-catch allowance was secured.

16.5 Lessons Learnt on the Journey

16.5.1 Reflection on the Projects' Achievements

With the successful implementation of two industry-led elasmobranch by-catch and discard research programmes, funded by UK Government, the Spurdog Programme ultimately realised its full potential, whereas the Shark, Skate and Ray Scientific By-catch Fishery did not. So why was this?

Both programmes had clear aims. The aim of the Shark, Skate and Ray Scientific By-catch Fishery was to '…provide fishery-dependent information in support of the assessment and management process, to design by-catch mitigation measures and thereby provide the foundation to support the development of a regional management plan for elasmobranchs…'. The aim of the Spurdog Programme was the following: 'An evidenced-based nominal by-catch allowance for dead spurdog that does not incentivise targeting of the stock to be investigated, together with options for potential management measures, such as spatial and temporal avoidance. Options will be presented by Defra to the Commission for their consideration'.

The Shark, Skate and Ray Scientific By-catch Fishery successfully engaged with the fishing industry in a way that probably had not been pursued before in the UK and certainly not in that particular fishery. The programme recorded data on elasmobranch by-catch and discards that would have otherwise gone unrecorded, and without this work, we would still be unsure as to the level of by-catch and discards that occurs (Ellis et al. 2015). However, the project stopped short of realising the aim to develop a regional management plan. In spite of this shortcoming, the Shark, Skate and Ray Scientific By-catch Fishery was of critical importance in underpinning the success of the Spurdog Programme, which utilised data on spurdog by-catch rates and on-deck vitality to provide evidence of current fisheries practices and estimates of dead discarding within the region.

Both the Shark, Skate and Ray Scientific By-catch Fishery programme and the preceeding FSP project laid the foundations of the Spurdog Programme and this latter's aim of an evidence-based dead by-catch allowance, with the European Commission amending Council Regulation (EU) 2016/72, granting a by-catch allowance for dead spurdog for vessels participating in the UK led Spurdog By-catch Avoidance Programme (Hetherington et al. 2018).

So back to the why? The Spurdog Programme had a clear goal, and a significant ongoing investment was made in defining a clear pathway to realising that goal. As well the effort involved in engaging all stakeholders in conducting the scientific research and in collecting the data, just as much effort was expended in turning the data into evidence and feeding the evidence into policy. Through targeting the appropriate policy chain, consideration could then be made as to how policy could then be implemented through the European Commission. The entire process required sustained effort and drive, way beyond robust scientific data and evidence collection alone. This ultimately was the difference between the Shark, Skate and Ray Scientific By-catch Fishery, which invested energy and drive into the creation of new industry data collection methods, but which invested less in when, how and by whom the data should be used to achieve its aim. Despite incentivisation through monthly payments, but with no visible impacts from the Shark, Skate and Ray Scientific By-catch Fishery, and a continuing burden on the skipper and crew to provide detailed information on multiple elasmobranch by-catch species, fishermen-fatigue set in. Fishers became unwilling to continue the data collection programme in the same format, ultimately becoming a sticking point for the Shark, Skate and Ray Scientific By-catch Fishery pilot study.

As reported by Mackinson et al. (2015), there are widely held views in science that fishery-dependent research is not appropriate for peer-reviewed publication. Thus, the publication-based reward system may to some extent deter scientists from engaging in participatory or industry-led research due to the length of time it can take to yield publications. For the Spurdog Programme, a fishery-dependent approach was the mainstay of the research, but the industry-generated data were in themselves insufficient. For example, to inform how by-catch could be avoided, the fishery-dependent data was complemented with fishery-independent electronic tagging data detailing spurdog movements and behaviour. It was the electronic tagging data reported in scientifically robust customer project reports (Righton et al. 2013; Hetherington et al. 2015b) that were used by the Spurdog Programme (Bendall et al. 2014) to inform Government policy and effect change in fisheries management, as these data had not yet reached peer-reviewed publication, the benchmark by which all scientists are measured.

In summary, the former projects paved the way for the success of the Spurdog Programme, understanding the fisheries problem, building a network of relationships and partnerships, developing data collection methods and applying lessons learnt along the way. The Spurdog Programme clearly identified and agreed on its overarching goal with all stakeholders at the outset, maintained focus and invested resource in identifying the political and governance pathways to realise success. Engagement and feedback provided a continuous loop, with all the stakeholders involved, mitigating stakeholder fatigue, concluding in industry-led research influencing UK Government and European policy and fisheries management of spurdog in the North-east Atlantic.

16.5.2 Key Ingredients for Successful Engagement with the Fishing Industry

The key ingredients for successfully engaging with the fishing industry and for fishery-dependent data to feed into effective management can sometimes be difficult to discern. However, through our experience, five key ingredients were evident:

1. A passionate and driven fishing industry representative (in this case, the Chief Executive of the CFPO), well respected by fishermen, able to distil emotive and complex positioning of fishermen into specific and realistic objectives.

2. A proactive, reasoned, centre-ground eNGO (in this case, the Shark Trust), willing to constructively engage and support fishery-dependent data collection to increase the evidence base to support sustainable management.

3. A dynamic Government (Defra) policy advisor, genuinely engaged with and valuing the worth of fishermen's involvement in data collection, able to effectively communicate with the fishing industry, and respected by fishermen.

4. Participatory research scientists at Cefas, building trust and professional relationships with fishermen at sea, understanding the worth and utility of fishermen providing information on their fisheries, to complement traditional fisheries science approaches.

5. The fifth magic ingredient, without which all else fails – 'belief'. Forging out an industry-led research pathway requires dedicated time to fully engage and maintain engagement with the fishing industry to collect policy-relevant data, while pursuing and further developing open collaborative partnerships. This path is not easy and should never be taken lightly: Failure to succeed is all too easy and can take a lifetime to undo, with a detrimental impact to reputation and future collaborations.

16.5.3 Stages in the Process

This research programme on elasmobranch by-catch and discards in net fisheries identified and demonstrated important stages and pathways in the engagement process (Table 16.2).

16.5.4 Mitigating Fatigue

The fishery-dependent approach to data collection can complement and add value to traditional methodologies. However, there is a commonly held perception of a shortfall in the rigour of industry-led research as described by Hetherington et al. 2017, and that industry-led science is not always appropriate for peer-reviewed publication (Mackinson et al. 2015). This perception must be challenged. Here we provide an example of how participatory research with mutual understanding and scientifically robust guidance and shared ownership of scientific goals can yield scientifically robust outcomes, which can ultimately achieve the desired goals through the appropriate stakeholder, political and governance pathways. Challenging perceptions can be exhausting, creating fatigue for all sides; however avoiding fatigue amongst fishing industry representatives should be seen here as a priority, as industry-led research is likely to play an increasingly prominent role in

Table 16.2 Key stages of what is required in industry-led research on elasmobranchs in the UK, from identifying the issue to obtaining a satisfactory outcome

Key stage	Action required
Collaboration & Co-specification	Engage with fishermen. Involve all relevant stakeholders. Communicate to the fishing industry at large. Identify the issue. Set a clear, specific and realistic goal, shared by all stakeholders. Define the role of each involved stakeholder. Agree what success will look like.
Co-design & Implementation	Assess the level of involvement required of fishermen in data collection to meet the goal. Work with fishermen to collect data, ensuring quality and consistency. Feedback the data to the fishermen, allowing them to review the preliminary results.
Outreach	Engage with fishermen. Involve all relevant stakeholders. Include those who are most likely to use the data. Mitigate participatory fatigue by providing feedback and communicating 'wins'. Communicate the results to all stakeholders.
Evaluation	Determine if the goal was met, project uptake and impact. Lessons learnt applied to forward planning with continuous learning.

Adapted from Johnson and van Densen (2007)

future studies. The engagement of fishermen in the lengthy process to secure a dead by-catch allowance for the Spurdog Programme was dutifully captured by the Chief Executive of the CFPO in a joint article.

'We need to improve scientific understanding, and the working relationship between Cornish fishermen and Cefas scientists is contributing to this… This relationship is based on working towards sensible management regimes that will ensure improved scientific understandings, sensible by-catch provisions and the reduction of wasteful discards…The challenge now is to deliver this!' (Hetherington et al. 2013).

As each week, month and year go by with little output or return for the effort and contribution given by fishermen, fatigue and disengagement become an uphill battle to prevent. Fortunately, through relentless efforts over the 6-year period of engagement, collaborative efforts secured a scientifically robust, pragmatic solution to the management dilemma of spurdog, which reinvigorated participation of the fishing industry. This allowed full implementation of the Spurdog By-catch Avoidance Programme at a crucial time when the support of the fishing industry might easily have been lost.

16.6 Concluding Remarks

Engagement and data collection are just two components of any fishery-dependent or industry-led research approach. Equal account needs to be taken at the outset of the processes and pathways that the data (or 'evidence') must follow in order to make a difference to management.

Finally, there is a frequently held perception amongst fisheries scientists that see industry-collected data as inferior to fishery-independent data, usually due to a perceived lack of quality assurance through the data not having been collected by scientists or having been collected outside the use of established scientific protocols. It is imperative therefore that collaborating scientists invest the time to work with fishermen so that studies are carefully designed to accommodate the skipper, crew, vessel, gear type and normal fish handling practices and therefore mitigate against the potential pit falls in fishery-dependent data. With the belief and buy-in of all stakeholders involved in the study, collectively working to fill data-gaps, the results can be scientifically robust and fit for purpose, can feed into policy and can result in significant management outcomes. All stakeholders involved within the study must believe that working in collaboration with the fishing industry to collect scientifically robust data is the best approach to help fill data gaps, to better inform policy and to achieve effective management.

Acknowledgements We thank the Department for Environment, Food and Rural Affairs (Defra) for their support and funding of the work. We sincerely thank Jamie Rendell (Defra), Kirsty McGregor (Defra) and Carl O'Brien (Cefas) for their professional representation of the Spurdog Programme to the European Commission and overall support of the Spurdog Programme. We are very grateful to and thank Ali Hood of the Shark Trust for her balanced critique and robust defence of the Shark Trust's position and for her support of working with the fishing industry. We thank all the fishermen who participated and engaged in the research projects, especially John Walsh of the fishing vessel Charisma, Phillip Mitchell of the fishing vessel Govenek of Ladram and Alan Dwan of the fishing vessel Ajax. We thank Barrie Deas and Dale Rodmell of the National Federation of Fishermen's Organisations (NFFO) for their involvement in taking the Spurdog Programme to a national and international audience during consultation. Our appreciation and thanks go to Dr. Steven Mackinson, Dr. Ewan Hunter and Joana Barker for their detailed editorial review and contributions to improving this chapter. Finally, to Paul Trebilcock of the Cornish Fish Producers Organisation (CFPO) who was instrumental in the positive direction and outcomes of the engagement events, with Paul seeing this process through from the very beginning, highlighting the issue to Government, navigating science, policy, conservation and fisheries, keeping fishermen engaged.

References

Bendall VA, Hetherington SJ, Ellis JR, Smith SF, Ives MJ, Gregson J, Riley AA (2012) Spurdog, porbeagle and common skate by-catch and discard reduction. Fisheries science partnership 2011–2012, Final report. Lowestoft, Cefas, 88 pp

Bendall VA, Carson N, Cragg A, Hetherington SJ, McGregor K, O'Brien C, Riley A, Rendell J (2014) Proposal for a UK pilot project to develop a real-time spurdog by-catch avoidance programme to mitigate a choke species and so minimise fishing induced mortality. Working

document submitted to the Scientific, Technical and Economic Committee for Fisheries (STECF) – 47th plenary meeting
Ellis JR, Bendall VA, Hetherington SJ, Silva JF, McCully Phillips SR (2015) National evaluation of populations of threatened and uncertain elasmobranchs (NEPTUsNE), Project report. Lowestof, Cefas, 105 pp
Hetherington S, Rendell J, Trebilcock P (2013) Fishermen have their say on shark and skate bycatch and discards. Fishing News, January 2013, 12–13
Hetherington SJ, Bendall VA, de Rozarieux NA, Rodmell D, Stromberg P (2015a) Stakeholder involvement in a UK led initiative to align spurdog management with the landing obligation under the reformed common fisheries policy. Fisheries science partnership 2014–2015, Final report. Lowestoft, Cefas, 46 pp
Hetherington S, Bendall V, Riley A, O'Brien C, Wright S, Hunter E, Forster R, Koch A (2015b) Common fisheries policy reform implementation: aligning zero quota species and improving fisheries management – a spurdog case study (MB0125), Project report. Lowestoft, Cefas, pp 27
Hetherington SJ, Nicholson RE, O'Brien CM (2016) Spurdog by-catch avoidance programme, Final report. Lowestoft, Cefas, 52 pp
Hetherington SJ, Nelson P, Searle A, Bendall VA, Barreau T, Nicholson RE, Smith SF, Sandeman LR (2017) Remote electronic monitoring (REM) of common skate by-catch (ELECTRA MF6001: work package task 1.3), Project report. Lowestoft, Cefas, 30 pp
Hetherington SJ, Nicholson RE, Bendall VA (2018) Spurdog by-catch avoidance programme 2: 12 month progress update. Interim evaluation (Nov 2016 – Oct 2017), Project report. Lowestoft, Cefas, 87 pp
Johnson TR, van Densen WLT (2007) The benefits and organisation of cooperative research for fisheries management. ICES J Mar Sci 64:834–840. https://doi.org/10.1093/icesjms/fsm014
Mackinson S, Raicevich S, Kraan M, Magudia R, Borrow K (eds) (2015) Good practice guide: participatory research in fisheries science. Retrieved from http://gap2.eu/outputs/gap-good-practice-guide/
Righton D, Bendall VA, Ellis JR, Hetherington SJ, McCully SR, Metcalfe J, Silva JF (2013) Assessing the survivability of bycaught porbeagle and spurdog and furthering our understanding of their movement patterns in UK marine waters. Defra final report for contract M5201, 99 pp
STECF (2014) 47th plenary meeting report (PLEN-14-03). Publications Office of the European Union, Luxembourg, 138 pp
STECF (2015) 50th plenary meeting report (PLEN-15-03). Publications Office of the European Union, Luxembourg, 90 pp

Chapter 17
Bridging Gaps, Reforming Fisheries

Petter Holm, Maria Hadjimichael, Steven Mackinson, and Sebastian Linke

Abstract Scientific debates often revolve around the issues of 'unbiased science' with the majority of scientists keeping themselves at arm's length from policy-making to ensure their credibility. Participatory research has been shifting these dynamics and has led to the emergence of research practices and advice frameworks that allow co-creation of common knowledge bases for management. This chapter, following the description of 14 cases of participatory research, places these cases alongside each other, compares and examines them as pieces in a larger puzzle to let us identify emergent patterns. In doing that, we draw on the analytical basis developed in Chap. 2. To understand what goes on in the transition zone between top-down management and participatory governance, we focus on (i) participation, (ii) knowledge inclusion and (iii) institutional reform. What we are seeing is that the case studies, instead of becoming arenas for negotiating knowledge gaps and removing false preconceptions, worked much more pragmatically, allowing fishermen access to the resources of science. With the ongoing institutional reform, emphasizing stakeholder participation and the need for broader sharing of responsibility for management processes, fisheries governance is changing. We explore this change process through the concept of the 'scientific fisherman' introduced in Chap. 2, a character who is actively involved in management decision-making and a competent

P. Holm (✉)
Norwegian College of Fishery Science, UiT The Arctic University of Norway, Tromsø, Norway
e-mail: petter.holm@uit.no

M. Hadjimichael
Innovative Fisheries Management, Department of Planning, Aalborg University, Aalborg, Denmark

University of Cyprus, Nicosia, Cyprus

S. Mackinson
Centre for Environment, Fisheries and Aquaculture Science (Cefas), Lowestoft, Suffolk, UK

Scottish Pelagic Fishermen's Association, Fraserburgh, UK

S. Linke
School of Global Studies, University of Gothenburg, Gothenburg, Sweden

P. Holm et al. (eds.), *Collaborative Research in Fisheries*, MARE Publication Series 22, https://doi.org/10.1007/978-3-030-26784-1_17

and acknowledged participant in the processes of mobilizing knowledge for management purposes.

Keywords Collaborative research · Fisheries governance · European fisheries · Knowledge · Stakeholder involvement

17.1 Introduction

The GAP project was set up to explore ongoing transitions in fisheries governance. A key premise for the overall project was that existing gaps in knowledge, communication and trust between fishermen and scientists can be bridged through collaborative research (Mackinson and Wilson 2014). Joint knowledge production practices are expected to improve transparency, trust and social robustness in fisheries governance, and thereby contribute to the sustainability of fisheries practices (Holm and Soma 2016). The case studies presented in this volume have investigated how collaborative research works out and performs in practice. Collectively, they address fundamental questions: *What are the knowledge gaps that the GAP case studies are constructed to fill? Why are they not addressed by conventional designs? Can they be bridged through collaborative research? What characterizes the relationship between scientists and stakeholders within the collaborative research projects? To what extent can collaborative research remedy the legitimacy deficits created by unresponsive management practices? Do the GAP case studies represent new modes of science-society relations, or do they reproduce a conventional and deferential relationship between science and lay clients? Are the case studies sites where scientists get access to new platforms for pursuing scientific research? Or are they arenas where fishermen get access to the resources of science for their own purposes?* To address these questions, we bring together a set of theoretical perspectives and debates introduced in Chap. 2, and the practical experiences of the individual case studies (CS) projects, presented in the previous 14 chapters.

In Chap. 2, we introduced three interrelated strands of research that have investigated knowledge issues and the prospects for co-creating knowledge between scientists and stakeholders under new modes of governance. These research strands, or pillars as we call them (see Chap. 2, Fig. 2.1), focus on participation, knowledge inclusion and institutional reform. In order to make these arguments directly relevant to the GAP experience, however, we must consider what type of insights the different CS projects can contribute. At the outset, of course, it is reasonable to expect that all the 14 GAP CS projects are relevant to all the three issues. This is also confirmed by the CS chapters. Given the great variety represented by the CS projects, however, the issues are framed in different ways across CS projects. In the following section, we therefore examine the main categories of the CS projects, and discuss the specific take they offer on the main governance issues at stake. In the final section, we summarize the main results and discuss their implications.

17.2 Sorting out the CS Projects

As indicated by the GAP CS projects, the variation among fisheries across Europe is staggering. What fisheries are – their resource basis, technology, structure, economic importance and place in society – varies, and so do the principles and procedures by which they are governed. In order to understand what the CS projects specifically can teach us, we first need to consider the range of lessons they may provide.

Since the CS projects were not explicitly designed to explore issues of theoretical interests, connecting theory and practice is not straightforward. While the design and development of each CS certainly were constructed as part of the overall GAP project, their specific identities and thrusts were in important ways framed by the specific conditions of their local context. The three pillars (participation, knowledge inclusion and institutional reform) remain at the heart of a changing fisheries governance (see Chap. 2). The CS projects, however, are the sites where we can examine how that change actually happens. How are the issues of participation, knowledge inclusion and institutional reform articulated given the nature of the particular case in question? To address this question, we need to sort out the CS projects.

As a start, we note that some of the CSs fall under the *Common Fisheries Policy* (CFP) while others do not. For example, the CS project reported in Chap. 9, fishery monitoring for coastal cod, located in Steigen, Norway, engages with fisheries that remain squarely outside the CFP. Moreover, a number of the case studies within the EU work with fisheries that are not managed under the CFP, at least not directly. For instance, the CS on selectivity in Lake Vättern (Chap. 4) focuses on white fish fisheries in a freshwater lake managed by the Swedish authorities. In addition, the CS projects within the EU that feature coastal fisheries remain within Member State jurisdiction, as in the CS on sustainability of brown crab in the UK (Chap. 3), the CS on rare Wadden Sea species (Chap. 7) and the CS on mapping habitat and fishing in Galicia, Spain (Chap. 5). In the Estonian CS, featuring marine spatial planning in the Baltic Sea (Chap. 13), the fisheries involved may be covered by the CFP. Nevertheless, the focus of this case study is spatial planning conducted under national jurisdiction. For practical purposes, we can therefore count the Baltic CS as non-CFP.

At the outset, it makes sense to distinguish between CS projects on the basis of whether they fall under the CFP or not. As pointed out by Doug C. Wilson, the CFP stands out as unique in its geographical scope:

> No other fisheries management system in the world seriously attempts to manage fisheries through such a huge, top-down system. On a continental scale, the complexity of the information needed simply cannot be handled (Wilson 2009: 267).

The discourse on the three pillars of transformation – participation, knowledge inclusion and institutional reform – is particularly relevant to the CFP and has to some extent been developed explicitly to address concerns arising in the context of the CFP and the CFP reform. This is confirmed by the CS projects on fisheries governed under CFP, for instance, in the CS projects on mixed fisheries in the North Sea (Chap. 11), on French and Spanish tuna fisheries (Chap. 12) and on discard sampling in Dutch flatfish fisheries (Chap. 14). As it is apparent from these chapters, the rigidities imposed by the CFP are framing the CS projects in different ways.

Nevertheless, the relevant object of analysis here is not the CFP as such, but the governance framework in operation at the level of the CS fishery in question. For instance, even though the Norwegian case, reported in Chap. 9, is outside the CFP, the fishery in question is managed under the centralized Norwegian governance system for fisheries. This makes the Norwegian case more similar to the CS projects within the EU that are governed under the CFP than those which are not. On the other hand, some of the CS projects feature fisheries that in principle fall under the CFP, but are strongly influenced by the specific local governance practices in operation. This is the case in the CS project on management of NW Mediterranean red shrimp (Chap. 10) and on the Maltese fisheries management zone (Chap. 15). It is hardly a coincidence that both these CS projects are located in the Mediterranean, where important features of the CFP have not been implemented in the same way as in the northern EU regions (Hadjimichael et al. 2010; Smith and Garcia 2014).

As this suggests, an important factor affecting the governance framing of the CS projects is whether the fisheries in question are managed under a centralized TAC machine system or not. As introduced in Chap. 2 (Sect. 2.1), the TAC machine concept captures the highly specialized and institutionalized system dedicated to the production, authorization and deployment of TACs as the key management instrument (Holm and Nielsen 2004; Schwach et al. 2007; Nielsen and Holm 2008). It is on the basis of the institutionalization of the TAC machine as the standardized management regime that the possibility of fisheries management on a 'continental scale' comes within reach, although, as Wilson (2009) has noted, the complexity of the task suggests that the success of such an accomplishment will be limited. Since the TAC machine requires standardization and centralization, it comes with strong restriction on participation in knowledge provision and management decision-making.

In Table 17.1, we have sorted the CS projects along the two dimensions discussed above. In the following, we discuss how the issues of participation, knowl-

Table 17.1 The CS projects sorted according to two dimensions of the governance framework of the fisheries they engage with: first, whether they are managed under the EU *Common Fisheries Policy* (CFP) or not; second, whether the management system has TAC machine features or not

	TAC machine	Non-TAC machine
CFP	Ch. 6 WBSS	Ch. 8 Adriatic Sea
	Ch. 11 North Sea mixed fisheries	Ch. 10 Mediterranean red shrimp
	Ch. 12 Tuna FADs	Ch. 15 Malta
	Ch. 14 Dutch Discards	
	Ch. 16 Elasmobranches	
Non-CFP	Ch. 9 Steigen Coastal Cod	Ch. 3 Devon Brown Crab
		Ch. 4 Vättern
		Ch. 5 Galicia TURFs
		Ch. 7 Wadden Sea brown shrimp
		Ch. 13 Baltic MSP

edge production and reform get articulated for each main type of CS, i.e. whether it involves a TAC-machine structure or not.

At the outset, CS projects tied to TAC-machine structures will provide lessons that are directly relevant to the key problems taken up by the GAP project. If things like weak participatory mechanisms, low legitimacy and gaps in understanding and communication are typical for fisheries, it is because of the dominance of the TAC-machine regime, and not an inherent trait of fisheries as such. In the same way, the thrust of the reform is about issues that primarily relate to limitations and rigidities of TAC-machine structures. While this is important to keep in mind in order to draw inferences from the GAP project, it does not mean that the CS projects that somehow have escaped from the TAC-machine are less relevant. As we shall see, the more open and sometimes more benign governance settings outside the TAC-machine provide better opportunities for exploring the capacities and limitations of collaborative models. Also, they open up for an exploration of the possibilities of a nested design, as suggested in Chap. 2. We return to this topic below.

17.2.1 CS Projects Under the TAC Machine

Six of the CS projects are closely tied to TAC-machine structures, within the EU and in Norway. While the variation among these projects may be more striking than their commonalities, they all focus on issues that can be expected to arise within TAC machine settings. TAC-machine frameworks are centralized structures characterized by limited direct access and influence for those affected (Nielsen and Holm 2008; Schwach et al. 2007). Since they tend to be large-scale and standardized, with decisions taken through pre-programmed procedures, they will often produce decisions that are insensitive to regional and local complexities (Degnbol 2003; Wilson 2009; Symes 2012). This trait does not affect all fisheries the same way, of course. While TAC-machine frameworks are rigid, they are not immune to political and social pressures. The capacity for mobilizing such pressures varies greatly among fisheries, depending on factors like the size and economic importance of the fishery, the social and cultural and political standing of the fisheries in the society in question, the organizational and economic organization of the sector and so on (Hallenstvedt 1982). This means, in short, that well-organized, economically important fisheries, like the Norwegian fishery for North East Arctic cod (Kolle et al. 2017), the North Sea pelagic fisheries (Coers et al. 2012) or the Icelandic cod fisheries (Pálsson 1991), usually will be able to bend the TAC machine to serve their respective purposes. For marginal fisheries and non-standard issues, however, the regulations imposed will sometimes be disruptive, with interventions perceived as uninformed and counterproductive at the local level (Wilson and Degnbol 2002; Symes et al. 2015).

This is a well-known and necessary consequence of large-scale systems, which must be constructed with a standardized set of problems in mind (Wilson 2009; Hadjimichael et al. 2010). The efficiency of the TAC-machine in dealing with the standard problems as seen from the perspective of the economically and politically dominant fishing interests is of course a strong justification of its existence and support. Nevertheless, this justification is challenged under the present transition of fisheries governance, as we argue in Chap. 2. With the acceptance of governance principles related to participation and knowledge inclusion, combined with the insights related to ecosystem interactions and increasing competition among interests in marine space, relevant management and knowledge objects are becoming more complex (Wilson 2009). This development is an important driver behind the efforts to make TAC-machine structures more dynamic, flexible and participatory. Nevertheless, participatory approaches may also be more vulnerable to the influence of well-organized and wealthy interest groups. An important argument in support of a centralized system is that it lends itself to democratic control and, hence, the will and values of the majority of civil society. While making fisheries management more dynamic and participatory is important in order to meet increasing complexity and local variation, this may come with greater exposition to the power play of well-organized interest groups.

The CS projects on FADs in tuna fisheries (Chap. 12) and discard sampling for flatfish fisheries (Chap. 14) are cases that illustrate, albeit in different ways, the types of strain generated by this transition. In the tuna FAD case, the CS project was initiated by scientists, building on the assumption that the industry stakeholders would be interested in and had the capacity for collaborative approaches to management and knowledge issues. The project focused on use of Fishing Aggregation Devices (FADs), a fishing technique that improved catch rates, but at the same time introduced by-catch and discard problems. The CS project was organized as a common arena where fishermen, scientists and managers could collaborate on defining research goals and approaches. Instead of fertile interaction, however, the project became a site where distrust and frustration spilled out in the open. Opening up a Pandora's Box of mutual suspicion, the project had to take a big step back, focusing on basic trust-building activities. In a dramatic fashion, Chap. 12 demonstrates some of the consequences in terms of lost trust and legitimacy imposed by TAC machine structures (though not necessarily only because of that). Moreover, it indicates the enormous investment required in order to repair this problem. Trust and collaborative spirit are not simply capacities that are available on demand, ready for action when a strategic deliberation suggests that it would be useful.

In the Dutch discard case (Chap. 14), in contrast, the new governance norms of participation and mutual interest already had resulted in a collaborative arrangement and growing trust between scientists, fishermen and managers. In particular, this materialized in a series of industry-government self-sampling projects, whereby fishermen collected data on by-catch and discards in the flatfish fisheries. The Dutch CS project was set up to explore this emerging practice, with a particular interest in improving collaborative research design. As reported in Chap. 14, however, this exploration, as well as the practices under examination, was disrupted by the implementation of the landing obligation, by which hard targets for discard reduction were introduced in the form of a directive thrust down from the EU Council. This case, of course, is a striking demonstration of the logic and limitation of standardized solutions implemented across a field of great complexity and variation (Wilson 2009; de Vos et al. 2016).

In the CS on tuna FADs and flatfish discards, the GAP projects were neither prepared for nor geared up to deal effectively with the anger and distrust that were revealed by the CS. In these instances, the GAP project could not really help building solid bridges, but had to retreat into the more modest task of charting the gaps in question. Nevertheless, the GAP portfolio also contains cases where the project's capabilities were more appropriate for the task at hand. The CS project on by-catch and discards of elasmobranches, reported in Chap. 16, is a case in point. As in the Dutch discard case, the landing obligation played an important role here. The case features the Cornish cod and hake fishery, where elasmobranches – sharks, skates and rays – are occasional but regular by-catch. The by-catch rates had not been recorded in a systematic way, however. In the combination of the new landing obligation, the red-listing of elasmobranches, and standard precautionary procedures for dealing with data-poor situations, the fishermen were hit by regulations that effectively would exclude them from some of their most productive fishing grounds. Since the CS project, involving fishermen collecting by-catch data, could supply the missing information and deliver it in an appropriate format at the right place, the crisis was averted. The regulation was changed and the fishery remained open. Collaborative research made a difference! In this case, we argue, the complexities of the Cornish fisheries were too local and specific to attract attention within the centralized management system. Nevertheless, the problem was framed in such a way that a solution primarily hinged on the provision of the missing data, which could be done through fishermen's participation in this CS project.

In the case on the Western Baltic Spring Spawning herring (WBSS), reported in Chap. 6, the CS project also made a difference. The purpose of the project was to develop an assessment model and harvest control rules that could improve the prediction and management of the WBSS herring fisheries. While this fishery clearly is subject to a TAC-machine framework, it did not fit in easily, due to a combination of the stock's migration pattern and a complex pattern of fisheries. As the spawning ground is located in the Baltic, it falls under the Baltic management framework, where the EU is in charge. Since its feeding migration brings it to the North Sea, it is also affected by the joint Norway/EU fisheries agreements. Due to such complications, the WBSS stock remained both misunderstood and mismanaged. The CS project sought to repair this, allowing the development of an appropriate Multi Annual Management Plan for the stock. In order to accomplish that, the stock behaviour needed to be modelled. This was challenging in part because of the lack of understanding of the stock structure and its migration pattern and in part by imprecise catch statistics and misreporting.

Initiated and led by a scientist holding key positions in the herring assessment system, the project also included stakeholder representatives from the industry and relevant Advisory Councils (ACs). While it perhaps can be characterized as 'science driven', the project was deeply committed to industry collaboration, both in the reconstruction of faulty catch statistics and during the modelling stage. By establishing a common knowledge platform, the project built a foundation for a better management plan that was worked out in the CS.

Similar to the CS on elasmobranches (Chap. 16), a critical challenge for the project on WBSS fisheries was to make project results count in management decisions. In both cases, the projects were able to identify the problems, collect and clean up data, and establish a common knowledge basis. Solving the modelling problem, though challenging, proved to be a success. Nevertheless, this was not sufficient for getting agreement on new management decisions. The WBSS CS collided with a 'management wall', although there were indications that this eventually could be climbed. In the elasmobranchs CS, the positive outcome was only achieved after several years of systematic effort.

In the CS project on coastal cod in Steigen, Norway (Chap. 9), the outcome was less encouraging. The original objective of the coastal cod project was to develop and test a model for monitoring coastal cod resources organized by fishermen themselves. The Norwegian coastal cod stock had been at a reasonably stable albeit poor state for the last 10 years, and it did not seem to respond well to the management measures. There was agreement among managers, scientists and fishermen that the knowledge basis for coastal cod is weak. The way stock assessment is carried out is an issue of debate due to the limited number of sample events over the course of the year, the surveys only covering areas where trawling is possible, and because the surveyed areas only constitute a minor part of the coastal zone. In addition, the catch statistics are not accurate enough to give a precise measure of the fishing pressure in the fjords. Such knowledge gaps have been recognized both by the fisheries authorities and the stakeholder groups including the fishermen.

In much the same way as the elasmobranchs and WBSS cases, the Steigen CS sought to fill recognized knowledge gaps through collaborative research. The idea for the Steigen project was to develop the prototype for a local, fishermen-operated data collection mechanism that could improve the data basis for coastal cod management. In its attempt to achieve this objective, an important requirement for the CS was that the data collected would be acceptable as input in the existing data collection regime in Norwegian fisheries. On the advice of the science partner in the project, the CS invested in a state-of-the art scientific data collection device (echo-sounders). While this solved the data quality problem, it trapped the Steigen project into a marginal position of a large data collection machinery. While most of the funds and effort in the project was spent in making the method work (producing quality data), less attention was paid to the utilization of the data and what difference they made. Although not a spectacular failure, perhaps, the disappointing outcome of this CS demonstrates some of the difficulties involved in trying to establish an independent role for fishermen in knowledge provision in the face of a well-established system. The CS was hence 'trapped in the TAC machine', unable to realize its objective.

All of the CSs tied to TAC-machine structures were in one way or another affected by institutional reform. In several cases, as in the projects on tuna FADs (Chap. 12), Dutch discards (Chap. 14), and elasmobranch by-catch (Chap. 16), the stronger emphasis on biodiversity issues, an important reform issue, is obvious. In the projects on WBSS herring and Norwegian coastal cod, the strategy of active stakeholder participation was tested out as a way to fill knowledge gaps. In the CS on multispecies and mixed fisheries in the North Sea (Chap. 11), moreover, the project was explicitly motivated to inform an ongoing fisheries reform in a more direct way. This project was designed to develop methodologies and tools adapted to the complex, recursive dynamic of an ecosystem-based approach. Set up as a collaboration between the scientists and fisheries stakeholders, the project aimed to build a decision support tool that would facilitate stakeholder interaction in long-term management planning. The idea was not specifically to find what would be the best management option for North Sea mixed fisheries, but to find the language, approach and knowledge required for identifying and evaluating management options. The ultimate knowledge product was a software decision support tool that would help focus the thinking when making an evaluation of the options, and improve the ability to make better-reasoned choices. The project utilized the North Sea Advisory Council (NSAC) as a platform, mobilizing the participants' knowledge and insights on the problems of mixed fisheries and multi-species interactions and on that basis developed options plausible for the industry. Originally, the Advisory Council was also intended as the main user of such a tool for framing and informing discussion when developing its advice. However, after the reform of the CFP, and the proposed regionalization structure, it was realized that other end-users of the tool (such as the Member States) are relevant and could have been considered. In practical terms, the project struggled with the day-to-day demand of the NSAC to give advice on plausible management options, compounded by the requirements generated by the ongoing reform process. Under the high strain from ongoing activities, it was hard for the project to mobilize NSAC partners' attention to project activities. While a 'slow-burn' process was the inevitable outcome, delays in CFP reform meant that the tool was ready in time and applied to evaluate the EC's proposed North Sea multiannual plan (Mackinson et al. 2018; see Chap. 11).

A common feature of the six TAC-machine CS projects was their strong framing by 'a management wall' (cf. Chap. 6) and the existence of a centralized arena for management decisions on which specific fisheries depend but from which they remained excluded. Exactly how the management wall problem defined the CS project varied. In some cases, the absence of meaningful ways of participation had left fishermen distraught, with huge gaps in understanding and trust revealed by the GAP project (cf. Chaps. 12 and 14). In other cases, the top-down features of the system had fewer devastating effects, allowing the possibility of mobilization of better information from the local fisheries to lead to improved management decisions. While 'scaling the management wall' in this way appears highly difficult, and sometimes beyond the scope of the GAP CS projects (Chap. 9), it sometimes proved possible, as in the cases of WBSS herring and elasmobranch by-catch (Chaps. 6 and 16).

17.2.2 CS Projects Outside the TAC Machine

While the CS projects under the TAC-machine formed a reasonably homogeneous category, this was not so for the remaining cases. Despite their differences, however, their escape from the constraints of the TAC machine sometimes allowed for easier and more direct access of stakeholders and their knowledge to the management and decision-making processes. In the absence of a management wall as a strong framing, these projects developed under a different logic than those reviewed above. Nevertheless, the variability of the governance conditions for these projects makes it challenging to understand how these projects speak to our overarching research questions. In pragmatic terms, we can distinguish between two different groups of CS projects in this category, depending on the type of management and knowledge issues they focus on. The first group was united by their focus on issues related to access and management of marine space rather than classical fisheries management issues like fishing mortality, quotas, by-catch and discards. The second group comprised projects that remain with classical fisheries management issues, but where the absence of TAC machine structures has opened up for local and participatory (co-)management.

One of the CS projects, Mapping habitats and fishing in Galicia, Spain (Chap. 5), exemplifies both issues. It started out with research goals in support of a specific local management arrangement under fishermen control, but ended up demonstrating how collaborative research can mobilize local knowledge as a political commodity for coastal people in the fight for their position within marine space. From the start, the Galician CS project was clearly management oriented, focusing on the documentation and local involvement required in order to get Galician authorities to establish an exclusive fishing zone controlled by local fishermen in thearea of Aguiño. At the time, this was a recognized, albeit not much used, area-based management instrument under Galician authorities (de Oliveira 2013). As the economic crisis made this instrument politically unavailable, however, the CS transformed, focusing instead on demonstrating the practical feasibility of participatory mapping as a method. This highlighted the trust and legitimacy gap, which exists in Galicia. Fishermen consider management regulations to have low legitimacy due to lack of participation. At the same time, the management does not trust the fishermen because their arguments are usually not supported by data. Thus, after the initial plan failed, the Galician CS project refocused on the development of a methodology for collaborative management. Through knowledge cartography and vessel monitoring, segments of the small-scale fisheries have been mapped, creating 'knowledge-based *metiérs*'. This empowered fishermen, providing research-based evidence using their knowledge of fishing grounds. Fishermen from different *metiérs* and of different ages participated in the project. In addition, scientists and students participated as well as technical assistants (as intermediaries between fishermen and scientists). Neither the official representatives of the fishermen nor administrative bodies participated in the project, since it was expected that this would make it more difficult for fishermen to participate. Knowledge produced in this CS has created a potential political commodity that may allow the fishermen more leverage in their negotiation with management authorities and other stakeholders in the struggle for marine space.

The CS projects on sustainability of brown crab fishery in Devon, UK (Chap. 3) and on rare Wadden Sea species, Germany (Chap. 7), although different in many ways, focus squarely on the struggle for marine space. An important contender in this struggle, sometimes in direct conflict with traditional small-scale fishing, is that of conservation and the establishment of Marine Protected Areas (MPAs) (Jentoft et al. 2007).

The Devon brown crab CS (Chap. 3) was organized in collaboration between scientists from the University of Leicester and eight fishers from the South Devon and Channel Shell Fishermen's Association and focused on the fisheries within an area managed under the Inshore Potting Agreement (IPA). This agreement is a voluntary management system, in place since 1978 to reduce conflict between static gear (trap and net) and towed gear (trawl and dredge) fishermen. The aim of the CS was to produce locality-specific stock assessments as official data on the status of the crab stock were unsatisfactory. This was not an immediate management problem, since the crab stock assessment indicated that it was fished sustainably and no new restrictions were on the agenda. In the light of the reopening of the discussion around the management of the British waters and the setting up of a network of MPAs, however, an area-specific stock assessment was deemed vital for the protection of the existing IPA and the right of the inshore crabbers to continue with their activities. Thus, the collaboration between the inshore crabbers and the scientists produced an Individual Based Model (IBM) of the South Devon crab fishery. This model was based on data collected at sea with the fishermen as well as with interviews collecting fishermen's experience-based knowledge. Though it is early days to identify if and how the model will be used, it seems that this collaboration has strengthened the political stature of the fishermen, lending them more leverage in their struggle to preserve the IPA. In addition, the collaboration also helped generate new ideas of how the data could be used, for example, to attain an 'environmental license', a sort of environmental accreditation, which would distinguish the small-scale fishermen from large-scale operators.

In Devon, the GAP project contributed to a reconfiguration of the societal standing and political effectiveness of fishermen as a group, making them able to stand up for themselves in the battle over priorities in the marine domain. In much the same way, the CS project on the Wadden Sea (Chap. 7) was motivated by the vulnerability of the brown shrimp fishermen in a political setting where the environmental discourse was predominant. The whole Wadden Sea area, including the traditional fishing grounds of the German brown shrimp fishermen, is protected, under the EU's Birds and Habitats Directive. While the fishery itself is not considered to be unsustainable by management authorities, it has a negative reputation as it is often claimed – without evidence – to have high by-catch rates. Collaborative research was seen as an opportunity to document actual by-catch rates, end hence disprove such notions. In practical terms, the project collected data on the occurrence of rare and migratory species in the shrimp catches. Project outputs included a species inventory of rare and migrant species for the area, valuable for monitoring Good Environmental Status. Nevertheless, since the project was not based on an organizational platform that could refine and bring this information to the relevant audiences, the CS struggled to realize its objectives.

The GAP CS projects in Galicia, Devon and the Wadden Sea focused on governance issues and frameworks that are relatively new in fisheries, reflecting emerging demands of the Blue Economy and the new realities at sea this agenda has created for traditional fishermen and fishing communities (Johnsen and Hersoug 2014; Arbo et al. 2018; Hadjimichael In press). In the larger picture, activities and interests like marine aquaculture, shipping, oil and gas extraction, seabed mining, renewable energy production, tourism and environmental protection are posed to expand their claims in marine space. While these activities continue to be governed under sectoral frameworks, with underdeveloped capacities for cross-sectoral dependencies (Raakjær et al. 2014), the need for greater integration and holistic planning is recognized. For instance, the adoption of the Marine Spatial Planning Directive by the EU in 2014 made zoning and planning at sea a requirement, building on important developments in many coastal states and regions before that (Flannery and Cinnéide 2012; Knol 2011; Smith and Brennan 2012). While such development surely recognizes the fact that fisheries increasingly must compete with other interests at sea, the planning instruments also provide new mechanisms for documenting, utilizing and authorizing fishing practices and fishermen's knowledge (Johnsen et al. 2014). Whereas the projects in Galicia, Devon and Wadden Sea illustrate this in general, the CS project on Baltic fisheries ties this explicitly to an MSP process (Chap. 13).

The Baltic GAP CS was primarily a collaboration between scientists from the Estonian Marine Institute at University of Tartu and small-scale fishing interests in the Estonian Pärnu county. Following an MSP process, this CS was established to improve the ability of small-scale fishermen to put their knowledge forward, formulate their views and impact on the MSP process. The aim was not to produce new knowledge but rather to find a way to make their knowledge available to the instruments used in the MSP process. During this process, the identification and the mapping of the actual or planned competing sea uses took place, as well as an assessment of the possible impact on the spatial and temporal allocation of fishing possibilities. This was done in two phases: first by putting fishermen's knowledge on maps and second by including other stakeholders, trying to understand how the different activities could coexist. Although the general requirements for an MSP plan are set out in the EU MSP directive, the specific governance process as to how the plan could be created is to a large extent up to the respective EU Member States. With the aid of the GAP project, Estonian small-scale fisheries interests were included in the MSP process.

Together, the CS projects from Galicia, Devon, Wadden Sea and Estonia forcefully demonstrate the new reality of marine space, with growing competition for space and societal recognition of worth. Fishermen cannot hide from public view, expecting to be carried through by the force of tradition. In order to be recognized as important and worthy of support, fishermen must struggle to make themselves visible. In doing that, knowledge is an important resource, particularly if it is collected, refined and disseminated in collaboration with scientists.

In the CS project in Galicia (Chap. 5), the initial plan was to establish a fisheries reserve, within which the fishermen themselves could take on management responsibilities. In the CS project on brown crab fisheries in Devon, the idea was to establish data collection procedures by which the fishermen could assess the sustainability of their fishery. Both these projects, as turned out, must be understood in terms of building societal capital useful in the power struggle over marine space as well as competencies and knowledge in support of fisheries management responsibilities. Indeed, the co-creation of policy-relevant knowledge is what unites the last group of the CS projects.

The CS project on selectivity in Vättern, Sweden (Chap. 4) is a case in point. As a part of a broader national initiative, a fisheries co-management group for Lake Vättern had been formed in 2004–2005, with membership of fishermen, regional authorities and scientists under leadership of the Lake Vättern Water Conservation Society. The co-management group works as an arena for management advice, conflict resolution, general discussions and information exchange between different groups. The co-management group has no formal authority for regulating the fisheries but has an advisory function. Nonetheless, its advice is in most cases taken seriously and implemented by the national authority. The purpose of the GAP project was to undertake research in support of the advisory work of the co-management group. This happened in the context of a crisis, in which a new set of fisheries restrictions were implemented in the lake, including large areas closed for all fishing, increased minimum landing sizes for Arctic charr and a substantial increase in the minimum mesh size of gill nets. These management efforts appeared to have the desired effect. At the same time, however, they also made it very difficult for commercial fishermen to target whitefish, traditionally a very important fishery in terms of economic benefits. Since Arctic charr and whitefish were exploited in a mixed gill net fishery, the strong protection of the weak stock (the charr) made it difficult to pursue the abundant and underexploited species (the whitefish).

In the Vättern CS, the measures taken to protect one type of fish threatened the economic viability of an entire fishery. Nevertheless, the existence of a co-management framework for Vättern allowed for a more orderly process for trying to sort out the problem. In keeping with the collaborative ideals of GAP as well as the Vättern management framework, the research goals of the CS project were developed in a collaborative process, anchored in and approved by the co-management group. The project focused on solving the selectivity problem of the whitefish fishery, exploring methods and fishing practices that would reduce the by-catch of Arctic charr. On the basis of this research, carried out with strong involvement of the fishermen, a set of recommendations were provided to the co-management group and, with the expectation that the regulations will be updated accordingly.

However, the account of the Vättern CS project in Chap. 4 does not tell the story to the end, leaving out whether the collaborative research made a difference on the regulations or not. What we do know is that the existence of a co-management framework has consequences for the existence of a 'management wall', as discussed previously. Since the collaborative research was anchored in, if not commissioned by, the co-management group, the salience and legitimacy of the research were already secured. Where the GAP CS projects under top-down TAC-machine frameworks have struggled mightily in order for their findings to be acknowledged and put to use by managers, as documented in the WBSS herring (Chap. 6) and elasmobranches (Chap. 16) cases, access to the decision-making arena is less restrictive under co-management conditions. Co-management is a governance arrangement, in which user groups are directly involved in and co-responsible for management decision-making (Symes 2006; Linke and Bruckmeier 2015; see Chap. 2).

In the CS project on the Maltese fisheries management zone, reported in Chap. 15, there were also close linkages between the research undertaken within GAP and the management arena. In Malta the management authority resides with government authorities, and there is no formal co-management arrangement in place. In practical terms, however, the smaller scale and close ties among stakeholders allowed for collaboration between scientists, managers and fishermen. This CS was a collaboration between fishermen from the Maltese trawling fleet and scientists from the Maltese fisheries department aiming for a better understanding of the demersal resources of the Maltese Fisheries Management Zone (which extends to 25 nm from the coast). The study was originally proposed by the fishermen themselves and the methodology was determined through continuous discussions between fishermen and scientists from the Maltese Department of Fisheries and Aquaculture so that fishermen's knowledge would be used to determine sampling locations while at the same time obtaining sound results. The research included mapping the nursery and spawning grounds targeted by the trawlers with a sharper seasonal and spatial analysis of how fish in Maltese waters change according to season throughout the year; where they spawn, and mature, and where is the greatest concentration of young fish throughout the year. Although fisheries management in Malta works through a top-down decision-making system, the CS allowed for the establishment of a collaborative platform, where fishermen, scientists and managers worked together collecting data relevant for management. The chapter does not report on the extent to which the new data actually influenced management decisions. Since the study was designed with direct involvement of management authorities, however, there seems to be no principled reasons why it will not do so, as demonstrated by similar projects elsewhere (Bjørkan 2011; Kraan et al. 2013; see also Chap. 14).

The CS project on fishing and habitat in the northern Adriatic Sea (Chap. 8) was designed to fill a gap in understanding of the sustainability status of the fishery. While the responsible managers and affiliated scientists, with support from some fishermen, argued that the health of the stocks was not in danger, other fishermen and scientists described the fisheries as in crisis, with stock collapse and ruination on the way. These groups saw the GAP project as an opportunity to establish new research that could help establish a common understanding and lead to more appropriate regulations. The CS included marine scientists from ISPRA and fishermen from the trawling fleet in Chioggia, Northern Adriatic. A main thread in the account in the chapter is the struggle to overcome division and distrust. This was apparent not only in the limited success in getting important stakeholder groups onboard, but also on the reluctance of the fishermen who were actually included. In this respect, the Italian CS project confronted the same type of problems reported in the tuna FAD CS (Chap. 12). In spite of these challenges, however, the project was reasonably successful in the collection of new data, both on the ecology and the state of commercially important stocks and on the viewpoints of the fishermen regarding the management interventions. In particular, the project results indicated that the summer fishing ban, the main management measure in place, should be extended and that this would be supported by fishermen. Whether such a proposal will be successful is far from clear, as a number of obstacles still remain. While the proposal, as reported in Chap. 12, was well received in some quarters, it does not seem to be well aligned with the perspectives and mandate of the relevant management authorities.

In particular, the effective implementation of the proposal seems to be outside the scope of the regional management authority, requiring agreement and coordination between all the relevant coastal states in the Northern Adriatic. As such frameworks are still not in place, the prospect of improved management seems remote.

An illuminating contrast to the Italian case, the CS project on management of NW Mediterranean red shrimp was more successful in terms of influencing management decisions. The purpose of the project, a collaboration between local red shrimp fishermen in Palamós and marine scientists in Barcelona, was to undertake research in support of a long-term management plan (LTMP) for the local fishery. Twenty-four provided daily data on catches and haul positions. This information was used to record the fishing grounds visited over a two-year period and identify the recruitment areas and juvenile capture season for the red shrimp. On this basis, the fishermen and scientists developed a management plan proposal, which included an extended period of no fishing during winter months. The CS team cooperated with regional fisheries managers who strongly supported the initiative. In the end, the implementation of the management plan hinged on the approval from central Spanish authorities. After 2 years of negotiation, an agreement with the Spanish Government was reached and the LTMP was approved in May 2013.

Unfolding like a text-book case on the conditions for successful resource governance (Ostrom 1990), the Mediterranean red shrimp case is truly unique. Here, the fishing grounds were clearly delimited, with recognized territorial user rights already secured for the Palamós fishermen. The red-shrimp fishery has virtually no by-catch, the presence of which easily could have introduced tricky boundary issues. There were well-established organizations of fishermen and crew, which allowed for an effective arena for coordination and agreement among stakeholders. There was the happy escape from centralized TAC machine structures, allowing national authorities to recognize a local management plan without infringing on established management systems. There was the presence of interested managers at the regional level, who saw the advantage in having the fishermen take the lead. And to kick it off from the start, there was the successful explanation by marine scientists of a stock collapse in the early 2000, completed with an accurate prediction of its recovery. This made the fishermen trust the scientists to the extent that they were willing to engage in a collaborative project within GAP, resulting in the development of a local management plan.

The red shrimp case is interesting not least because it illustrates how many different resources and conditions must be aligned in order for local fishermen to take control and manage their own fishery. In this respect, the red-shrimp case stands in contrast, as already suggested, to the CS on the northern Adriatic fisheries. While most of the key institutional conditions for successful resource management were already in place in Palamós, even before the GAP team started its work, the fishermen in Chioggia could not even agree whether their fisheries were in crisis or not. This points to the pivotal importance of a 'step zero' of joint problem framing, on which successful collaboration always depends upon. While the (successful) collaboration on the red shrimp management was based on such a solid basis of trust between scientists and fishermen, the main challenge in Chioggia was to establish and consolidate such trust.

17.3 GAP and the Pillars of Transformation

An important premise for the GAP project, as indicated by its title, was the persistence of gaps in understanding, participation and influence in the fisheries, and that such gaps affect the legitimacy and effectiveness of fisheries governance. The CS in this volume confirm and qualify this. Some fisheries are infested with mutual suspicion and legitimacy problems, requiring substantial investment in trust-building activities in preparation for collaborative work. In the FAD tuna case (Chap. 12), the GAP project became an arena for venting frustration that seems to have been building up in a system without appropriate participatory mechanisms. In Chioggia (Chap. 8), fishermen worried that the research would be used against their interests. In the Galician case (Chap. 5), the CS project retreated into basic trust-building activities through mapping of fishing practices and knowledge, with no immediate plans for using such maps for management purposes. As these examples indicate, the availability of mechanisms for meaningful stakeholder participation and influence remain a problem in many fisheries across Europe (cf. Chap. 2). How this may affect issues of legitimacy and trust becomes evidently clear in comparison with the CS projects where more appropriate participatory mechanisms are in place, be it under co-management frameworks as in Vättern (Chap. 4) or under conventional government direction as in Malta (Chap. 15).

As we pointed out in the beginning of this chapter, the governance framework – including participatory arrangements – for the fishery in questions must be taken as given from the perspective of a specific GAP CS project, with only limited prospects of contributing to its change. Accepting such limitations, what can we reasonably expect a GAP CS project to accomplish in terms of improved participation? The general answer, of course, is that the GAP CS aim to improve conditions for participation and influence through participatory research. With access to solid, agreed and certified knowledge, and better capacities to produce and promote such knowledge, fishermen will be in a better position to improve the knowledge basis for management, and to yield influence and protect their interests. In Galicia (Chap. 5), the CS team accepted this as the primary task, bringing scientists and fishermen together as a first step to establish trust. In Chioggia (Chap. 8), the project went one step further, and using that hard-won trust and the shared-knowledge basis, it allowed to promote an improved management solution, even though no appropriate management authority seems to exist for acting on the proposal. In Palamós (Chap. 10), like a dream come true, the construction of the management plan and the knowledge to justify it went hand in hand, allowing fishermen to take charge of the management of the fishery.

In these cases, the CS projects were surely informed, but not strongly framed, by existing management structures. Collaborative research served as a stepping stone towards more appropriate and participatory management practices, rather than aiming at changing specific management measures. In the Vättern case (Chap. 4), this was different. Here, a co-management framework was already in place, allowing fishermen and other stakeholders access to and real influence over management decisions. While the collaborative approach towards project identification and data collection was emphasized in support of co-management ideals, the focus of the research was to solve a typical fishery problem relating to selectivity in a mixed fishery. In a similar way, although within a very different management setting, the WBSS herring (Chap. 6) and the elasmobranches (Chap. 16) cases maintained a strong focus on the technical issue at hand. In all these cases, the GAP projects were able to carry out research that could justify different and more sustainable management measures. In contrast to the relatively small-scale Vättern case, where the road from advice to implementation is rather short, the CS projects on WBSS herring and elasmobranches encountered the 'management wall' problem, typical for larger-scale management systems.

The 'management wall' problem, we argue, is a specific version of a participation problem, where a relevant and effective management arena exists but where access to and influence on decisions are highly restricted. Large and centralized TAC-machine structures will by their nature include restrictive management walls, since they work by deploying a limited set of standardized solutions, fit to cover a set of standardized problems. In some cases, the solutions available in the existing TAC machine repertoire do not fit, for some reason or another, the local situation in the fisheries. The repertoire of interventions available from the centre does not allow the fine-grained resolution required to capture the local issue. Or if it does, the intervention tools at hand do not have the required precision to fix the problem. To the extent fisheries governance comprises a range of specific and dynamic local problems, in addition to the brief catalogue of standard issues serviced by the TAC machine, the management wall problem will be re-occurring as a regular feature.

Following Stange (2016), we can apply the concept of a 'boundary object' to come to grips with the 'management wall' problem. This concept was introduced by Star and Griesemer (1989) in order to understand how collaboration is possible among actors that come from different social worlds. Boundary objects, according to Star and Griesemer (1989: 393) are 'both plastic enough to adapt to local needs and the constraints of several parties employing them, yet robust enough to maintain a common identity across sites'. In a fisheries management context, this concept is useful in order to understand collaborative efforts across the divides among fishermen, scientists and managers.

Depending on the context, boundary objects can take a variety of forms, and may be abstract or concrete (Star 2010). In Stange's (2016) analysis of the collaborative approach to management of boarfish, both an acoustic survey and a management plan served as boundary objects for different collaborative purposes. This can be confirmed and qualified in light of the GAP CS projects. We notice that specific knowledge pursuits, like acoustic or trawl surveys, GIS mapping, discard sampling or any other relevant research method, can be found at the heart of the CS project. Usually, this is a boundary object bounced back and forth between fishermen and scientists, with management concerns in the background. Typically, the management implications of these activities come up towards the end of the project. Here, the boundary object is no longer the primary knowledge item in itself or the research method, but becomes part of the management decisions towards which it speaks, which sometimes are embedded in a management plan, as Stange suggests. The management wall problem as formulated above may indicate that it is easier to organize collaborative research than to make it count. This makes sense since the boundary issue involved becomes more complex when it involves managers as well as fishermen and scientists.

In Stange's formulation, the management plan itself is designated as a boundary object. We suggest a slightly different usage, indicating that the management plan proposal becomes the boundary object. It is not the plan itself, but the process of working out submitting, reviewing, rejecting, revising and resubmitting the plan proposal that should be in focus here. It is in this back-and-forth process of 'boundary work' that the boundaries between the different social worlds are negotiated and sometimes transformed. An important factor in deciding how this work can proceed is the boundary infrastructure facilitating such exchanges (Bowker and Star 1999: 313). As pointed out by Stange (2016), a common vocabulary among the actors is important for transferring experiences and shared meaning, and that takes time and practical collaboration to establish. Such a vocabulary, together with the standards and shared meanings that support it, is part of the boundary infrastructure. In addition, we argue, the framework defining what a management plan is and what it can do constitutes an important part of the boundary infrastructure in a fisheries management context.

Now, the focus of the GAP CS projects has been to organize local processes of co-creating common knowledge and management proposals as boundary objects, flexible enough to contain local needs and viewpoints, yet robust enough to survive the passage across the management wall and to make a difference on management decisions. In this way, the GAP CS method is a prototype for improved local participation in knowledge mobilization and management decision-making. As we have seen, the success of this approach is highly variable, depending on a number of factors within and outside the respective CS project. One important dimension here concerns the state of the boundary infrastructure facilitating the development and utilization of the co-created local boundary objects.

As already noted, the 'GAP method', common across all the case studies, was collaborative research. If the major problem to be addressed was connected to weak participatory mechanisms, as discussed above, the approach to fix that relied on the promises of collaborative research and co-creation of knowledge. The CS chapters confirm that this was the common approach, although its modes of deployment and effectiveness depend on the particularities of the case-study setting. Some of the cases were motivated by specific management problems, conducting research that could justify retraction or modification of punitive regulations, as in the CS projects on Lake Vättern (Chap. 4) and on elasmobranches (Chap. 16) or allow better management plans, as in the CS projects on WBSS herring (Chap. 6); red shrimp (Chap. 10), or in Chioggia (Chap. 8). In other cases, the focus was broader, involving the assessment of key stocks in a fisheries zone in Malta (Chap. 15) or mapping the fishing practices and resources of a local area in Galicia (Chap. 5). Some of the case studies focused on method development, like the use echo-sounders for monitoring coastal resources, as in Steigen (Chap. 9) or developing decision tools to help stakeholders make sense of complex management options as for the North Sea mixed fisheries (Chap. 11). Some of the cases started out with a management-related focus but were overwhelmed by legitimacy issues, as in the case of FAD tuna (Chap. 12) and Dutch discard (Chap. 14). Some cases were not primarily motivated to provide a new type of knowledge-based advice relevant for fisheries management decisions. Instead, they worked to strengthen the fishermen's capacity for knowledge mobilization, improving their status and effectiveness in the tightening struggle for space in the coastal zone, as in the CS projects of Devon brown crab (Chap. 3), Wadden Sea brown shrimp (Chap. 7) and Baltic MSP (Chap. 13).

The CS projects in this volume seem to comprise the same range of variation in themes and approaches that is generally observed in collaborative fisheries research, be it covered under the heading of Fisheries Dependent Information (FDI) (Graham et al. 2011; Dörner et al. 2015; Mangi et al. 2018) or Fishermens Knowledge (FK) (Hind 2015; Stephenson et al. 2016). Nevertheless, FDI and FK include approaches where the participation of fishermen and other stakeholders is limited to practical tasks, usually data collection, but where strategically important issues like project design and data analysis remain the prerogative of the science partner. In line with recent shifts towards more complete engagement of stakeholders (Dörner et al. 2015; Mangi et al. 2018), the GAP CS projects stand out by a commitment to active stakeholder participation beyond that (Johnson and van Densen 2007) with particular emphasis on joint problem definition. In functional terms, we want to argue, an important feature of the GAP approach was on how the project allowed systematic research efforts to be dedicated to knowledge gaps and problems as defined by the particular context and complexities of the local situation. In the design of the specific research approach to such problems, i.e. the choice of appropriate methods and the analysis and packaging of the results, the stakeholders in general have been happy to stand back and let the science partner take the lead (see also Stange 2016).

An important practical reason for this, of course, is the realization of the key role of fisheries scientists as gatekeepers for knowledge and advice for management purposes. In order to penetrate the management wall, then, you need fisheries scientists firmly in charge of formatting and pushing the message, as demonstrated most clearly in CS projects on WBSS herring (Chap. 6) and elasmobranches (Chap.16). In cases like Vättern (Chap. 4) and Palamós (Chap. 10), where strong fishermen participation in management functions was already secured, the collaborative organization of the research itself (i.e. the legitimacy issue) seems to be less important than the credibility and management relevance (saliency) of the knowledge it produces.

The GAP CS projects indicate that a range of different configurations are possible in the relationship between stakeholders, scientists and managers. We have previously suggested the term 'scientific fisherman' as a handle for the new agency afforded by this configuration (Dubois et al. 2016; see Chap. 2). The scientific fisherman is constituted by a network of scientists, fishermen and managers, aligned in a common effort to maintain the social, economic and environmental sustainability of the fisheries. In that struggle, the involved actors certainly have different roles and responsibilities. Nevertheless, as indicated by the qualifier 'scientific', the fishermen can no longer afford to leave knowledge and management issues in the hands of other actors. In the information age, fishermen must have ready access to the language and resources of science. In order for that to be possible, however, fishermen and other stakeholders must engage actively in knowledge and management work, and not leave this to the outside experts. This is important in order for fishermen to explain and defend their own interests. As noted in Chap. 2, this new actor constellation comes about from the transformation of fisheries management discussed as a 'communicative turnaround' resulting from sharing the burden of proof (Linke and Jentoft 2013). The GAP CS projects not only demonstrate how this is possible, but also how useful such projects can be in order to identify and contain the overflows arising from the transition taking place in the governance of fisheries, whether the case studies explored are from within the CFP or not.

It is worth noting that the knowledge and competencies of the scientific fisherman in this sense are of a different order than that imputed to fishermen in classical work on Fishermen's Ecological Knowledge (FEK; cf. Holm 2003; see also Hind 2015). In the FEK discourse, the knowledge and skills of fishermen are portrayed as existing unrelated and prior to their interaction with scientists and managers. FEK could complement scientific knowledge because it existed as a separate knowledge formation, independent of scientific knowledge. The scientific fisherman, in contrast, becomes knowledgeable through interaction with scientists and managers. The most useful aspect of such knowledge, moreover, is not about fish stocks or marine ecology, since these are the home domains of scientific experts. Instead, it is knowledge about the dynamic interaction between ecology, exploitation and management intervention, mediated by technology, economic motives and social responsibilities. For such inter- and trans-disciplinary knowledge, the scientific fisherman is clearly at home in a new political domain.

In this perspective, then, the GAP CS projects can be seen as sites for forging the networks among fishermen, scientists and managers out of which the scientific fisherman emerges. In this sense, the scientific fisherman is characterized by its command of a common vocabulary across social worlds (cf. Stange 2016). While the GAP project in itself has contributed to the creation and stabilization of such networks and infrastructure, its realization is part of a more general movement towards more inclusive governance approaches, as noted in Chap. 2.The scientific fisherman, then, is the result of a broad acceptance of active stakeholder participation in fisheries management and marine research (see Stephenson et al. 2016; Holm and Soma 2016).

As already noted, the GAP project was conceived as a research programme and a reform agenda, set up to explore and contribute to the development towards more participatory and inclusive fisheries science and management. The individual CS projects, as we have seen, were informed by institutional reform issues in different ways. Some of the projects became exemplary models for stakeholder-driven research and co-management (Chap. 10 Mediterranean red shrimp; Chap. 4 Selectivity in Lake Vättern). Some of the projects demonstrated the potentials and pitfalls of collaborative solutions for collecting fisheries-dependent information (Chap. 15 Maltese Fisheries Management Zone; Chap. 5 Mapping habitats and fishing in Galicia; Chap. 9 Fishery monitoring for coastal cod). Some of the projects brought out the challenges of making a collaborative knowledge count, of making a difference in management (Chap. 16 By-catch and discards of elasmobranchs; Chap. 6 on Management of herring). Some of the projects became entangled in ongoing reform processes in Europe (Chap. 14 Discard sampling for flatfish fisheries; Chap. 11 Multispecies and mixed fisheries in the North Sea), while others were swamped by distrust and legitimacy issues generated by weak mechanisms for participation and influence (Chap. 12 Tuna FADs; Chap. 8 Fishing and habitat in the northern Adriatic Sea). Some of the projects pointed towards the growing importance of broader processes of marine management, in which the fisheries are challenged by other interests in the coastal zone (Chap. 3 Sustainability of brown crab fisheries; Chap. 7 Rare Wadden Sea species; Chap. 13 Baltic fisheries and Marine Spatial Planning).

One of the CS projects, on Multispecies and mixed fisheries in the North Sea (Chap. 11), engaged with the ongoing CFP reform in a more direct way than any of the others. Using the North Sea Advisory Council as a platform, its objective was to develop a decision tool needed to support effective collaboration among fishermen, scientists and managers in the development of a multi-annual management plan for North Sea mixed fisheries. Sticking to the terminology introduced above, we could say that such decision tools, together with the shared vocabulary they require and disseminate, form part of the boundary infrastructure that facilitates collaboration across different social worlds.

In the same way, an important output of the GAP project as a whole was a set of guidelines for participatory research between fishermen and scientists (Mackinson et al. 2008; Mackinson et al. 2015). It is of course difficult to assess the extent to which such guidelines are picked up and put to use. Nevertheless, we can note that their impact is strengthened by an alignment with the discourse on Fisheries Dependent Information (Graham et al. 2011; Dörner et al. 2015; Hind 2015; Stephenson et al. 2016), suggesting that it is approaching an accepted normative model for involving stakeholders in research. Based on the emerging practices in the UK, consistent with international trends, Mangi et al. (2018: 622) conclude, 'There is considerable evidence of a paradigm shift from the conventional practice of asking fishermen to provide data for scientific analysis towards full engagement of key stakeholders in data collection'.

The work to establish a common framework and vocabulary for research collaboration forms an important part of the process of creating a boundary infrastructure that facilitates co-creation of knowledge. The GAP project has contributed to the development and institutionalization of such an infrastructure, building a foundation for reform. In addition, however, the GAP CS projects also explored the prospects of more specific reform approaches. In particular, the GAP project speaks to the possibilities and challenges of effective local participation in making co-created knowledge count in management decisions, as in Chap. 4 on selectivity in Lake Vättern and Chap. 10 on Management of Mediterranean red shrimp.

In both of these CS project, the fisheries in questions are managed under co-management frameworks. As pointed out in Chap. 2, co-management comes with strong participatory ideals, and is as such an important model in the context of fisheries reform. Nevertheless, the co-management model cannot be a complete alternative to top-down management, replacing it altogether. While management issues sometimes can be contained locally, as in the red shrimp case, this is often not the case, as demonstrated in Chap. 8 on fishing and habitat in the Northern Adriatic Sea. With the increasing complexity and interconnectivity acknowledged in the ecosystem approach to fisheries management, any functional governance format must include strong elements of centralization. The GAP experience indicates the possibility of a nested design, by which the structures and processes of top-down systems can be adjusted in order to allow co-management modules to operate within the larger framework. This in itself is of course not a new idea. One of Ostrom's design principles for successful local resource management is nested design, highlighting that the local management institutions must be supported, or at least not undermined, by higher authorities (Ostrom 1990). The GAP project indicates the potential of one specific format of nested design, namely one where the management plan proposal serves as boundary object for communication and coordination between local co-management units and the higher decision-making levels. It is in the process of bouncing the proposal back and forth among different stakeholders that the possibility of alignment and mutual adjustment occur, as we have seen in some of the CS projects (Chaps. 6, 10 and 16). This is also consistent with research in EcoFishMan project, where a model for management planning along these lines was developed and tested (Nielsen et al. 2015, 2018).

While the GAP project was not specifically designed to explore a nested design, it clearly demonstrates the feasibility of a nested model for knowledge construction. Whereas the GAP experience surely indicates that this may work, it has also made clear the need for developing an appropriate boundary infrastructure in support of such a model. We already see the traces of such an infrastructure, in constructions of a shared vocabulary, best practice guidelines for collaborative research and decision support tools. Needless to say, realizing a complete governance model for fisheries based on a nested design as indicated here would require a more systematic process to establish an appropriate management plan framework, with the appropriate legal, organizational and economic standards for a division of responsibilities and the streamlining of processes. While this seems to be far off at the moment, we note that the recent CFP reform includes a legal definition of a management plan process that may open up for stronger regional involvement in management plan processes (Nielsen et al. 2018), within a system where the key values and principles remain under democratic control in a centralized system.

The power of a nested systems model along these lines lies in its capacity for opening up for more active stakeholder participation at the same time as it can handle the problem of increasing complexity that follows from the ecosystem approach. To realize such a model would require considerable investment in appropriate boundary infrastructure. Nevertheless, the GAP experience also indicates that modest approaches towards such a model are possible and do not have to wait for such infrastructure to be put in place. In several of the GAP CS projects, stakeholder-led management was realized within the established structures, as demonstrated in Chap. 6 on WBSS herring, in Chap. 10 on Mediterranean red shrimp and in Chap.

16 on by-catch of elasmobranchs. Management plan proposals, it hence seems, can function effectively as boundary objects even in the absence of a clear boundary structure to streamline the process.

References

Arbo P, Knol M, Linke S, St. Martin K (2018) The transformation of the oceans and the future of marine social science. Maritime Stud 17(117). https://doi.org/10.1007/s40152-018-0117-5

Bjørkan M (2011) Fishing for advice: the case of the Norwegian reference fleet. PhD. Thesis. University of Tromsø, Tromsø

Bowker GC, Star SL (1999) Sorting things out: classification and its consequences. The MIT press, Cambridge

Coers A, Raakjær J, Olesen C (2012) Stakeholder participation in the management of North East Atlantic pelagic fish stocks: the future role of the Pelagic regional advisory Council in a reformed CFP. Mar Policy 36(3):689–695

de Oliveira LP (2013) Fishers as advocates of marine protected areas: a case study from Galicia (NW Spain). Mar Policy 41:95–102

de Vos BI, Döring R, Buisman FC, Frangoudes K, Goti LL, Macher C, Maravelias CD, Murillas-Maza A, van der Valk O, Vasilakopous P (2016) New modes of fisheries governance: implementation of the landing obligation in four European countries. Mar Policy 64:1–8

Degnbol P (2003) Science and the user perspective: the gap co-management must address. In: Wilson DC, Nielsen JR, Degnbol P (eds) The fisheries co-management experience. Accomplishments, challenges and prospects. Kluwer Academic Publishers, Dordrecht, pp 31–49

Dörner H, Graham N, Bianchi G, Bjordal Å, Frederiksen M, Karp WA, Kennelly SJ, Martinsohn JT, Murray K, Pastoors M, Håkon Gudbrandsen N (2015) From cooperative data collection to full collaboration and co-management: a synthesis of the 2014 ICES symposium on fishery-dependent information. ICES J Mar Sci 72(4):1133–1139. https://doi.org/10.1093/icesjms/fsu222

Dubois M, Hadjimichael M, Raakjær J (2016) The rise of the scientific fisherman: Mobilising knowledge and negotiating user rights in the Devon inshore brown crab fishery, UK. Mar Policy 65:48–55

Flannery W, Cinnéide MÓ (2012) Stakeholder participation in marine spatial planning: lessons from the Channel Islands National Marine Sanctuary. Soc Nat Resour 25(8):727–742

Graham N, Grainger R, Karp WA, MacLennan DN, MacMullen P, Nedraas K (2011) An introduction to the proceedings and a synthesis of the 2010 ICES Symposium on Fishery-Dependent Information. ICES J Mar Sci 68(8):1593–1597. https://doi.org/10.1093/icesms/fsr36

Hadjimichael M (In press) Loosing space: the rise of the Blue Economy and the marginalisation of the small-scale fishers in Cyprus. In: Status and resilience of small-scale fisheries in Europe: in the shade of the CFP reform and economic crisis. Springer, Cham

Hadjimichael M, Edwards-Jones G, Kaiser MJ (2010) Distribution of the burden of fisheries regulations in Europe: The north/south divide. Mar Policy 34(4):795–802

Hallenstvedt A (1982) Med lov og organisasjon. Organisering av næringsinteresser og markeder i norsk fiskerinæring. Universitetsforlaget, Tromsø

Hind EJ (2015) A review of the past, the present, and the future of fishers' knowledge research: a challenge to established fisheries science. ICES J Mar Sci 72(2):341–358

Holm P (2003) Crossing the border: on the relationship between science and fishermen's knowledge in a resource management context. Maritime Stud 2(1):5–33

Holm P, Nielsen KN (2004) The TAC machine. In: Report of the working group on fishery systems. WGFS annual report. ICES, Copenhagen, pp 40–51

Holm P, Soma K (2016) Fishers' information in governance: a matter of trust. Curr Opin Environ Sustain 18:115–121
Jentoft S, van Son TC, Bjørkan M (2007) Marine protected areas: a governance systems analysis. Hum Ecol 35(5):611–622
Johnsen JP, Hersoug B (2014) Local empowerment through the creation of coastal space? Ecol Soc 19(2):60. https://doi.org/10.5751/ES-06465-190260
Johnsen JP, Hersoug B, Solås AM (2014) The creation of coastal space – how local ecological knowledge becomes relevant. Maritime Stud 13:2
Johnson TR, van Densen WLT (2007) The benefits and organization of cooperative research for fisheries management. ICES J Mar Sci 64:834–840
Knol M (2011) Mapping ocean governance: from ecological values to policy instrumentation. J Environ Plan Manag 54(7):979–995
Kolle N, Nilsen AR, Døssland A, Christensen P (2017) Fish, coast and communities: a history of Norway. Fagbokforlaget, Bergen
Kraan M, Uhlmann S, Steenbergen J, van Helmond AT, van Hoof L (2013) The optimal process of self—sampling in fisheries: lessons learned from the Netherlands. J Fish Biol 83:963–973. https://doi.org/10.1111/jfb.12192.
Linke S, Bruckmeier K (2015) Co-management in fisheries – experiences and changing approaches in Europe. Ocean Coast Manag 104:170–181
Linke S, Jentoft S (2013) A communicative turnaround: shifting the burden of proof in European fisheries governance. Mar Policy 38:337–345
Mackinson S, Wilson DC (2014) Building bridges among scientists and fishermen with participatory action research. In: Urquhart J, Acott T, Symes D, Zhao M (eds) Social issues in sustainable fisheries management. Springer, Dordrecht, pp 121–137
Mackinson S, Neville S, Raichevich S, Clausen LW (2008) Good practice guide to participatory research between fisheries stakeholders and scientists. GAP Project deliverable 1:23
Mackinson S, Raichevich S, Kraan M, Magudia R, Borrow K (eds) (2015) Good practice guide: participatory research in fisheries science. Retrieved from http://gap2.eu/outputs/pr-handbook/
Mackinson S, Platts M, Garcia C, Lynam C (2018) Evaluating the fishery and ecological consequences of the proposed North Sea multi-annual plan. PLoS One 13(1):e0190015. https://doi.org/10.1371/journal.pone.0190015
Mangi SC, Kupschus S, Mackinson S, Rodmell D, Lee A, Bourke E, Rossiter T, Masters J, Hetherington S, Ctachpole T, Righton D (2018) Progress in designing and delivering effective fishing industry – science data collection in the UK. Fish Fish 19:622–642. https://doi.org/10.1111/faf.12279
Nielsen KN, Holm P (2008) The TAC Machine: on the institutionalization of sustainable fisheries resource management. In Science|politics: boundary construction in mandated science. The case of CIES' advice on fisheries management. K.N. Nielsen. PhD thesis, NCFS, University of Tromsø, Tromsø, Norway
Nielsen KN, Holm P, Aschan M (2015) Results based management in fisheries: delegating responsibility to resource users. Mar Policy 51:442–451
Nielsen KN, Aschan M, Agnarsson S, Ballesteros M, Baudron A, Borges M, Campos A, Chapela R, Daníelsdóttir AK, Erzini K, Gregersen Ó, Holm P, Lucchetti A, Margeirsson S, Mendes HV, Olsen P, Rangel M, Sala A, Santiago JL, Sigurðardóttir S, Silva C, Sykes D, Viðarsson JR, Virgili M, Wise L, Fernandes PG (2018) A framework for results-based management in fisheries. Fish Fish 19(2):363–376
Ostrom E (1990) Governing the commons: the evolution of institutions for collective action. Cambridge University Press, New York
Pálsson G (1991) Coastal economies, cultural accounts. Human ecology and Icelandic discourse. Manchester University Press, Manchester
Raakjær J, van Leeuven J, van Tatenhove J, Hadjimichael M (2014) Ecosystem-based marine management in European regional seas calls for nested governance structures and coordination—a policy brief. Mar Policy 50:373–381

Schwach V, Bailly D, Christensen AS, Delaney AE, Degnbol P, van Densen WL, Holm P, McLay HA, Nolde Nielsen K, Pastoors MA, Reeves SA, Wilson DC (2007) Policy and knowledge in fisheries management: a policy brief. ICES J Mar Sci 64(4):789–803
Smith G, Brennan RE (2012) Losing our way with mapping: thinking critically about marine spatial planning in Scotland. Ocean Coast Manag 69:210–216
Smith ADM, Garcia SM (2014) Fishery management: contrast in the Mediterranean and the Atlantic. Curr Biol 24(17):R810–R812
Stange K (2016) Building a knowledge base for management if a new fishery: boarfish (Capros aper) in the Northeast Atlantic. Fish Res 174:94–102
Star SL (2010) This is not a boundary object: reflection on the origin of a concept. Sci Technol Hum Values 35(5):601–617
Star SL, Griesemer JR (1989) Institutional ecology, 'translations' and boundary objects: amateurs and professionals in Berkely's Museum of vertebrate Zoology, 1907–39. Soc Stud Sci 19:387–420
Stephenson RL, Paul S, Pastoors M, Kraan M, Holm P, Wiber M, Mackinson S, Dankel D, Brooks K, Benson A (2016) Integrating fishers' knowledge research in science and management. 2016 ICES J Mar Sci 73(6):1459–1465. https://doi.org/10.1093/icesjms/fsw025.
Symes D (2006) Fisheries governance: a coming of age for fisheries social science? Fish Res 81:113–117
Symes D (2012) Regionalizing the common fisheries policy: context, content and controversy. Maritime Stud 11:6
Symes D, Phillipson J, Salmi P (2015) Europe's coastal fisheries: instability and the impacts of fisheries policy. Sociol Rural 55(3):245–257
Wilson DC (2009) The paradoxes of transparency: science and the ecosystem approach to fisheries Management in Europe. Amsterdam University Press, Amsterdam
Wilson DC, Degnbol P (2002) The effects of legal mandates on fisheries management deliberations: the case of Atlantic bluefish in the United States. Fish Res 58:1–14

Chapter 18
Conclusion

Steven Mackinson, Petter Holm, and Maria Hadjimichael

Abstract This chapter summarizes the arguments and discusses the results of the GAP project in the context of the ongoing reform in fisheries governance.

Keywords Participatory research · Governance · Co-management · Stakeholder engagement · Knowledge brokering

18.1 Introduction – Exploring the Transition Zone

A prominent feature of global marine environmental governance and the management of fisheries during the last decade has been the building of opportunities for scientists, stakeholders, policy-makers and non-governmental organizations to communicate, negotiate and work together. Progress has been made. The GAP experience has demonstrated that collaborative research does work. Nevertheless, much work remains to be done. While GAP has shown that collaborative research may succeed at the level of individual projects, things get more complicated if the approaches deployed in the GAP project are going to form part of the standard institutional set-up for fisheries governance. There are unresolved issues in relation to how multi-actor collaboration in research and innovation can provide a vehicle for

---

S. Mackinson (✉)
Centre for Environment, Fisheries and Aquaculture Science (Cefas), Lowestoft, Suffolk, UK

Scottish Pelagic Fishermen's Association, Fraserburgh, UK
e-mail: steve.mackinson@scottishpelagic.co.uk

P. Holm
Norwegian College of Fishery Science, UiT The Arctic University of Norway, Tromsø, Norway

M. Hadjimichael
Innovative Fisheries Management, Department of Planning, Aalborg University, Aalborg, Denmark

University of Cyprus, Nicosia, Cyprus

P. Holm et al. (eds.), *Collaborative Research in Fisheries*, MARE Publication Series 22, https://doi.org/10.1007/978-3-030-26784-1_18

inclusive governance. At the same time, much work needs to be done in order to evaluate what happens at the margins of collaborative approaches. Even collaborative approaches cannot include all stakeholders and accommodate all interests, and we need to understand more about how such approaches distribute and redistribute power and influence (c.f. Griffin 2013).

Despite being perhaps the most science-driven policy area in Europe, the Common Fisheries Policy (CFP) has suffered from a legitimacy crisis that also focused on the knowledge base for management and decision-making (Schwach et al. 2007; Wilson 2009; see Chap. 2). The foundation of the legitimacy crisis has centred on how the science-policy system makes informed management decisions when (i) it is known (or believed) that the underpinning science is uncertain, (ii) the impact of science on the policy process can be unclear and (iii) there are differing perspectives on what we should do with our seas. Understandably, these questions of legitimacy also undermine the credibility of the institutions responsible for scientific assessment and advice.

The command and control management paradigm of the CFP has meant that until recently, conditions have not been favourable for catalysing the type of participatory research initiatives necessary to rebuild trust and credibility. But things are changing. Reflections on failings of the CFP have led to a tangible change in attitude among stakeholders and scientists, and created opportunities for developing inclusive governance approaches in the 2014 reform of the CFP (see Chap. 2). This backdrop means that efforts to bring together the knowledge and know-how of scientists, fishermen, policy managers and civil society organizations are becoming more prevalent; a key motivation being that the knowledge base for management plans and how they are implemented is accepted by society and those whose livelihoods depend upon them.

The shift towards the ecosystem approach adds force to this movement because its added complexity and diversity of interests that need to be taken account of has contributed to the broadening of the knowledge and expertise required for management as well as the need for active user participation (Mackinson and Middleton 2018). At the same time, new Information and Communication Technologies (ICT) have drastically improved the capacity for collecting, combining and utilizing data and information from different sources (Holm and Soma 2016). As a result of these shifts – and the EU reform initiatives seeking to take advantage of them – the contours of a decentred and participatory governance regime are visible, built on co-management, active stakeholder engagement, and co-creation of common knowledge bases for management.

However, because participation and inclusive approaches are rather new in many areas of the EU fisheries research agenda, there is a desperate need to learn quickly how to do this well and thus help avert tensions that arise between society, policy and science when environmental sustainability concerns appear in conflict with maintaining livelihoods. Arguably, the organization and authorization of knowledge for management forms one of the most dynamic parts of the evolution of European fisheries policy and management.

In the case studies presented in this book, we have explored what goes on in the transition zone between top-down management and participatory governance, in particular in the ongoing transition from expert-dominated towards collaborative arenas for knowledge creation. The chapters expose the bare bones of experiments in participatory research, showing how the knowledge frameworks and relationships shape the chances of success and provide personal critical accounts of what it takes to do better.

18.2 Putting 'Participatory Research' on the European Fisheries Research Agenda

As a collective, the CS papers describe the complexity of European fisheries in some detail. As practical field experiments, they are significant in demonstrating applied participatory methods and their utility across a range of research issues. By having to confront the mechanisms of the established order, they reveal how participatory research can generate new insights and produce a wealth of new knowledge to contribute to the world of science and management. One particularly successful example is the red shrimp case study in Palamós, Spain (Chap. 10). Here, excellent collaborative working between local fishermen, scientists and the regional Government of Catalonia has produced well-accepted, environmentally sound management measures – many promulgated by local fishermen themselves – which have prevented the collapse of an extremely valuable fishery. The red shrimp fishery is now thriving, and used as a best-practice example along the Mediterranean coast. Not only has each of the GAP case studies produced useful new fisheries data, the personal accounts described in each chapter testify their contribution to understanding the challenges and benefits of the participatory approach.

In addition to the in-depth case studies, GAP served to facilitate and promote inclusive dialogue on emerging issues of the reform of the CFP and implementation of the Marine Strategy Framework Directive (MSFD). Two pivotal workshops during CFP reform provided a stage for policy-makers, industry, managers and scientists to discuss 'Putting the Science into Regionalisation'. Other pioneering workshops included a fisherman-focused interactive workshop on collaborative management of octopus fisheries in Northern Spain, UK–French collaboration on the Channel scallop fishery, multi-stakeholder workshops on Irish Sea herring management plans, discussions on the process for developing long-term management plans for North Sea demersal fish and Western Baltic herring and Spanish and French collaboration on sustainable FAD fishing for tuna in the Indian Ocean.

A re-occurring theme across all case studies and workshops has been the need for effective communication between all partners. This has meant working together to develop a 'common language', an exemplar being the international symposium on the theme of 'Participatory Research and Co-Management in Fisheries' (GAP 2015a), where innovative approaches were used to help participants share their experiences of participatory research and discuss its evolution in the EU and beyond. As with the development of the 'common language', it has been important to tailor

GAP's outputs to those who will be using them, such as in the final policy briefing (GAP 2015b) that summarizes the key impacts. Open access to the Participatory Research Toolbox (GAP 2015c) and all briefings and reports is available via the GAP website (www.gap2.eu). In making the knowledge gained from the project accessible in such ways, we hope that GAP lives beyond the natural conclusion of the project and helps to evolve collaboration in other areas of research. By continuing to build strong working relationships between partners in such diverse fisheries as the UK crab, Danish herring, Norwegian cod, Dutch flatfish, and Italian cuttlefish, it has been the aim of the project to leave a footprint of – or rather a blueprint for – collaboration. Signs of its application are now visible across a range of national (e.g. Pastoors and Quirijns 2017; Mangi et al. 2018) and EU projects (e.g. Discardless (www.discardless.eu) and Pandora (http://pandora-fisheries-project.eu/)), and have been used to develop detailed guidelines to support them (Mackinson et al. 2017).

18.3 GAP Lessons

Active inclusion of stakeholders in research from conception has been crucial to the development and success of the case studies. Rather than a tokenistic involvement in data collection, GAP processes attempted to bring all stakeholders together in the design of research at the outset, as well as offering ownership of data collected at the conclusion of the research. This is perhaps nowhere more clearly illustrated than in GAP's Swedish case study in Lake Vättern (Chap. 4), where members of a co-management group proposed, and undertook their own research experiments, studied the population structure and the selective fishing methods for Whitefish. This process continues beyond GAP with the methods being proposed to enter into local legislation. Similarly, and following the example set by the Red Shrimp case study (Chap. 10), the Maltese case study resulted in fishers and the Maltese Department of Fisheries and Aquaculture collecting data, which will potentially be used to complement the country's management plan for trawl fisheries (Chap. 15). At the same time, some of the case studies, for example, the Western Baltic spring spawning herring case study (Chap. 6) and Steigen coastal cod (Chap. 9), highlight the thin line between success and failure of collaborative research, particularly when the 'management wall' is reached.

Taken together, the CS chapters bring out the great variability in fisheries across Europe. This is important to keep in mind when we are extracting lessons from the GAP experience. Since the fisheries of Europe are confronted with a range of different problems, and work under a variety of technical, economic and societal conditions, it is not likely that we can find one model to fix it all. In line with this, we want to warn against one simplistic interpretation of the GAP experience, namely that developing the knowledge base for effective management points to local co-management as a general solution, even though the obvious success stories in GAP, be it the Red Shrimp in the Mediterranean or the White Fish of lake Vättern, indeed seem to promote co-management as an effective mechanism. While these case stories are fabulous and inspiring, it is important to emphasize the unique set of preconditions that have made these cases work. In this sense, the CS from Chioggia, Italy, and Steigen, Norway, and FADs in Spain and France are equally important, despite the multiple obstacles encountered. In general, we want to argue that the conditions that these CS projects have to deal with are more typical across European fisheries; thus Chioggia might serve as a better test case for collaborative research than Palamos. Or rather, a robust inclusive approach to fisheries governance must be able to withstand the challenges of Chioggia, not only enjoy the fruits of Palamos. To be sure, under the right conditions, co-management remains an ideal allowing active stakeholder participation and collaborative approaches on knowledge production as well as decision-making, but it is not a requirement. Nevertheless, because of the nested scales and complexities of contemporary fisheries, governance often necessitates linking local, regional and global issues in increasingly intricate ways. Fisheries management cannot be solved primarily at the local level. Instead, the crucial lesson from GAP is about how the top-down structures of the TAC machine can be modified in order to accommodate co-management component as integrated elements. Instead of either top-down control or local co-management, we need both within a nested systems approach (Ostrom 1990; Wilson 2009). The GAP project has demonstrated that this is indeed possible, even without massive investment and explicit reform efforts. Within the emerging structures, based on collaborative platforms of the ACs and the emerging institution of Long Term Management Plans, we can see the shape of a nested system operating in practice, as in the cases of WBSS herring (Chap. 6) and Elasmobranches bycatch (Chap. 16).

The top-down system of fisheries management is already under heavy pressure to handle the complexities of fisheries management. With the acceptance of the ecosystem approach, and the general movement towards including a more heterogeneous set of interests as stakeholders, this pressure will continue to grow. A nested approach along the lines suggested here seems to be an obvious approach to meet this challenge: by systematic mobilization of knowledge and management effort at the local level, within a generalized framework that takes care of higher-order problems. The main challenge, from this perspective, is to strengthen the boundary infrastructure that allows for appropriate division of responsibilities and interaction between the levels in such a nested system.

18.4 Responsible Research and Innovation (RRI): A Knowledge Framework for the Ecosystem Approach

The principles of inclusion, participation and transparency upheld in GAP resonate with the wider EU research agenda for adopting Responsible Research and Innovation (RRI, see Box 18.1) as a framework for directing and delivering research

Box 18.1 Responsible Research and Innovation
RRI means societal actors (researchers & innovators, citizens, policy makers, business, research funders, NGOs etc.) working together during the whole research and innovation process to better align both the process and its outcomes with the values, needs and expectations of society. Active engagement and participatory research are cornerstones of RRI (EU 2019).

and its outcomes in a way that society wants and accepts. We see the RRI framework as an enabler of opportunities to create an environment for collective action initiatives to emerge and be sustained.

GAP is just one example of the broader range of Responsible Research and Innovation (RRI) projects funded by the European Commission. It is necessary now that the lessons learnt from this contribute to helping research policy-makers and funders understand how they can build collaborative approaches into future projects – both within fisheries and beyond.

With the experience gained from participatory research case studies around the world, it has become clear that the basic idea of collaboration in research and innovation is deeply connected with the principles of inclusive governance, which are embodied within the ecosystem approach to fisheries (EAF). Indeed, 50 years' experience on the principles and operational guidance for ecosystem management-shows that elements of inclusive governance form part of the foundations of EAF:

> Involve all stakeholders in knowledge-sharing, decision-making and management; Ensure coordination, consultation and cooperation, including joint decision-making, between fisheries and other sectors; Recognize that management objectives are a matter of societal choice; Decentralize decision and action to the lowest appropriate level. (FAO 2001–2019).

This perspective highlights that getting the science right is not always enough. Having more and better data only goes part way to address issues of sustainability in managing fisheries. In itself it cannot substitute for stakeholders' non-compliance with management measures, their feelings that science misrepresents what they see on (or under) the water and their lack of trust in the data and how it is used. Getting it right requires generating the knowledge base on par with developing the trust and confidence among stakeholders, researchers and managers, so that solutions are understood and fit for purpose (Holm and Soma 2016). By evolving what it takes to carry out RRI in practice and to ensure its utility in management, future work needs to focus on embedding collaborative approaches in a systematic way.

But why is this more important now than ever?

The regional approach to fisheries management, established in the 2014 reform of the Common Fisheries Policy (CFP), strived to provide the conditions that EU Member Countries now need to collaborate with stakeholders and scientists when deciding how to best manage fisheries in regions where they share fishing interests. While the regionalized approach may take some time to produce mature collaborative partnerships (Eliasen et al. 2015; Linke and Jentoft 2016), it is important to begin moving in the right direction. European citizens today expect their seafood to come from sustainable, responsible, and ethical fisheries, which puts fishermen, managers, and scientists in the spotlight. Indeed, societal acceptance, or 'the social licence' to fish has never been more important given the public awareness of the wasteful discarding of fish and the environmental imperative for healthy ecosystems.

Without taking too great a leap then, it is reasonable to expect that the CFP's focus on an ecosystem approach in the context of regionalization has the potential to lead to the proliferation of participatory research practices, the rise of the 'scientific fisherman' (Dubois et al. 2016), and the development of more collaborative management arrangements.

18.5 Conclusion

Stakeholders may frequently challenge the validity or interpretation of scientific advice because of the negative impact that policy decisions arising from it can have on their lives. This 'tension' between society, policy and science is plainly evident when environmental sustainability concerns appear in conflict with maintaining livelihoods. This is why research seeking to integrate the experiences of stakeholders in the knowledge base for management is a rapidly developing field. As an example of this phenomenon, GAP made apparent the disparity between the political desire to actively engage a broad range of stakeholders and the practical means by which to achieve it (Mackinson et al. 2011). It challenged the barriers and promoted ideas to better enable the participation of stakeholders in research (Mackinson and Wilson 2014). The case studies presented in this volume have been the vanguard of a transition of management approaches and the changing knowledge requirements to support it. Active participation in research has been used as a way to reduce tension by focusing on creating knowledge that is both scientifically credible and legitimate. Today there are many projects operating in Europe and around the world, where this is 'par for the course'.

After seven years at the forefront of participatory research in Europe's fisheries, the GAP team and all those who have shared in the projects work are moving forward with co-created new knowledge, and adding momentum to the ever-growing enthusiasm for the value of collaboration in research and policy-making. Our sense is that in the slow transition to an ecosystem approach – and the inclusive governance it demands – research and management approaches are at a tipping point in their readiness to make it work. Policy bodies are more sensitized than ever to the idea that society should be actively involved in research that underpins policy 'with and for' EU citizens.

There have been positive signs from a range of international organizations with regard to a growth in research participation – the UN FAO's 'Voluntary Guidelines on Securing Sustainable Small-Scale Fisheries' recently highlighted the importance of co-management (an outcome of participatory research) in ensuring long-term sustainability (Chuenpagdee and Jentoft 2019). The Directorates General for Research and Innovation and Maritime Affairs and Fisheries have both closely followed GAPs work and are committed to spreading 'responsible research and innovation' more widely across all forms of policy-making. This policy interest is reflected in the scientific arena, exemplified by the ICES conference in 2016 entitled 'Understanding marine socio-ecological systems: including the human dimension in Integrated Ecosystem Assessments', where discussions about stakeholder collaboration and participatory research were in the spotlight.

Finally, the whole GAP team would like to thank all those who have been involved in, or simply followed, the GAP journey. We hope to have imparted some useful knowledge, challenged perspectives and inspired you.

References

Chuenpagdee R, Jentoft S (eds) (2019) Transdisciplinarity for small-scale fisheries governance: analysis and practice. Springer, Dordrecht

Dubois M, Hadjimichael M, Raakjær J (2016) The rise of the scientific fisherman: Mobilising knowledge and negotiating user rights in the Devon inshore brown crab fishery, UK. Marine Pol 65:48–55

Eliasen SQ, Hegland TJ, Raakjær J (2015) Decentralising: The implementation of regionalisation and co-management under the post-2013 Common Fisheries Policy. Marine Pol 62:224–232

EU (2019) Responsible research and innovation, H2020 programme. Retrieved from https://ec.europa.eu/programmes/horizon2020/en/h2020-section/responsible-research-innovation

FAO (2001–2019). Fisheries and Aquaculture topics. Fisheries and aquaculture governance. Topics Fact Sheets. In FAO fisheries and aquaculture department [online]. Rome. Updated 7 January 2016. Retrieved from http://www.fao.org/fishery/governance/en and http://www.fao.org/fishery/topic/16034/en

GAP (2015a) Policy briefing 4: GAP2 international symposium. Retrieved from (http://gap2.eu/gap2general/gap2-international-symposium-post-event-brief/)

GAP (2015b) Policy briefing 5: building bridges through participation – concrete outcomes & cementing change. Retrieved from (http://gap2.eu/news/policy-briefing-5-building-bridges-through-participation-concrete-outcomes-cementing-change/)

GAP (2015c) Participatory research toolbox. Retrieved from (http://gap2.eu/methodological-toolbox/)

Griffin L (2013) Good governance, scale and power: a case study of North Sea fisheries. Routledge, New York

Holm P, Soma K (2016) Fishers' information in governance: A matter of trust. Curr Opin Environ Sustain 18:115–121

Linke S, Jentoft S (2016) Ideals, realities and paradoxes of stakeholder participation in EU fisheries governance. Environ Sociol 2(2):144–154

Mackinson S, Middleton D (2018) Evolving the ecosystem approach in European fisheries: Transferable lessons from New Zealand's experience in strengthening stakeholder involvement. Mar Pol 90:194–122

Mackinson S, Wilson DCK (2014) Building bridges among scientists and fishermen with participatory action research. In: Urquhart J, Acott T, Symens D, Zhao M (eds) Social issues in sustainable marine fisheries management. Springer, Dordrecht, pp 121–139

Mackinson S, Wilson DC, Galiay P, Deas P (2011) Engaging stakeholders in marine research. Mar Policy 35:18–24

Mackinson S, Mangi S, Hetherington S, Catchpole T, Masters J (2017) Guidelines for Industry-Science Data Collection: step-by-step guidance to gathering useful and useable scientific information. Fishing into the Future report to Seafish. 65p. June 2017

Mangi S, Kupschus S, Mackinson S, Rodmell D, Lee A, Bourke E, Rossiter T, Masters J, Hetherington S, Catchpole T, Righton D (2018) Progress in designing and delivering effective fishing industry-science data collection in the UK. Fish Fish, Wiley 2018:1–21. https://doi.org/10.1111/faf.12279

Ostrom E (1990) Governing the commons: the evolution of institutions for collective action. Cambridge University Press, New York

Pastoors MA, Quirijns FJ (2017) PFA Self-sampling report 2015–2016. PFA report 017/02, 2017

Schwach V, Bailly D, Christensen AS, Delaney AE, Degnbol P, van Densen WL, Holm P, McLay HA, Nolde Nielsen K, Pastoors MA, Reeves SA, Wilson DC (2007) Policy and knowledge in fisheries management: a policy brief. ICES J Mar Sci 64(4):789–803

Wilson DC (2009) The paradoxes of transparency: science and the ecosystem approach to fisheries management in Europe. Amsterdam University Press, Amsterdam

Index

P. Holm et al. (eds.), *Collaborative Research in Fisheries*, MARE Publication Series 22, https://doi.org/10.1007/978-3-030-26784-1

MIX
Papier aus verantwortungsvollen Quellen
Paper from responsible sources
FSC® C105338

If you have any concerns about our products,
you can contact us on
ProductSafety@springernature.com

In case Publisher is established outside the EU,
the EU authorized representative is:
Springer Nature Customer Service Center GmbH
Europaplatz 3, 69115 Heidelberg, Germany

Printed by Libri Plureos GmbH
in Hamburg, Germany